Some of the most exciting recent advances in animal behaviour have
occurred at the interface between that subject and the study of
evolution. This book is written by experts in this area, and illustrates
how the profound changes in our understanding of evolution have
influenced behavioural research. Its chapters cover studies of how
behaviour itself has evolved, dealing with topics such as comparative
studies, the genetics of behaviour, speciation and the evolution of
sociality and of intelligence. They also deal with the adaptiveness which
this evolution has brought about, with mating and fighting strategies,
and with theories of kinship and altruism.

   Behaviour and Evolution will be invaluable to senior undergraduates
and graduate students of biology and psychology, especially those
in the fields of animal behaviour, behavioural ecology, sociobiology,
evolution, ecology and environmental biology.

D0139733

BEHAVIOUR AND EVOLUTION

BEHAVIOUR AND EVOLUTION

Edited by

P. J. B. SLATER

Professor of Natural History and Head of the School of Biological and Medical Sciences, University of St Andrews, Fife, UK

T. R. HALLIDAY

Professor of Biology and Head of the Department of Biology, The Open University, Milton Keynes, UK

Pen-and-ink illustrations by Priscilla Barrett, S.W.L.A.

CAMBRIDGE
UNIVERSITY PRESS

PUBLISHED BY THE PRESS SYNDICATE OF THE UNIVERSITY OF CAMBRIDGE
The Pitt Building, Trumpington Street, Cambridge, CB2 1RP, United Kingdom
CAMBRIDGE UNIVERSITY PRESS
The Edinburgh Building, Cambridge CB2 2RU, UK http://www.cup.cam.ac.uk
40 West 20th Street, New York, NY 10011-4211, USA http://www.cup.org
10 Stamford Road, Oakleigh, Melbourne 3166, Australia

First published 1994
Reprinted 1998

Printed in the United Kingdom at the University Press, Cambridge

Typeset in Linotron Times 10/13pt [KW]

A catalogue record for this book is available from the British Library

Library of Congress Cataloguing in Publication data
Slater, P. J. B. (Peter James Bramwell), 1942–
    Behaviour and Evolution / P.J.B. Slater, T.R. Halliday
      p. cm
    Includes bibliographical references (p.) and index.
    ISBN 0 521 41858 5 (hardback).
    ISBN 0 521 42923 4 (paperback).
    1. Animal behavior.   2. Behavior evolution.   I. Halliday, Tim.
    1945–   II. Title.
    QL751.S615   1994
    591.51 – dc20   94-6536 CIP

Contents

Contributors

R. K. Butlin
Department of Genetics, University of Leeds, Leeds, UK

R. W. Byrne
Scottish Primate Research Group, Department of Psychology, University of St Andrews, St Andrews, Fife, UK

D. M. Decker
Department of Zoology, University of Tennessee, Knoxville, Tennessee, USA

J. L. Gittleman
Department of Zoology, University of Tennessee, Knoxville, Tennessee, USA

T. R. Halliday
Department of Biology, The Open University, Milton Keynes, UK

I. F. Harvey
Department of Zoology, University of Liverpool, Liverpool, UK

A. A. Hoffmann
Department of Genetics and Human Variation, La Trobe University, Bundoora, Victoria, Australia

P. C. Lee
Department of Biological Anthropology, University of Cambridge, Cambridge, UK

M. G. Ritchie
School of Biological and Medical Sciences, University of St Andrews, St Andrews, Fife, UK

P. J. B. Slater
School of Biological and Medical Sciences, University of St Andrews, St Andrews, Fife, UK

1
Introduction

Evolution is a subject of central importance to biologists interested in the study of behaviour. Darwin himself devoted a chapter in *The Origin of Species* to behaviour and, like so much else in biology, the growth of interest in behaviour during the past century took much of its impetus from the framework that he provided. Yet the study of behavioural evolution itself has lagged behind other aspects of behavioural work because, for various reasons, it is not an easy subject in which to achieve insights. Set against this, there are reasons to regard behaviour as an especially important field to study from the evolutionary viewpoint. What is both difficult and important about behaviour?

The most fundamental problem in the study of behavioural evolution,

as compared to that of morphology, is the lack of a fossil record. To say that 'behaviour leaves no fossils' is almost a cliche, but it is not entirely true. The physical structures that we find fossilised can sometimes give a clue to what the behaviour of their possessors must have been like. That *Archaeopteryx* had wings but no keel suggests that it glided rather than flew. Who could doubt that the sabre-toothed tiger was a carnivore? In some cases the fossil record has even allowed a closer examination of behavioural evolution. For example, marine invertebrates can sometimes leave traces of their burrows in the fossil record. The shape of these burrows indicates how they foraged so as to avoid covering the same ground twice. Furthermore, the particular way in which they refilled their tubes with material that they had processed gave a laminated structure of characteristic form to the burrows they left behind. These burrows are products of behaviour, but their form is just as characteristic of the species as are aspects of its morphology; tracing the changes in form through the fossil record shows how their foraging patterns changed as their behaviour evolved (Seilacher, 1986).

Gastropods can also leave fossil evidence of their behaviour, through the holes that they bore in the shells of their prey. Examination of prey can indicate whether or not a predatory attempt was made, whether it succeeded, where on the shell the hole was made and, by the size of the hole, the size of the predator. Changes in these features over hundreds of millions of years indicate how predators and their prey evolved (Kitchell, 1986).

Despite these examples, there is no doubt that our direct knowledge of how behaviour evolved is poor, for we can only infer it from structures that were left behind. In most cases we lack these and must rely on more indirect knowledge, comparing present-day species to determine how their common ancestor must have behaved and what changes have taken place since they split. The classic study by Konrad Lorenz (1941a, b) on the displays of ducks blazed a trail in this direction, but for many years there were few further developments. The field of comparative studies is a methodologically complex and difficult one, calling for rigorous statistical techniques: only recently have these been placed on a sound footing. As John Gittleman and Denise Decker make clear in Chapter 4, a start has been made in studies that make use of these new methods, and prospects for the future in this area look bright.

The early ethologists, such as Lorenz with his study of ducks, concentrated on displays; there was good reason for these to diverge between species and little reason for convergence. Species are the fundamental units within the animal kingdom. We can, rather arbitrarily, split them up into races or subspecies. We can equally group them into genera, families, and so on. But of all these groupings, the species is the most clearly defined (Mayr, 1963). One of the major reasons for this relates to the behaviour of the individuals that make them up, especially their displays. Behavioural changes accompany speciation, and may indeed lead to it. Differences in behaviour between species are a major reason why individuals of different species seldom even attempt to mate with one another. The topic of behaviour and speciation is considered by Roger Butlin and Michael Ritchie in Chapter 3.

A major difficulty in studying the evolution of behaviour lies in its very flexibility. While displays may be rather constant and fixed within a species, and have to be to ensure their recognition, this is far from true for many other aspects of behaviour. Taxonomists draw the distinction between homology, where two structures are the same because of common ancestry, and analogy, where they merely look the same through convergence. In tracing evolution, the distinction is clearly a very important one. Detailed studies of the bones of the forelimb allow us to conclude that the pectoral fin of a fish, the wing of a bird and the human arm are homologous: all are examples of the tetrapod forelimb. But the wing of a bird and that of an insect, while performing the same function, are simply analogous, as they are based on quite different structures. Application of the idea of homology to behaviour is fraught with difficulty (Atz, 1970). While the behaviour of closely related species is often very similar, the potential for convergence between non-relatives following a similar way of life is tremendous. Furthermore, such responses can occur within the lifetime of the individual. Thorpe (1963) defined learning as the process manifested by 'adaptive changes in individual behaviour as a result of experience'. The word 'adaptive' is there, just as it is in evolutionary arguments. When we see an animal performing an apparently well-adapted action, such as washing a potato before it eats it, we cannot tell whether the behaviour has a long evolutionary history or was learned for the first time a few moments before. That is, we cannot tell without further study. For the flexibility of behaviour, and the role that experience plays in shaping it, is not just a problem for studying its evolution; it is one of the prime reasons why the evolution of behaviour is of interest. Learning and flexibility are, of course, themselves the products of evolution, and are of particular relevance to ourselves. Dick Byrne takes up this theme in Chapter 8.

The flexibility of behaviour leads to a complex interaction between it and evolution. Animals can, by their behaviour, move to different places, choose new foods or modify the world around them in various ways. Any such actions will change the selective forces operating on them. So behaviour can modify the rules of the game. Indeed, it has been argued that such changes may be of fundamental importance in evolution, behavioural change coming first and genetic adaptation in line with it following on behind, the so-called 'Baldwin effect'. To take an example, an animal that eats only red berries will have a problem in a year when the harvest of these is poor. Only those flexible enough to switch to blue ones may survive. Learning to switch may be involved, and those that learn quickly do not starve, whereas the slow learners go to the wall. Over a series of disasters, selection may favour faster and faster learners until, ultimately, the survivors will not have to learn to eat blue berries at all but will do so on their first encounter. As Darwin would have put it, a habit has become an instinct. This may sound Lamarckian, but it is not; it is simply indicative of the complex and fascinating interplay between behaviour and evolution, which is very much a two-way process.

One problem one might expect in studying the evolution of behaviour, which is not really a problem at all, is whether or not the behaviour pattern under study has a genetic basis. Put more strictly, evolution can only take place where *variation* in a feature is at least partly genetic, so that selection can act upon it and changes are carried over from one generation to the next. Some of the arguments in this area have smacked of genetic determinism, and have rightly irritated those who thought the nature–nurture controversy long resolved in favour of an interactionist stance. Genes, as such, do not cause behaviour patterns, and it is not necessary for them to do so for selection to affect behaviour. To stress the point, all that is required is for variations in the behaviour pattern to have some genetic component on which selection can act. It is hard to imagine any aspect of behaviour of which this will not be true. Selection experiments may lead to change more or less quickly, depending on the extent to which individuals differ genetically in ways that affect the behaviour, but it is hardly likely that such an experiment would not work at all. All behaviour is genetically based, and affected by many genes; at least some of the variation in it is bound to stem from differences between individuals in their exact genotype. Ary Hoffmann explores the ways in which genes affect behaviour in more detail in Chapter 2.

As mentioned earlier, developments in the comparative method, allowing us to reconstruct the course of evolution from studies of extant species, have done a great deal to enhance our understanding of behavioural evolution in the past few years. Comparisons are the essence of science, and here we are talking of comparisons between species or 'taxa'. A quite different approach to comparison has had even more impact, and this is the comparison between the way animals behave and evolutionary models of how they would be predicted to behave. Since Maynard Smith and Price (1973) applied game theory to the fighting behaviour of animals, and coined the phrase 'evolutionary stable strategy' (ESS), models relying on this approach have burgeoned. Some of them are highly mathematical and not all of them have their feet as firmly in the real world as the more innumerate of us might like. But they have had a strong impact, both on our thinking about behaviour and on our understanding of it, as the frequent references to ESS theories in this book will testify. Chapter 5 is specifically devoted to models of behaviour such as these: in it Ian Harvey skilfully covers this ground without assuming that we all have a degree in mathematics.

Darwin (1871) was aware that sex was something rather special, so he devoted much of *The Descent of Man* to it. A century and more later, it is still a central topic for evolutionary discussion. Indeed, where 20 years ago behavioural conferences were dominated by talks on foraging strategies and the application of optimality theory in that area, today breeding systems and mate choice seem the focal subjects. The diversity of animal mating systems and understanding them in terms of natural selection are in themselves intriguing. So is the role of sexual selection in leading to the lavish displays, ornaments and weapons shown by many species. But the importance of sex in evolution runs deeper than these. There is no doubt that sexual reproduction, with its capacity to generate variety through recombination, is a potent force for change in evolution. What remains in some doubt is why sex evolved. Just what advantage is it to a female to throw half her eggs away by producing males rather than producing entirely asexual daughters? This and other perplexing aspects of sexual reproduction are dealt with by Tim Halliday in Chapter 6.

Darwin's theory was delightfully simple; T. H. Huxley remarked on how stupid he was not to have thought of it himself. But it has been, and remains, subject to a great many misunderstandings. There have also been changes in thinking. The neo-Darwinian will agree with much in *The Origin of Species*, but he also has ideas of his own. Perhaps the most significant difference comes from the development of kin selection theory, largely dating from the two classic papers by Hamilton (1964a, 1964b). The level at which natural selection is best viewed as acting, whether on individuals or on genes, remains a subject of debate, but few would doubt that the theory of kin selection gives the right perspective on the issues. If mate choice is the most popular seminar topic in behaviour today, kin recognition must run it a close second. Realisation of the importance of kinship in behaviour has led to many studies in this field; these are amongst the topics explored by Peter Slater in Chapter 7.

If, as kin selection theory predicts, animals should behave in such a way as to benefit their genes rather than, for example, the group or species to which they belong, we can account simply and elegantly for much of the great diversity of behaviour that we observe in the natural world. But a few interesting problems remain. Altruism is one, although there are remarkably few well-documented examples of animals behaving apparently in the interests of others rather than themselves, as the account of the subject in Chapter 7 shows. Another problem area is social structure and organisation, which Phyllis Lee discusses in Chapter 9. How do such phenomena as complex social structures emerge from groups of individuals, if each is acting purely to enhance its own inclusive fitness? Are kinship and reciprocal altruism the sole keys we need to find a solution to this question, or do animal societies have emergent properties that we must address at a different level? Herbert Spencer viewed an orchestra as an immoral organisation because it forced the individal to become subordinate to the group. Few would take such an extreme stance when it comes to human organisations, but do animal groups consist of more than just collections of individuals playing their own tunes?

The evolution of behaviour, how natural selection acts on it, and the adaptations that have been produced as a result are a particularly exciting research area today. Not only is there the superb diversity of nature to be accounted for, but there are also some clear and ingenious theories against which to test it. Finally, a battery of tools has been developed which puts us in a powerful position to carry out such tests. If Charles Darwin were alive today, surely this would be his field – and he would certainly be in his element!

2
Behaviour genetics and evolution

A. A. HOFFMANN

2.1 Introduction

Behaviour genetics is concerned with the genetic analysis of individual differences in behavioural traits. These differences may occur naturally or they may be induced by a researcher in the form of mutations.

Naturally occurring variation forms the basis of traditional behaviour genetic approaches. Much of the early literature in this area is concerned with demonstrating a genetic basis for individual differences. Studies tended to focus on simple behaviour patterns that could be easily and rapidly measured under laboratory conditions with a high degree of repeatability. Organisms were used that could be easily maintained in laboratory cultures, in particular fruit flies and mice. This enabled researchers to accumulate observations on a large number of individuals, which is a prerequisite for any genetic analysis.

Examples of these early studies include geotaxis in *Drosophila melanogaster*, in which flies were forced to make a series of up or down movements in a vertical maze (e.g. Hirsch, 1963), and 'emotionality' and 'open field behaviour' in rats and mice, which was measured as the rate of defaecation and urination when rodents were placed in a new environment (e.g. Broadhurst, 1960). By selecting over successive generations for individuals with high and low emotionality scores or negative and positive geotactic responses, lines were created which differed markedly for these activities. Such studies illustrated that genetic differences existed among individuals and that behavioural traits, just like morphological and physiological traits, could be subjected to a genetical analysis.

Naturally occurring variation also forms the basis of human behaviour genetics, which has traditionally been concerned with testing whether individual variation in human behaviour has a genetic component. In particular, much effort has been devoted to the study of intelligence as measured by IQ tests, and personality as measured by responses to questionnaires. More recently, human behaviour genetics has moved beyond simply determining whether or not behavioural variation has a genetic component. The impetus for this development has come from psychology in which behaviour genetics is seen as a tool for further understanding behavioural variation rather than as a means of studying the functioning of genes underlying variation. For example, recent genetic research on intelligence has focused on the extent to which the same genes influence different intellectual skills, and on the roles of genes and the environment in different stages of intellectual development (Hay, 1986; Plomin, DeFries & McClearn, 1990).

Induced behavioural mutations, particularly those altering the nervous system, have been used extensively to understand how nervous systems are assembled and how their components function. This work comprises the relatively new field of neurogenetics, and has been largely carried out using protozoans, nematodes and *Drosophila* (Hall, Greenspan & Harris, 1982). Neurogenetics is expanding rapidly with the availability of techniques from molecular biology to study specific genes and their products.

This chapter is concerned with aspects of behaviour genetics relevant to evolutionary processes. Because neurogenetics and human behaviour genetics have generally not addressed evolutionary questions, these areas will only be discussed briefly. Early behaviour genetic studies will not be considered in detail because the ecological significance of many traits used in these studies is not clear. For example, it is difficult to imagine how the response of fruit flies in a tube to light or gravity is relevant to their ability to locate resources in nature (Rockwell & Seiger, 1973). Similarly, it is not known how defaecation and urination rates of mice in an open field experiment relate to their exploratory behaviour in nature (Hay, 1986). The discussion will therefore focus on recent research using traits relevant to the ecology of animals.

The chapter starts with a brief look at the way genes can influence behaviour. This is followed by examples illustrating the genetic analysis of variation for ecologically relevant behaviour. Some complications and limitations evident from these examples are discussed. The genetic analysis of differences between populations is then considered to illustrate behavioural adaptation to different environmental conditions. A final section considers the genetic analysis of behavioural differences between closely related species.

2.2 How genes affect behaviour patterns

Mutations which block or alter behaviour patterns provide a useful tool in understanding how genes influence behaviour (Hall *et al.*, 1982). Two types of information can be obtained from comparisons of mutant and normal individuals. First, mutations provide a way of disrupting behaviour, similar to disruptions obtained with direct surgical intervention or drug treatment. For example, the importance of the antennae in the behavioural response of an insect can be investigated by surgically removing the antennae of a normal individual or by isolating mutant insects which lack antennae. Mutations are particularly useful when disruptions cannot be readily obtained by experimental manipulation. For example, in the analysis of the visual behaviour of flies, extensive use has been made of mutants which lack or change particular cell types in the eye (Hall *et al.*, 1982), alterations that would have been difficult to obtain by surgical intervention. However, a limitation of mutations is that they are often less specific than direct intervention techniques because a single mutation may cause disruptions in several different traits.

Second, mutations provide a way of isolating genes and gene products affecting behavioural patterns. Enzymes and other proteins involved in the development and control of the nervous system can be identified. This approach has been particularly useful in isolating genes and gene products involved in complex processes such as learning and memory (Dudai, 1988).

Most neurogenetic studies have been carried out with micro-organisms and invertebrates. This is partly because the effects of mutations at the biochemical and physiological levels are easier to determine in simpler organisms. It is hoped that an understanding of gene products influencing behavioural processes such as learning and memory in invertebrates will provide insights into these processes in higher organisms. In addition, mutations with drastic effects can readily be isolated in micro-organisms and invertebrates by 'screening' for individuals with radically altered behaviour patterns. Screening is normally carried out after individuals are exposed to a treatment inducing mutations, such as exposure to a mutagenic chemical. Mutations can also arise without mutagenic treatments, but such 'spontaneous' mutations occur at a much lower frequency than induced mutations and they are therefore not easy to detect.

The following studies illustrate some of the ways mutations have been used to understand behaviour patterns. The example of defaecation in nematodes shows how mutants are isolated and the type of information they readily provide. Research on learning in *Drosophila* indicates how mutations can lead to the isolation of gene products involved in a complex process. Finally, the shiverer mutations in mice illustrate how the effects of specific genes can be studied in mammals with the help of molecular biological techniques.

2.2.1 *Defaecation in* Caenorhabditis elegans

The nematode *Caenorhabditis elegans* has been used extensively for the genetic analysis of behaviour. The nervous system of this nematode is relatively simple and well known, and can be altered substantially without causing death. These features make *C. elegans* particularly suitable for understanding the neurobiological basis of behaviour patterns. Many mutations have been isolated which influence behaviour such as egg laying, osmotic avoidance, locomotion and touch sensitivity. More than 200 genes are now known that have specific effects on the nervous system of *C. elegans*.

An example of a screening for behavioural mutations is a study by Thomas (1990) on defaecation. As outlined in Figure 2.1, defaecation is achieved when nematodes undergo a cycle of muscle contractions, starting with the contraction of the posterior body muscles which causes the gut contents to be pushed forwards. The muscles relax and the gut contents accumulate near the anal region. The intestinal contents are then pressurized and the anus is opened to enable expulsion. Each cycle therefore consists of three phases, the posterior body contraction, the anterior body contraction, and finally the expulsion of faeces.

Thomas (1990) obtained mutations by feeding nematodes a mutagenic chemical (ethylmethanesulphonate). To isolate the mutations, adults were kept individually and produced F1 progeny (the *C. elegans* strain used was a self-fertilizing hermaphrodite). Most newly arisen mutations are recessive rather than dominant and will therefore not be expressed in a diploid organism unless they are in the homozygous form. Any recessive mutation arising from the mutagenic treatment should be in a heterozygous condition in the F1 generation and not recognizable. To identify mutations, F2 worms were produced from the F1 generation. Following the basic rules of Mendelian genetics, some recessive mutations will be expressed in the F2 generation because $\frac{1}{4}$ of the progeny should be homozygous for a newly arisen mutation. Mutants could therefore be identified by observing the defaecation behaviour of F2 individuals under a microscope.

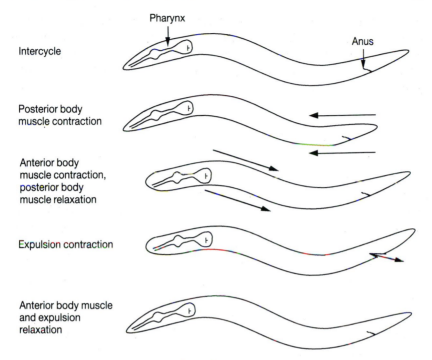

Fig. 2.1. Steps involved in the defaecation cycle of the nematode *Caenorhabditis elegans*. The top diagram shows a nematode in its typical configuration between cycles. The posterior body muscles then contract, followed by their relaxation, and contraction of the anterior muscles. This is followed by expulsion via contraction of the anal muscles, and muscle relaxation to return to the intercycle phase. (After Thomas, 1990.)

In this way, Thomas (1990) isolated 24 behavioural mutations with abnormal defaecation cycles. Some mutants were defective in the expulsion phase, whereas other mutants were defective in either of the two contraction phases. Some mutants were defective in both the anterior contraction phase and the expulsion phase, but seemed normal for the posterior contraction phase. Finally, two mutants had normal behaviour, but their defaecation cycle was much longer than normal as well as being more variable in duration.

These mutations demonstrate a number of important aspects of defaecation behaviour. First, different steps in the defaecation cycle appear to be under independent control because mutants could be isolated for each phase. However, expulsion and anterior contraction may partially be under a common control mechanism because some mutants affected both these phases. Second, locomotion and defaecation appear to be controlled by different mechanisms because mutations did not affect both behaviour patterns, despite the fact that they used the same muscles. Third, there appears to be a timer influencing defaecation because mutants with long expulsion cycles could be isolated. Further studies could be undertaken to isolate gene products associated with these mutations and to confirm the control pathways by direct manipulation.

2.2.2 *Learning in* **Drosophila**

A number of mutations affect learning in *Drosophila melanogaster*. These were first isolated by testing flies in a conditioning procedure. Flies were exposed to an odour cue and simultaneously given an electric shock. Individuals learned to avoid the olfactory cue when subsequently exposed to it in the absence of a shock. Normal flies retained this avoidance response for several hours afterwards, but learning mutants rapidly lost this ability and therefore seemed to be defective in remembering to associate the olfactory cue with the electric shock (Tully & Quinn, 1985).

The best known of the learning mutations are alleles mapping to the *dunce* locus, which is on the X chromosome of *Drosophila melanogaster*. Dunce mutants forget the avoidance response extremely rapidly (Fig. 2.2), and are also defective on other learning tasks not involving olfactory cues. This suggests that dunce mutants generally have a poor learning ability.

Dunce and other mutations have been used to investigate the biochemical basis of learning (Davis & Dauwalder, 1991). Dunce flies lack activity in a phosphodiesterase enzyme and the dunce locus appears to code for this enzyme. Phosphodiesterase breaks down the compound cyclic adenosinemonophosphate (cAMP), resulting in an elevated level of this compound in mutant flies. Levels of cAMP therefore seem to be important in learning and memory, which is consistent with other evidence associating cAMP levels with neurotransmission. Phosphodiesterase activity is high in parts of the fly brain where neuronal connections and synaptic processes are concentrated, particularly in structures known as 'mushroom bodies' where information is believed to be stored and processed.

In the case of dunce, modification of one part of the machinery controlling learning has allowed an important aspect of this process to be identified. Several other genes influencing learning have also been isolated and some have been associated with specific proteins (Dudai, 1988). It is hoped that detailed studies of such mutations will provide clues about learning in higher organisms. Genes closely related to dunce have been

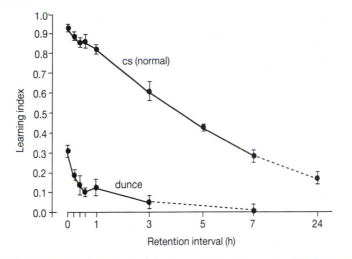

Fig. 2.2. Retention of a learning task in *Drosophila melanogaster* flies from a normal strain (cs) and from the mutant strain *dunce*. Bars represent standard errors. Dunce flies have a lower learning ability and forget the task more rapidly than normal flies. (After Tully, T. & Quinn, W. G. (1985). *Journal of Comparative Physiology*, 157, 271.)

isolated in mammals and these seem to code for enzymes related to phosphodiesterase (Davis & Dauwalder, 1991).

2.2.3 Shivering and myelin deficiency in mice

Although the isolation of behavioural mutants is more difficult in mammals than in insects and nematodes, several mutations have been identified that provide insight into the biochemical and physiological basis of some behaviour patterns. Mutations that cause shivering in mice represent one such example. Two recessive mutations have been isolated (Readhead & Hood, 1990). The first of these, known as shiverer (*shi*), causes mice to suffer seizures, tremors and a reduction in lifespan. Tremors develop when mice are 12 days old. Seizures are induced by several stimuli such as handling, light and sound. The *shi* mutation is associated with changes in myelin, the sheath laid down along selected nerve fibres which consists of lipids placed between two layers of protein. Shiverer mice almost completely lack myelin in their brains, as well as having subtle abnormalities in the myelin sheaths of their peripheral nervous systems. This sheath facilitates the transmission of signals along nerve cells.

The second mutation is known as the myelin-deficient mutation and is designated as *shi^{mld}*. This mutation is allelic to shiverer because it maps to the same locus. It is characterized by a shivering motion which becomes evident when mice are 12 days old. Mice with this mutation also have a reduced lifespan. Myelin formation in mutant mice is reduced in the central nervous system but the peripheral nervous system appears to be normal.

These mutations are associated with the protein component of myelin sheath. In the central nervous system, 60–80% of the protein component consists of a proteolipid protein (PLP) and myelin basic protein (MBP), while these proteins are less important in myelin from the peripheral nervous system. Mice homozygous for the shiverer mutation lack MBP, while mice homozygous for the myelin-deficient mutation have very low levels of this protein. Molecular characterization of the gene coding for MBP has identified a complex gene with seven 'exons' separated by 'introns'. Exons are lengths of DNA which are transcribed into messenger RNA, and these in turn can be translated into proteins, in contrast to introns which are not transcribed. By comparing DNA from normal and mutant individuals, it has been shown that the *shi* mutation has a deletion in the gene coding for the myelin basic protein, while in the *shi^{mld}* mutation the MBP gene has been duplicated and then inverted (Fig. 2.3).

A recent development in this work is the use of 'transgenic' animals,

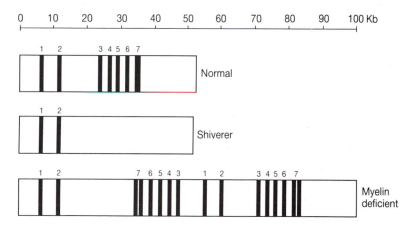

Fig. 2.3. Structure of the myelin base protein gene in normal mice (wild-type) and in shiverer and myelin-deficient mutants. Numbers show position of exons; unshaded areas show introns. The shiverer allele has a deletion of exons 3–7. The myelin-deficient allele consists of a duplication of exons 1–7 and an inversion of exons 3–7. (After Readhead & Hood, 1990.)

constructed by injecting DNA from a foreign source into eggs. These can be used to examine the effects of a single gene product on a behaviour pattern. DNA consisting of wildtype MBP genes was introduced into the eggs of shiverer mice (Readhead *et al.*, 1987), and the transgenic mice were found to have normal behaviour and normal myelin in the central nervous system. This demonstrated conclusively that the mutant behaviour was associated with the product of the shiverer gene.

2.3 The genetic analysis of discontinuous variation

Naturally occurring variation in a behavioural trait is discontinuous when animals can be assigned to two or more discrete categories, such as when one male defends a territory and another intercepts females before they move into a territory. Variation may also be continuous so that differences between individuals are quantitative rather than qualitative, such as when males differ in their tendency to defend a territory. A first step in the genetic analysis of both types of variation is to determine the extent to which differences between individuals are due to environmental factors as opposed to genetic factors. For continuous variation, this involves the techniques of quantitative genetics to be described later.

The genetic analysis of discrete differences normally proceeds by breeding animals in the same environment to see if differences between parents are also evident in progeny. Heritable factors are suggested when differences are passed on to progeny. However, it is also necessary to carry out crosses between individuals to ensure that differences are not due to maternal influences. These often arise because a trait is affected by an organism's prenatal or postnatal environment. For example, an animal's size at birth may be influenced by its mother's nutritional state, or an animal's size after birth may depend on its mother's ability to obtain food. To detect maternal influences, reciprocal crosses are usually carried out between males and females with different behaviour patterns, to see if progeny resemble their mothers more closely than their fathers. If maternal influences are indicated in organisms with parental care, cross-fostering studies can be undertaken to distinguish prenatal and postnatal factors.

Genetic analyses of discontinuous variation for traits relevant to an animal's ecology have rarely been carried out, but provide some of the best evidence that behavioural differences between individuals can be under genetic control. The following two examples illustrate this approach and the information that can be obtained from it.

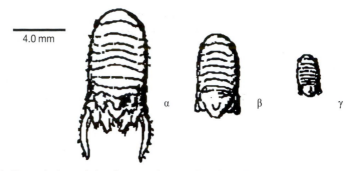

Fig. 2.4. Dorsal view of the three male morphs of the isopod *Paracerceis sculpta*. (After Shuster, S. M. (1989). *Evolution*, 43, 1685.)

2.3.1 Mating strategies in a marine isopod

The marine isopod, *Paracerceis sculpta*, breeds in sponges in the gulf of California (Shuster, 1989; Shuster & Wade, 1991). There are three discrete male morphs (Fig. 2.4) which coexist in the same population. The large alpha males are ornamented and defend harems within the central area of sponges known as the spongocoel. Smaller beta males resemble sexually mature females in behaviour and morphology and enter spongocoels by deception. Finally, the tiny gamma males invade spongocoels by stealth and place themselves inside the spongocoels and out of reach of the alpha males.

Breeding experiments indicate that these discrete differences in male morphology have a genetic basis (Shuster, personal communication). The results of crosses between males and randomly selected females suggest that a single autosomal locus is responsible for the size variation. Alpha males appear to be homozygous for a common recessive allele (*a*). Gamma and beta males usually carry the *a* allele in a heterozygous form. Gamma males have another allele (*b*) which is dominant over *a*. Beta males have a third allele which is dominant over the *a* and *b* alleles.

One of the interesting questions raised by the presence of these discrete genetic morphs is why all three persist in natural populations. Different morphs should only coexist if they have a similar mating success, because the morph with the highest mating success would otherwise displace the other morphs. It turns out that the mating success of the morphs of *P. sculpta* varies considerably depending on prevailing conditions (Shuster & Wade, 1991). Alpha males sire most offspring when there is only one female per spongocoel, but beta males sire about 60% of all offspring when there is more than one female. Gamma males become relatively more successful as the size of the harem held by alpha males increases. Shuster and Wade (1991) argue that the three male morphs have a similar mating success overall, which may help to explain why the three strategies coexist in the same population.

Alternative mating tactics have been observed in many animals other than isopods. In particular, large males often tend to court and defend females whereas small males tend to engage in 'sneak' matings in or outside a territory. A genetic basis for such tactics has been established in only a few cases other than in isopods, such as in swordtail fish (Zimmerer & Kallman, 1989). Nevertheless, many alternative tactics are likely to be associated with age or environmental variation rather than having a genetic basis (Caro & Bateson, 1986).

2.3.2 Larval foraging in Drosophila

The foraging behaviour of *Drosophila melanogaster* larvae represents another case where discrete genetic strategies appear to coexist within a population. Larvae that are placed on a surface with food move around as they forage, and individuals can be divided into two classes depending on the length of the foraging path they take. Larvae that crawl a long way and have a long path length are known as 'rovers', whereas those with a short path length are known as 'sitters' (Sokolowski, 1980).

Whether larvae act as rovers or sitters seems to depend largely on their genotype. In the laboratory, strains can be isolated where most larvae from the same strain act as either rovers or sitters. Crosses between such strains indicate that differences in foraging behaviour can largely be explained by a single gene model with the allele for the rover strategy dominant over the allele for the sitter strategy (de Belle & Sokolowski, 1987). Further work has led to the localization of a gene controlling these strategies to one particular region of the *Drosophila melanogaster* genome, the left arm of chromosome II (de Belle, Hilliker & Sokolowski, 1989).

By taking advantage of an association between foraging behaviour and pupation site preference, it has been shown that behavioural differences between the rover and sitter strategies are expressed under field conditions (Sokolowski *et al.*, 1986). In the laboratory, rover larvae tend to pupate further away from food than sitter larvae. Pupae were therefore collected from the field at different distances from rotting pears, and the progeny of eclosing flies were tested for larval foraging. Flies emerging from pupae away from the fruit produced larval offspring with long path lengths, whereas those emerging from pupae on the fruit produced larvae with

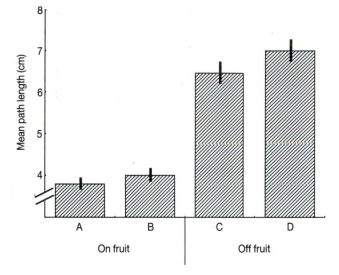

Fig. 2.5. Path lengths of larval offspring from *Drosophila melanogaster* collected at the pupal stage from different microhabitats associated with rotting pears. Pupae were collected from the top of the pears (A), near the bottom of the pears (B), in soil near the pears (C), or some distance away from them (D). (After Sokolowski, M. B. *et al.* (1986). *Animal Behaviour*, 34, 405.)

short path lengths (Fig. 2.5). This is consistent with the laboratory pupation behaviour of rover and sitter larvae.

As in the case of isopod mating strategies, these findings raise questions about the persistence of the different foraging genotypes in the same population. The fitness of rovers and sitters depends on environmental conditions. Sitters pupating on food may have higher survival than rovers when the soil moisture content is low because pupae become desiccated in dry soil. In wet conditions, however, pupae in soil may have higher survival than those on fruit because they are less likely to rot. In addition, sitters are parasitized less than rovers by wasps that use larval movement to detect their hosts. Both strategies may therefore persist in a population because different larval foraging strategies are favoured under different environmental conditions, although no estimate has yet been made of the fitness of rovers and sitters under field conditions.

Both this example and that of isopod mating strategies suggest that discontinuous genetic variation may often have a simple genetic basis. However, genes with minor effects will often modify the behavioural phenotype determined by a major gene. For example, there are genes with small effects that modify the expression of the foraging gene in *D. melanogaster* (de Belle *et al.*, 1989). Moreover, many discrete behavioural differences may represent thresholds where the expression of discrete phenotypes depends on an underlying variable that is continuously distributed. For example, animals may adopt a different behavioural strategy depending on whether or not they exceed a particular threshold value for size, and size is often influenced (in contrast with the isopod *P. sculpta*) by the effects of several genes acting in combination with environmental factors. Threshold models for behavioural traits are well known in human behaviour genetics, where the expression of psychiatric disorders such as schizophrenia is often assumed to depend on the combined effects of many genes and environmental factors (Hay, 1986). However, threshold genetic models have rarely been applied to discontinuous variation in animal populations.

2.4 The genetic analysis of quantitative traits

Naturally occurring variation in many behavioural traits is continuous rather than discrete, and continuous variation is usually assumed to be influenced by the effects of many genes, each of which is assumed to have a small effect on the quantitative trait. This makes it difficult to study such genes individually. Statistical techniques have therefore been devised to study quantitative variation and these techniques comprise the area of quantitative genetics.

Methods for analysing quantitative traits are complex and are discussed in textbooks such as that by Falconer (1989). Briefly, variation in a trait is measured by its phenotypic variance (i.e. the variance measured by the researcher). This variance is partitioned into two components, the variance due to alleles segregating in the population (V_G) and the variance due to the environment (V_E). The genetic component can be further subdivided into a component due to 'additive' genetic variance (V_A) and a component due to 'dominance' variance (V_D). The additive variance is the chief cause of the similarity between relatives. The dominance variance arises because some alleles of a gene influencing a quantitative trait are dominant over other alleles. This component of the genetic variance contributes to similarity among siblings but not to similarity between parents and their offspring. The partitioning of the phenotypic variance (V_P) in a trait is therefore written as $V_P = V_G + V_E$, where V_G is defined as $V_A + V_D$.

Several techniques are available for carrying out this partitioning procedure, and their application to behavioural traits is discussed in Ehrman and Parsons (1981) and Hay (1986). Once variances have been estimated, they are normally used to estimate the heritability of a trait. The narrow-sense heritability (h_n^2) is defined as the ratio $V_A : V_P$, while the broad-sense heritability (h_b^2) is defined as the ratio $V_G : V_P$. These heritabilities vary between 0 (phenotypic variance determined only by the environment) and 1 (phenotypic variance solely due to genetic factors).

The narrow-sense heritability provides an estimate of how similar progeny are to their parents because only V_A contributes to this similarity. It also determines what the effects of selection on a trait will be. For example, if a population is subjected to selection for increased aggression, the narrow-sense heritability provides an indication of how much of an increase occurs in each generation of selection. On the other hand, the broad-sense heritability provides a measure of the overall genetic determination of variation in a trait.

These genetic parameters have been used extensively in behaviour genetics to make inferences about the history of natural selection on behavioural traits. Selection may be 'directional' when extreme values of a trait are favoured. Alternatively, selection may be 'stabilizing' when the mean value of a trait is favoured. If a trait is under intense directional selection, alleles increasing or decreasing the value of a trait are expected to change in frequency and those alleles that are favoured are expected to become fixed. This means that V_A is expected to decline in a population, and traits with a history of directional selection are expected to have a low V_A. The narrow heritability will also decline if V_E does not change much during selection.

In addition, it has been argued that traits under directional selection should have a high V_D because favoured alleles are likely to be dominant over unfavoured alleles. This results in 'directional dominance' where the direction of past selection on a trait is reflected in its pattern of dominance. Arguments predicting directional dominance are contentious and too complex to be covered here. Basically, favoured alleles spread more rapidly in populations when they are dominant, and it has been proposed that selection pressures should lead to favoured alleles evolving dominance (see Broadhurst (1979) for an outline of this argument as applied to behaviour, and Kacser and Burns (1981) for an opposing view).

There are many genetic analyses of ecologically relevant traits, and some examples are discussed below. As in the case of discontinuous variation, these examples raise issues that extend beyond simply looking at the genetic component of the phenotypic variance. In particular, they raise questions about why a population exhibits a behavioural pattern and why genetic variance persists in a population.

2.4.1 Calling-bout length in crickets

Hedrick (1988) examined the heritability of calling-bout length in males from a wild population of crickets (*Gryllus integer*). This trait is used by females to select mates, females preferring males with longer calling bouts. Females were collected from the field and the calls of male offspring were recorded. These males comprised the parental generation and were mated randomly to females in order to produce a progeny generation whose call length was also recorded.

A comparison between offspring and their fathers provided an estimate of 0.75 for the narrow heritability of call length, while a comparison of siblings from the same family provided an estimate of 0.76 for the broad heritabiltiy. Standard errors were 0.19 and 0.24 respectively for these estimates. The heritability values are high, indicating that variation in calling-bout length is largely controlled by genes. The similarity between the narrow and broad heritabilities indicates that there is little genetic variance due to dominance. Calling-bout length should respond rapidly to selection if similar heritability values apply to cricket populations under field conditions.

These findings raise the question of why calling bouts are no longer because female preference for long calling bouts should rapidly increase the mean of this trait. This selection process would also decrease the genetic variance because alleles for long calling bouts would continually be favoured over those for short calling bouts. An unknown process acting in the opposite direction is presumably responsible for the persistence of genes for short calling bouts in the population. For example, it is possible that males with long calling bouts are more susceptible to predation or parasitism than those with short bouts because predators and parasites are attracted to cricket songs (Cade, 1975).

2.4.2 Nest building in mice

Lynch and co-workers have carried out several genetic experiments on nest building in mice. In the laboratory, mice will build nests from cotton, and an estimate of nest building can be obtained by measuring the amount of cotton they use. Lynch (1980) selected lines for high and low nest-building ability and obtained a response to selection in both directions. Narrow heritabilities were estimated as 0.15 for the high lines and 0.23 for the low lines, based on the magnitude of the selection response, suggesting a fairly low degree of genetic influence. As well

as building larger nests, mice from the high lines had bigger litters, suggesting that mice with large nests may enjoy increased reproductive success.

Lynch and Sulzbach (1984) used a different approach to examine genetic variation in nest building. Four inbred strains were crossed to themselves and to each other in all possible combinations. By crossing males and females reciprocally, 16 combinations were generated. This design is often used in quantitative genetics and is known as a 'diallel' cross. Diallels can be used to estimate the dominance and additive components of the genetic variance as described in Mather and Jinks (1971). Lynch and Sulzbach (1984) found significant additive genetic variance among the strains and obtained a narrow heritability estimate similar to that obtained from the selection experiment carried out by Lynch (1980). There was also evidence for genetic variance due to dominance. This tended to be directional for building large nests, because F1s from crosses between strains with high and low scores for nest building tended to be more similar to the high strains than to the low strains. This finding and the low narrow heritability estimate were interpreted as evidence for past directional selection for building large nests.

Nest building in mice can also be used to illustrate the problem of genotype–environment (GE) interactions often encountered in behavioural studies of quantitative traits. These interactions occur when behavioural differences between genotypes depend on the environment in which they are measured. GE interactions are best illustrated by comparisons of different inbred strains over a range of environments. For example, Lynch and Hegmann (1973) examined nest building in two inbred strains of mice at two temperatures (5°C and 26°C). The results (Fig. 2.6) indicate that strain divergence was greater at 5°C than at 26°C for both sexes, even though one strain consistently had higher scores than the other strain. A genetic difference between these strains was therefore more likely to be detected at one temperature than at the other temperature.

Behavioural traits often show large GE interactions (Ehrman & Parsons, 1981), and this can mean that alleles increasing a trait in one environment may not have the same effect on a trait in a different environment. Alleles favoured by artificial selection under one set of conditions might therefore not necessarily be selected under different conditions. The effect of genes on a quantitative trait should be evaluated in a range of environments if findings are to be extrapolated to natural conditions. Because genetic differences between strains depend on the environment, estimates of heritability and other genetic parameters can change as environmental conditions change, so that they should not be regarded as constant.

2.4.3 Aggression in sticklebacks

Sticklebacks (*Gasterosteus aculeatus*) are aggressive at the juvenile stage and aggressors will bite or bump opponents. Mature male sticklebacks engage in fights when attempting to obtain a territory in competition with other males. Once males obtain a territory they will defend it against rival males. Sexually mature females aggregate in groups and also behave aggressively towards one another. Aggression in sticklebacks therefore occurs in both sexes and at different developmental stages.

Bakker (1986) selected for high and low aggression levels in sticklebacks. He selected males and females separately at both the juvenile and adult stages. Selection for both increased and decreased aggression was successful in adult females, and narrow heritabilities for selection responses in both directions were estimated to be around 0.3 (Fig. 2.7). Decreased aggression was also obtained at the juvenile stage and for territorial males, but in these cases there was no response to selection for increased

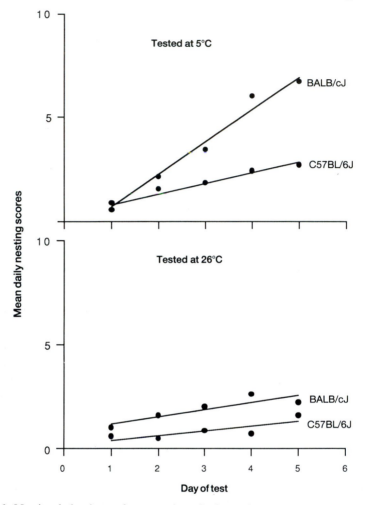

Fig. 2.6. Nesting behaviour of two strains of mice (BALB/cJ and C57BL/6J) at two temperatures. Nesting scores represent the grams of cotton mice used each day to build their nests. Individuals were tested over several days. (After Lynch, C. B. & Hegmann, J. P. (1973). *Behavior Genetics*, 3, 151.)

aggression. These experiments show that there is genetic variation for aggression in adult and juvenile sticklebacks, but that the population may be at a selection limit for high levels of aggression in juveniles and males. Bakker suggests that a limit may reflect a history of directional selection for increased aggression so that alleles increasing aggression would already have been at a high frequency in the population. In support of this argument, he found that the ratio of V_D to V_A was higher for male

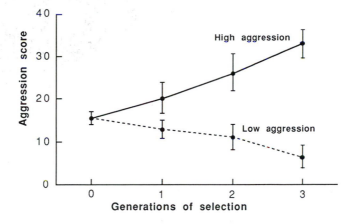

Fig. 2.7. Response to selection in female sticklebacks for increased and decreased levels of aggression. Bars indicate standard errors. The response was successful in both directions. (After Bakker, T. C. M. (1986). *Behaviour*, 98, 82.)

territorial aggression than for female aggression, although the direction of dominance was not determined.

One of the advantages of generating selected lines is that they can be used to look at other traits that may have changed as a consequence of selection. These 'correlated' selection responses can provide information about interactions between traits and about the physiological basis of a selected trait. By examining different types of aggression in the selected lines, Bakker (1986) found that selection for juvenile aggression also altered female aggression, whereas selection for male aggression did not lead to correlated responses in female and juvenile aggression. This suggests that the same genes influenced juvenile and female aggression but not male aggression. To look for a possible physiological basis of the selection response, the sizes of kidneys of males from the selected lines were compared because kidney size in reproductively mature males is determined by androgens. Males from lines selected for low territorial aggression had smaller kidneys than other lines, suggesting that aggression levels were mediated by androgen levels. In contrast, there was no difference in kidney size between lines selected for high and low juvenile aggression. However, gonadotropic hormones may have been involved in the selection response of the juvenile lines. These hormones affect sexual maturation, and males and females from the high aggression line became sexually mature earlier (as measured by courtship activity) than those from the low aggression line.

2.4.4 *Territoriality in* Drosophila

Males of *Drosophila melanogaster* defend small areas of food and so increase their access to females attracted to the food for feeding and oviposition. Lines from two populations of *D. melanogaster* were selected for increased territorial success by breeding from males holding territories over successive generations (Hoffmann, 1988). After 15 generations of selection, males from the selected lines held almost all the territories when they competed with males from unselected lines, indicating heritable variation for territorial success within both populations. Selected males were more successful than males from unselected lines because they had greater fighting ability and showed an increased tendency to establish territories. The selection response was not associated with a change in the size of the flies. Crosses between selected and control lines indicated that alleles for high levels of territorial success were dominant over those with low levels of success.

As in the stickleback study, a comparison of selected and unselected lines was undertaken to look for correlated selection responses. In this case, correlated responses were used to ask questions about why populations showed a particular level of territorial aggression. If territorial males have a higher mating success than non-territorial males, then the level of territorial aggression should increase in a population because of heritable variation for territorial success. However, there may be costs associated with a high level of territorial aggression because of a decrease in mating success in some situations or because aggressive males perform poorly for another fitness-related trait.

To look for costs associated with mating success, the number of matings obtained by males from the lines was examined in different situations (Hoffmann & Cacoyianni, 1989). Males from selected lines had an enormous mating advantage over control males in some situations. However, they did not have an advantage in other situations, such as when there were many food areas available for territorial defence. In addition, when males from the selected lines were very common, they had a lower mating success than males from control lines, suggesting that mating success is frequency dependent. Presumably, when most males have a high level of aggression, they spend much of their time and resources engaged in territorial encounters, which leaves little time for the courtship of females.

To look for other costs, selected and control lines were compared for a number of fitness components including development time, variability and longevity (Hoffmann & Cacoyianni, 1989). The only difference detected was for longevity. Males from the selected lines had a decreased lifespan compared to control males when flies were held in cages with defensible resources. This laboratory study illustrates how selected lines can be used to examine costs at the genetic level. The relatively low level of territorial aggression in natural populations of *D. melanogaster* may reflect such costs, although these should ultimately be evaluated under field conditions.

2.4.5 *Migratory tendency in sand crickets*

Behaviour genetic studies can also provide information about genetic interactions between different traits that contribute to a complex behaviour. Such traits may be independently inherited or they may be controlled together during development.

Migratory tendency in insects may be affected by a number of traits, including the propensity of an insect to fly, wing morphology and the condition of flight muscles. To see whether or not these traits are controlled by a similar developmental system, Fairbairn and Roff (1990) looked at genetic variation in the migratory tendency of the sand cricket, *Gryllus firmus*. This species consists of two types of individual, a 'micropterous' type which has reduced hindwings and cannot fly, and a 'macropterous' type which has normal functional wings. Artificial selection was used to increase the frequency of macropterous individuals in a population from 50% to 90% and to decrease it from 50% to 5%.

To test for genetic associations among the three traits, lines were scored for flight propensity and retention of flight muscles after selection. Flight propensity was measured in macropterous individuals from both lines by attaching the crickets to a cotton thread and observing their tendency to fly when they were no longer in contact with a surface. Crickets from lines with a high incidence of macropters showed an increased flight propensity, indicating that alleles increasing the frequency of macropters also increased flight activity. Macropters from each line were also dissected to see if flight muscle tissue had become disintegrated, which is a process known as histolysis. Tissue histolysis was greater in lines with a low incidence of macropters than in unselected lines, while lines with a high incidence of macropters showed decreased histolysis (Fig. 2.8). Genes increasing the tendency to produce macropters therefore also increased the tendency for crickets to retain muscles needed for flight.

These results indicate that some genes influenced variation in all three

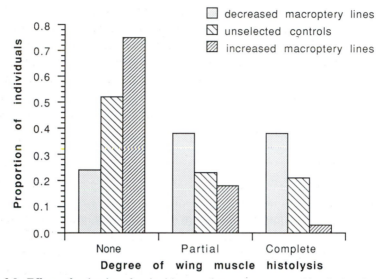

Fig. 2.8. Effect of selection for incidence of macropters on histolysis of wing muscles in sand crickets (*Grylus firmus*). Crickets came from lines selected for an increased or decreased incidence of macropters, or from unselected lines. Individuals were scored as having intact flight muscles (no histolysis), muscles reduced in size (partial histolysis), or without muscle fibres (complete histolysis). (After Fairbairn, D. J. & Roff, D. A. (1990). *Evolution*, 44, 1790.)

traits, so that different components of migratory ability are at least partly inherited together. Fairbairn and Roff (1990) suggest that genes influencing levels of juvenile growth hormone might be involved because this hormone is known to control wing reduction as well as the loss of flight muscles.

2.4.6 Concluding remarks

While the above studies provide examples of heritable variation in ecologically relevant behaviour patterns, they are limited in that they have been carried out under laboratory conditions. The presence of genotype–environment interactions for many behavioural traits, as illustrated by nest building in mice, means that estimates of heritability and other genetic parameters will be sensitive to environmental conditions. This restricts the extent to which estimates in one situation can be extrapolated to other situations.

Ideally, quantitative genetic studies should be carried out under field conditions, but this is extremely difficult when the parents and their offspring need to be identified individually. Nevertheless there are ways of obtaining information relevant to field conditions by collecting parents from the field and testing their offspring under laboratory conditions. For example, Singer and Thomas (1988) collected adults of the butterfly, *Euphydryas editha*, from a natural population and tested the oviposition response of females to two host plants. Offspring from these females were bred in a laboratory environment and tested for host preference in a glasshouse. By comparing offspring preference with parental preference, Singer and Thomas (1988) obtained a heritability estimate of 0.90, indicating that much of the variation in oviposition preference among females reared in the field had a genetic basis.

The above examples illustrate ways in which behaviour genetic studies have moved beyond simply estimating the extent to which variation is genetically determined. By changing the mean of a behavioural trait away from its normal value, selection experiments can be used to ask questions about the costs associated with different behaviour patterns. Selection experiments and other types of genetic analyses can also be used to examine correlations among traits due to interactions at the genetic level, as illustrated by migration in crickets. These correlations may help to elucidate common physiological mechanisms underlying variation in different behavioural traits. Correlations are also useful in testing for genetic interactions that may constrain the evolution of some behavioural traits. For example, the study of aggression in sticklebacks provided evidence for an interaction between sexual maturation and aggression, and it is possible that selection for increased or decreased aggression levels may be constrained because of changes in sexual maturation time.

While the heritable variation demonstrated for many behavioural traits might give the impression that behaviour patterns can evolve rapidly in populations, it is becoming clear that there are factors constraining evolutionary change. A particularly striking case is the use of host plants by phytophagous insects (Jermy, Labos & Molnar, 1990). The adult females of many insect species simply fail to lay eggs on hosts that are perfectly suitable for development of larvae. In some cases, insects lay eggs on host plants that are less suitable for larval development than plants that are not utilized (e.g. Wiklund, 1975; Karowe, 1990). The nature of such evolutionary constraints should become clearer as more information becomes available on interactions among genes and on the expression of genetic variation under natural conditions.

2.5 Identifying genes influencing quantitative traits

Several attempts have been made to isolate genes involved in quantitative behavioural traits. This may seem an impossible task if numerous 'minor' genes with small effects contribute to the genetic variance. However, it is possible that variation in quantitative traits is mostly determined by a few genes with large effects, and such 'major' genes could be studied individually.

One way of searching for major genes is to focus on genes likely to determine the behavioural trait. These may be identified by considering the underlying physiological and biochemical basis of a trait. For example, to study variation in phototaxis in *Drosophila*, a researcher might focus on genes producing proteins controlling the synthesis of eye pigments or proteins directly involved in photoreception. Genes likely to contribute to variation in a quantitative trait are referred to as 'candidate' genes.

Alternatively, a 'shotgun' approach may be used in which variants at a large number of randomly selected loci are surveyed to see if any are associated with variation in a trait. This technique depends on genetic linkage between the genes that are screened (the 'marker' genes) and genes that actually contribute to the variance in a trait. When two genes are tightly linked (i.e. close together on the same chromosome) and one allele at the behavioural locus is non-randomly associated with an allele at the marker locus, these alleles will tend to be passed on together to the next generation. By examining the association between the marker locus and the behavioural trait, evidence can be obtained for a major gene around the area of the chromosome where the marker locus is located. Ideally, marker loci should be scattered throughout an organism's genome to maximize the chances of detecting associations with a behavioural trait.

2.5.1 Shell banding patterns and resting site selection

The behavioural response of snails differing in their shell banding patterns provides an example of the candidate gene approach. Some snails of the species *Cepaea nemoralis* have numerous dark bands whereas others are effectively unbanded. These different morphs act as if they are determined by a single gene, although they are actually influenced by a few genes very closely linked to each on a chromosome (Jones, Leith & Rawlings, 1977).

Differences in banding patterns influence a snail's thermal properties by affecting the amount of sunlight absorbed by the shell. By controlling

solar absorption, the banding patterns influence the internal thermal environment of the snail. This makes the banding polymorphism a good candidate gene for behavioural variation in the resting sites selected by snails because snails partly control their body temperature by selecting exposed or shaded sites.

To test this, Jones (1982) marked snails with a paint which fades at a measurable rate when exposed to daylight. Marked snails were placed in enclosures in the field which had been planted with nettles and grass. After 60 days, the paint on banded snails had faded more than the paint on unbanded snails, indicating that a snail's banding pattern had influenced its exposure to daylight. Jones suggests that the dark-banded snails selected more exposed resting sites higher up in the vegetation than unbanded snails. This may have enabled banded morphs to avoid heat stress which is greater near the ground.

Banding patterns have also been related to behavioural responses in another snail, *Theba pisana* (Hazel & Johnson, 1990). Dark-banded morphs of *T. pisana* rested more often under the canopy of a wattle thicket where air temperatures were lower, whereas unbanded morphs tended to select exposed areas. Banded snails heated up more rapidly in sunlight than unbanded ones, and Hazel and Johnson suggest that the resting sites selected by these morphs enabled them to evade high temperature stress.

2.5.2 *Kin recognition and the major histocompatibility complex*

Kin recognition is important because it enables animals to behave altruistically towards their relatives and also enables them to avoid the deleterious effects of inbreeding which can arise when they mate with a close relative (see Chapter 7). In mice, it seems that genes comprising the major histocompatibility complex (MHC) are involved in kin recognition. MHC is associated with immunological functions and is best known for its role in tissue rejection. Because there are numerous alleles at the loci comprising the MHC complex, two unrelated animals are unlikely to have the same MHC types, which is why it is difficult to match organ recipients with donors. This makes MHC an ideal candidate gene complex for kin recognition because only closely related family members have similar MHC types.

To test for an association between kin recognition and MHC, mouse strains were constructed differing only in the region of the MHC complex, which is known as the H-2 region in mice. These strains were used to test the mating preference of males for females from the same strain or a different strain (Beauchamp, Yamazaki & Boyse, 1985). Males had a clear preference for females with a different MHC type, implicating MHC in mate recognition.

Learning tests were used to show that this behavioural recognition was associated with olfactory cues. Mice can be trained to distinguish between pairs of odours by rewarding thirsty mice with a drop of water when they respond to a particular odour. Using this assay it was possible to train mice to distinguish between the odours from the urine of mice with different MHC types.

Cross-fostering experiments have shown that early experience plays an important role in the recognition of females with different MHC types (Beauchamp *et al.*, 1988). Males from one strain (B6) were reared by females from the same strain or by females from a different strain (B6-H-2^k). These strains were identical except for their H-2 region. B6 males which were not fostered or which were fostered with B6 females showed a preference for females from the B6-H-2^k strain, while B6 males fostered to B6-H-2^k females preferred to mate with B6 females. This suggests that kin recognition is mediated by some form of imprinting.

There is some evidence that MHC is associated with kin recognition in other animals. However, the importance of the MHC complex in kin recognition under field conditions remains to be demonstrated.

2.5.3 *Associations between genetic markers and psychiatric disorders*

The search for marker loci for behavioural traits has been pursued vigorously in the study of human mental disorders. Family studies are used to associate the disorder with marker genes. Disorders often segregate in families so that some members are affected and others are not. Genetic differences at marker loci can be examined in affected and unaffected individuals from such families.

An example of this approach is a study by Egeland *et al.* (1987) of the disorder manic depression in a group from the Amish religious order in Pennsylvania. These workers identified a pedigree within this genetically isolated group where some members developed manic depression. To look for marker loci associated with the disorder, molecular biological techniques were used to examine variation at the DNA level directly. DNA was cut with enzymes known as restriction endonucleases which recognize particular sequences of bases. The resulting DNA fragments can then be separated by electrophoresis. When the DNA from two individuals differs at a particular site cut by the enzyme, fragments of a different size result, and genetic variation at this site on a chromosome can be identified. Genetic variation at a site detected by this technique is known as a 'restriction fragment length polymorphism' (RFLP).

By using a number of restriction endonucleases recognizing different sites, Egeland *et al.* (1987) were able to test for an association between manic depression and several locations on human chromosome 11. They found that the disorder was inherited along with variation at two marker loci which were closely linked to one another. A major gene influencing this disorder is therefore located on chromosome 11 near these markers.

Unfortunately, several attempts have failed to find an association between depression and the same genetic markers in pedigrees from other human populations. Instead, manic depression seems to be associated with different chromosomes in other pedigrees. This suggests that different major genes influence manic depression in various populations. Inconsistent associations with genetic markers have also been found for other psychiatric disorders such as schizophrenia (Owen & Mullan, 1990).

2.5.4 Concluding remarks

The identification of genes influencing variation in behavioural traits will undoubtedly be an important research area in behaviour genetics over the next few years. A greater understanding of genes underlying behaviour from studies of mutants should help in the identification of candidate genes. The study of DNA polymorphisms is being increasingly applied to the identification of genes underlying morphological traits and it is only a matter of time before these techniques are applied to behavioural traits in non-human populations.

Once specific genes have been identified, techniques are available for determining the proportion of the variance accounted for by each gene. Such studies should help to resolve the issue of whether variation in behavioural traits is influenced by a few major genes or numerous genes with small effects. This is an important evolutionary question because it determines the relevance of various genetic models that have been proposed to account for the persistence of genetic variance in populations (Barton & Turelli, 1989). For example, a common explanation for the persistence of genetic variance in traits under stabilizing selection is that there is a balance between mutation introducing new variation and selection removing it, but models show that mutation does not maintain much variation when only a few genes are involved.

2.6 Behavioural adaptation

If variation in a behavioural trait is heritable and different mean values of the trait are selected in two populations, this should eventually result in genetic divergence between the populations as they adapt to their environments. Adaptive divergence has been demonstrated for many morphological and physiological traits, but there is surprisingly little information on behavioural traits. The three examples below represent cases where genetic divergence between populations has been related to environmental variation.

2.6.1 Flight behaviour in milkweed bugs

The milkweed bug, *Oncopeltus fasciatus*, has a widespread distribution. Dingle (1981) describes a series of experiments on the migratory ability of bugs from different populations. In Iowa, milkweed bugs have one generation per year, and there is an influx of adult migrants from southern areas in spring and an efflux of offspring in autumn. Migration enables the bugs to evade the stressful winters in Iowa. In contrast, milkweed bugs breed throughout the year in the warm climate of Puerto Rico. There is only localized movement in Puerto Rico as bugs track the changing suitability of host plants for breeding.

To look for genetic differences between the populations in flight ability, bugs were reared under the same laboratory conditions. Bugs were tested for flight ability by attaching them to a stick and lifting them free from a surface. Flight performance was measured as the length of time the bugs flew. Dingle found that 44 of the 185 bugs (24%) that were tested from the Iowa population flew for more than 30 minutes, while only 3 of the 62 bugs (5%) from the Puerto Rico population flew for this period. Iowa bugs therefore had a higher migratory tendency. This is consistent with the long distance movement undertaken by bugs from this population and the relatively sedentary nature of the Puerto Rico population.

Flight activity has been examined in several other migratory insects and variation within populations often seems to have a large genetic component, as illustrated by the sand cricket example discussed earlier. This trait should therefore respond readily to natural selection and there are likely to be many other cases of adaptive divergence in flight activity between populations.

2.6.2 Feeding in garter snakes

Arnold (1980) carried out a genetic analysis of population differences in the feeding behaviour of garter snakes (*Thamnophis elegans*). Two Californian populations were compared, one from an inland area where the snakes feed on fish and frogs, and another from a coastal area where they feed mainly on slugs. Arnold took newly born snakes from both habitats to eliminate possible effects of experience on feeding behaviour, and gave each snake the opportunity to eat a slug. While most coastal snakes readily fed on the slug, inland snakes generally did not, although there was some overlap between the populations. This suggests genetic differences between the populations in their tendency to feed on slugs consistent with the presence/absence of slugs in their habitats.

Differences between the populations were further examined by crossing the populations reciprocally (Arnold, 1981). This allows genetic and maternal influences to be distinguished. If maternal effects are involved, then progeny will be more similar to their mother than to their father. In Arnold's study, the F1s had a similar incidence of slug eating regardless of whether their mother was from the coastal or inland population, indicating that maternal effects were not important. The F1s were more similar to the inland population than to the coastal population, indicating dominance of slug-refusal alleles over slug-acceptance alleles.

The behavioural basis of the population difference was further examined by testing the response of snakes to slug extract presented to them on a cotton swab. Snakes respond to odour by flicking out their tongue, and the extent to which snakes are stimulated by an odour can be scored by their rate of tongue flicking. Snakes from the coastal population had a higher rate than those from the inland population, suggesting that odour responses contributed to differences in the acceptance of prey items.

It is believed that coastal populations in California are more recent in origin than inland populations, and directional selection has probably favoured slug-response alleles in coastal areas. Arnold (1980) suggested that the low frequency of slug-response alleles in inland populations may be due to the low fitness of snakes with these alleles in the inland habitat. It is possible that an increased tendency to feed on slugs also increases feeding on leeches found in an inland area. Leech feeding may be detrimental to the snakes because leeches persist in the gut after they are consumed. Directional selection for slug refusal may therefore have occurred in inland populations.

2.6.3 *Territorial behaviour in a spider*

Riechert and her collaborators have investigated population differences in the territorial behaviour of the spider *Agelenopsis aperta*. One of the populations comes from a desert grassland area in New Mexico, and the other is from a desert riparian habitat in Arizona. Individuals of this species defend areas around their webs. Fights between grassland spiders are more costly than those between riparian spiders in terms of their duration, the number of acts performed in an encounter, and the risk of injury. These differences are expected because favourable sites for constructing webs are rarer in the grassland habitat than in the riparian habitat and grassland sites are therefore more valuable for territory owners. In addition, spiders defend larger areas in the grassland habitat than in the riparian habitat, which is consistent with the lower availability of food in the grassland habitat.

To see if these differences had a heritable basis, spiders from the two populations and from crosses between the populations were reared under the same controlled conditions (Riechert & Maynard-Smith, 1989). To measure contest intensity, two spiders were marked and released at the edge of a web, and interactions between them were recorded. To measure territory size, two spiders were left to build webs in separate plastic boxes. These boxes were placed at opposite ends of a large cage, and were moved closer together every day until a distance was reached where one spider moved away or was killed. This distance was used as a measure of territory size.

Spiders from the grassland population had more intense encounters and defended larger territories than those from the riparian population, indicating a heritable basis for the population differences. Moreover F1 individuals interacted more intensely and defended larger territories than spiders from either of the parental populations. Riechert and Maynard Smith used these findings to suggest that the population differences were determined by two motivational tendencies related to 'fear' and 'aggression' (Maynard Smith & Riechert, 1984; Riechert & Maynard Smith, 1989). Spiders from the grassland are assumed to have high aggression and high fear levels, while those from the riparian habitat are assumed to have low aggression and low fear levels. These two tendencies are assumed to be independently inherited, with alleles for high aggression and low fear dominant to alleles for low aggression and high fear. Crosses between the populations should therefore produce spiders with high aggression and low fear levels, accounting for the more intense encounters observed in the F1s. This model was further confirmed with contests involving spiders from the F2 generation and from backcrosses. It remains to be seen if 'fear' and 'aggression' tendencies underlie other population differences in territorial behaviour.

2.6.4 Concluding remarks

These studies illustrate that adaptive genetic divergence can be demonstrated for behavioural traits, and raise several interesting issues for future work. In particular, the question of whether genetic variation between populations and within populations is based on the same genes or different genes needs to be addressed. This will determine whether evolutionary processes influencing variation within populations are relevant to processes acting between populations. Genes that contribute to variation at the two levels therefore need to be identified.

In addition, it is not known if interactions between genes are the same at the inter- and intra-population levels. It is widely believed that genetic differences between populations are associated with 'co-adapted' gene complexes, which are combinations of genes that have been selected together to produce optimal phenotypes in a population. The genes that interact to produce these co-adapted complexes are expected to differ between populations. Evidence for co-adaptation is normally obtained by crossing populations and obtaining an F1 and F2 generation. If co-adaptation exists, it should break down at the F2 level because favourable combinations of genes can be broken up by recombination. This has been demonstrated for several traits that are closely related to fitness, such as viability and fertility, but there is little information on co-adaptation for behavioural traits (Wichman & Lynch, 1991).

2.7 Behaviour genetics of species differences

While most behaviour genetic analyses have been concerned with variation within species, a few studies have addressed behavioural differences between closely related species. This becomes feasible when species can be crossed to obtain viable F1 and F2 progeny. An advantage of using behavioural differences between species is that they are usually much larger than differences within species. Interspecific differences can often be treated like discontinuous variation, and crosses can provide information on the genetic basis of a trait.

Interspecific crosses will eventually help to determine if genes influencing behavioural variation within populations and between populations are involved at the interspecific level. This is an important evolutionary question because it has often been argued that evolutionary processes operating at the species level are different from those acting within species, involving genes with different types of effects. The identification of genes involved in species differences provides a way of helping to resolve this issue.

In both the examples given below, some information has been obtained about the genetic basis of behavioural differences between the species. Only simple crosses have been carried out because of the fertility and viability problems normally encountered in offspring from interspecific crosses.

2.7.1 Interspecific variation in burrowing behaviour in Peromyscus

The oldfield mouse, *Peromyscus polionotus*, constructs a complex burrow system in dunes and sandhills. These burrows have a nest chamber, escape tunnel and entrance tunnel, and the mice plug the entrance when they

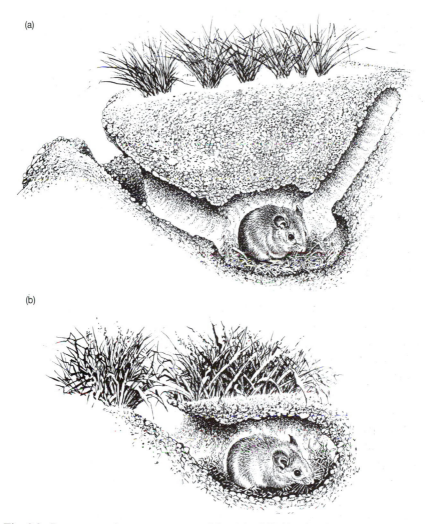

Fig. 2.9. Burrows and nests constructed by (*a*) oldfield mice (*Peromyscus polio-notus*) and (*b*) deermice (*P. maniculatus bairdii*). (After Dawson *et al.*, 1988.)

are in the nest (Fig. 2.9a). In contrast, the prairie deermouse, *Peromyscus maniculatus bairdii*, constructs a simple burrow consisting of a small cavity a few centimetres beneath a clump of grass (Fig. 2.9b).

In captivity, female deermice readily hybridize with male oldfield mice to produce viable and vigorous F1 offspring, enabling a genetic analysis of burrowing behaviour. Dawson, Lake and Schumpert (1988) tested populations of both species that had been reared for at least 20 generations in captivity. Burrowing performance was measured in a chamber filled with sandy soil. The soil was formed into a slope and a clump of grass was placed at the top of the slope to simulate natural conditions. Oldfield mice bred in captivity constructed elaborate burrows in this chamber similar to those constructed by their species in the wild. Likewise, prairie deermice constructed shallow cavities similar to their nests in the wild. Differences between species were therefore heritable because they persisted when animals were cultured under the same conditions.

Dawson *et al.* (1988) found that the F1 hybrids bred from deermouse mothers constructed burrows like those of oldfield mice, complete with escape tunnels and long entrance tunnels. This suggests that alleles controlling the type of burrow constructed by oldfield mice are dominant over those controlling the deermouse type. The offspring of F1 mice backcrossed to deermice were also tested, and these contructed a range of burrows. Some burrows were of the deermouse type and some were of the oldfield type, but most burrows had elements of both types. Individuals from this backcross generation have a range of genotypes consisting of different combinations of genes from the parental species. Because back-cross individuals have phenotypes other than those of the parents and F1 hybrids, one can conclude that burrowing in *Peromyscus* is not controlled by a single major gene, and that different genes control various aspects of this behaviour.

2.7.2 Oviposition preference in butterflies

Closely related species of butterfly often lay eggs on different host plants. Thompson (1988) examined the oviposition behaviour of two species of swallowtail butterfly, *Papilio oregonius* and *P. zelicaon*. *P. oregonius* feeds exclusively on a plant from the composite family (*Artemesia dracunculus*), whereas *P. zelicaon* feeds on two species of Umbellifer (*Lomatium grayi* and *Cymopterus terebinthinus*).

The oviposition behaviour of these species was compared after they had been reared in the same environment by presenting butterflies with the three species of host plant in a cage, along with a novel plant from another environment (*Foeniculum vulgare*). *P. oregonius* laid most of its eggs on the composite, whereas *P. zelicaon* laid most of its eggs on *L. grayi* and *C. terebinthinus* (Fig. 2.10). These heritable differences in oviposition behaviour were therefore consistent with the host plants the species used in the field.

Males and females from the two species were crossed reciprocally to

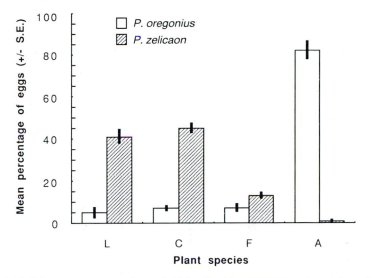

Fig. 2.10. Mean percentage of eggs laid by *Papilio oregonius* and *Papilio zelicaon* on four plant species. $L = Lomatium\ grayi$, $C = Cymopterus\ terebinthinus$, $F = Foeniculum\ vulgare$, and $A = Artemisia\ dracunculus$. (After Thompson, J. N. (1988). *Evolution*, 42, 1227.)

produce F1 hybrids and the oviposition behaviour of the hybrids was tested. Reciprocal F1s had different oviposition patterns. F1 females from the cross between female *P. oregonius* and male *P. zelicaon* laid most eggs on host plants used by *P. zelicaon*. In contrast, F1 females from the cross with a *P. zelicaon* female parent laid a similar number of eggs on the three hosts and fewer eggs on the novel plant. An attempt was made to obtain F2 butterflies but this was largely unsuccessful because of infertility problems.

This result can be explained by considering sex determination in butterflies. Male butterflies have two X chromosomes while female butterflies have one X. In the F1 generation, females obtain their sole X chromosome from the male parent rather than from the female parent. The paternal effect observed in this cross is therefore due to genes on the X chromosome that have a major influence on oviposition preference.

2.8 Conclusions

Early research in behaviour genetics demonstrated that variation in behaviour patterns which could be easily measured in the laboratory often had a genetic basis. More recently, it has been shown that genetic factors also contribute to individual differences in behaviour relevant to the ecology of animals, although the importance of genetic factors has rarely been examined for animals tested under field conditions.

Behaviour genetics is no longer concerned with simply testing for a genetic component to variation. Genetic studies of discrete behavioural variants within populations examine processes that maintain different strategies within a population. Genetic analysis of quantitative traits can lead to speculation about the history of selection on behavioural traits. The study of genetic interactions between traits indicates possible constraints on evolutionary change and provides information on the physiological and developmental basis of behavioural variation.

There is good evidence for adaptive behavioural divergence between populations and species, but it is not known if evolutionary processes acting within populations are important between populations and species. A comparison is needed of genes controlling variation within populations and those contributing to variation between populations and species. This should be facilitated by the identification of specific genes underlying behavioural variation.

3

Behaviour and speciation

R. K. BUTLIN and M. G. RITCHIE

3.1   What is a species?

The animals found at any one locality are not a continuum of forms but
fall into morphologically, behaviourally and genetically discrete clusters.
These clusters are called species. Species of sexually reproducing animals
remain distinct, primarily because they are genetically independent; genes
that are present within an individual of one species can usually only be
combined in future generations with genes from other members of the
same species. In the parlance of evolutionary genetics, the members of a
species are said to 'share a common gene pool' and the species is separated
from other species by 'barriers to gene exchange'.

These points are embodied in the most widely used species definition
introduced by Ernst Mayr (1942): 'Species are groups of actually or
potentially interbreeding populations, which are reproductively isolated
from other such groups'.

This definition includes both the property of sharing genes within a
species and the property of isolation from other species. These are difficult
criteria to apply in practice but they emphasize the underlying nature of
species and their significance to evolution. Since evolution is fundamentally
the accumulation of genetic modifications, and the boundaries of species
impose limits on the spread of genetic variation, species have a central
place in evolutionary biology.

This chapter considers the way in which behaviour patterns influence
species boundaries. The evolutionary forces which may lead to changes
in these behaviour patterns, especially those which comprise mating signal
systems, play a major role in the creation of new barriers to gene exchange
and thus in the origin of new species. This process is called speciation, or
cladogenesis (the splitting of lineages) – as opposed to anagenesis
(evolutionary change within lineages).

3.1.1 Barriers to gene exchange

In order for genes to be exchanged between two populations, it is necessary for individuals of the two populations to:

1. be present in the same locality
2. be sexually active at the same time
3. recognize each other as potential mates
4. stimulate sexual receptivity in one another
5. achieve intromission (internal fertilization) or co-ordinate gamete release (external fertilization).

It is then necessary that:

1. fertilization takes place
2. the zygote develops normally
3. the offspring survive to maturity
4. the offspring can obtain mates
5. the offspring are fertile.

An interruption at any stage in this sequence constitutes a barrier to gene exchange. Interruptions in the two groups of requirements listed above are normally referred to as 'premating' and 'postmating' barriers, respectively. Another possible classification is to separate intrinsic from extrinsic barriers. Intrinsic barriers are due to biological features of the animals and are often genetically determined, while extrinsic ones are those imposed by features of the environment, primarily geographic distance. This distinction is important because there is a barrier to gene exchange between, for example, a population of the meadow grasshopper (*Chorthippus parallelus*) in southern England and a population in northern France. The barrier is the English Channel. This is an extrinsic barrier and evolutionary biologists would usually define two such populations as conspecific. On the other hand, populations of the treehopper *Enchenopa binotata*, feeding on *Juglans* (walnut), are also unlikely to exchange genes freely with a population feeding on *Viburnum*. Here the barrier to gene exchange is intrinsic, the animals' host plant preference, and the populations could be considered separate species. Intrinsic barriers are more important in evolution. Many will show a high degree of genetic control and are unlikely to be reversed. In contrast, changes in species distributions may overcome extrinsic barriers.

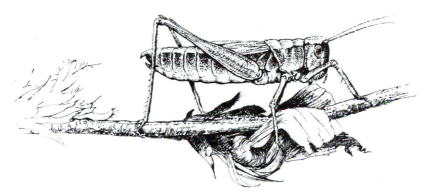

3.1.2 The role of behaviour

Behaviour is involved in the above sequence at all stages up to the release of gametes. Habitat choice and the timing of reproductive activity influence the chances of meetings between potential sexual partners; sexual signalling systems have a major role in the recognition of potential mating partners, bringing partners together and stimulating sexual receptivity, and copulatory behaviour is involved in ensuring meeting of gametes. Functional sexual behaviour of the offspring is also essential for effective gene exchange. Divergence between populations in any of these behaviour patterns can disrupt the sequence and act as a barrier to gene exchange.

Observing closely related, and therefore recently diverged, species provides examples of the critical involvement of each of these different types of behaviour in speciation. For example, in many phytophagous insect genera the species are highly host specific but, if forced together, will mate and produce viable and fertile offspring (the treehopper *Enchenopa* and the fruit fly *Rhagoletis* are good case studies that are discussed further below). Differences in mating signals between species are often much more marked than morphological differences, as in the light flashes of fireflies, the songs of grasshoppers or the visual displays of *Anolis* lizards (Fig. 3.1). As a result, mating signals are increasingly being used to identify species. The grasshopper songs shown in Figure 3.1 are from a field guide to species identification, and are much more reliable for this purpose than morphological or ecological differences.

Female grasshoppers use these characters when choosing a mate. If the differences are circumvented (e.g. by playing to a female the song of her own species while she is courted by a male of another species), hybrids are

Fig. 3.1. Examples of variability of sexual signals among closely related species. (a) Male fireflies use coded flashes of light, combined with distinctive flight patterns to elicit response signals from females. (Reproduced, with permission, from Lloyd, 1966.) (*continued*)

much more likely to be produced (Perdeck, 1958). This is not an isolated example. In the spider genus *Schizocosa*, interspecific hybrids can be produced by anaesthetizing females (Stratton & Uetz, 1986), otherwise they will only allow matings with males that produce the correct pattern of signals (in this case vibrations transmitted through the substrate). Thus behavioural characters may provide the major barrier between some

(b)

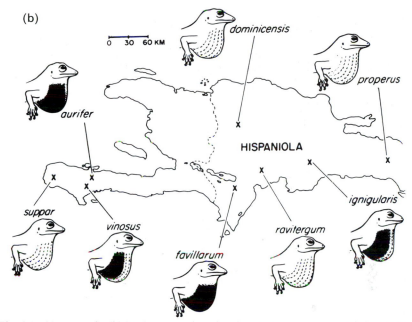

Fig. 3.1. (*Continued*) (b) Male *Anolis distichus* lizards display coloured flaps of skin, the dewlaps, to females. Subspecies vary in dewlap coloration (light stippling = white/pale yellow; heavy shading = orange. (Reproduced, with permission, from Crews & Williams, 1977.)

species pairs, apparently evolving before postmating barriers are complete, or in some cases before they are detectable at all.

3.2 Habitat choice and temporal isolation

The first requirement for gene exchange between two populations is that they should meet. Individuals in mating condition must have the opportunity to exchange mating signals. It is not enough for the populations simply to have overlapping ranges. For example, two species of blackfly may live together as larvae in one pond, but if they require different microhabitats in which to form mating swarms, they will not exchange genes.

Many species of phytophagous insects are highly host specific. In the small ermine moth, *Yponomeuta*, this seems to be controlled primarily by the oviposition behaviour of the females. Several species of this moth feed exclusively on different host bushes (*Prunus*, *Crataegus* and *Euonymus*) in the wild. However, there is very little difference im survival when the caterpillars of the different species are artificially placed on hosts they do

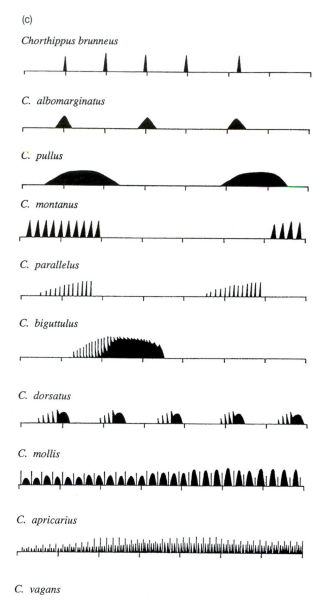

Fig. 3.1. (*Continued*) (c) Acoustic signals of male grasshoppers (genus *Chorthippus*) both attract and stimulate females. The patterns of amplitude modulation represented in these schematic oscillograms are species specific and highly distinctive, despite the morphological similarity of the species. (Adapted, with permission, from Bellman, 1988.).

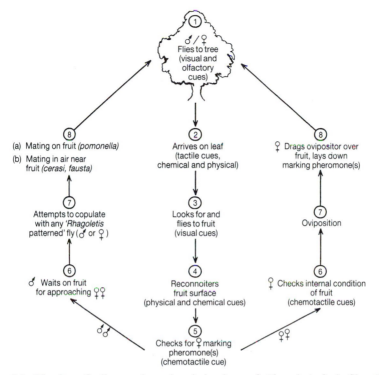

Fig. 3.2. The host-finding and mating behaviour of *Rhagoletis* fruit flies. Note that host cues are responsible for bringing mating partners together and that there is little or no exchange of signals following meeting. (Reproduced, with permission, from Bush, 1974.)

not normally use. The change in oviposition behaviour may have played a key role in speciation, but *Yponomeuta* mating signals must also have diverged since females preferentially attract conspecific males from long distances with pheromones (Menken, Herrebout & Wiebes, 1992).

In other cases there is no long-distance mate attraction; instead males and females are both attracted to the oviposition site and find each other there using short-range signals. Hence, even if different host races have similar mating signals, assortative mating (like mating with like) will be promoted by host plant preferences. A potential example of this is provided by the fruit fly genus *Rhagoletis* (Fig. 3.2), in which the females lay their eggs in developing fruit on the host plant. They are attracted to the host from long distances by its shape and colour, and then to the individual fruits by their size, shape, colour, and surface chemicals. Males search for fruit in similar ways and mating takes place on the fruit (Bush, 1974). Different species in the genus use one or a few host species and differ in their host-finding responses. Therefore sexually active males and females of different species are unlikely to encounter one another, especially as some of the different hosts (e.g. introduced apples and cherries) have different peak fruiting seasons.

How do divergent host preferences arise? One possibility is that geographically isolated populations are exposed to very different frequencies of alternative hosts and develop efficient search tactics for the most common species. If ranges subsequently expand so that the populations become sympatric, they may be able to coexist without gene exchange if the host preferences are sufficiently distinct. However, this potential history is difficult to reconcile with present insect and host distributions in well-studied groups such as *Rhagoletis*, *Yponomeuta* or *Enchenopa*.

The alternative is that some form of selection favoured either specialization of a generalist species onto two or more hosts, or a 'host shift' of a specialist species onto a new host. This selection may arise from many sources, such as adaptation to the host's biology, predator–prey co-evolution with the host or even with other predators or parasitoids associated with the host. In any case, the association between different host preferences and assortative mating that arises from mating on the host tends to promote divergence. This is because it increases the chance that different genes associated with adaptation to the host will be inherited together and occur in individuals on the appropriate host (Diehl & Bush, 1989).

Mating not only requires that individuals be in the same place but also that they are sexually active at the same time. Barriers to gene exchange may result from divergence between populations in the time of day when mating takes place, the time of year when individuals are sexually mature and active or, in cases like the periodical cicadas (see below), the year in which adults emerge. Isolation due to the time of breeding is called allochronic.

Allochronic isolation may be an incidental consequence of habitat choice. Eggs of the treehopper *Enchenopa binotata* hatch in response to the rise in sap in the host (Wood, 1993). This occurs at different times in different host species and, since the duration of development is very similar on each host, the link leads to separation in the emergence times of adults. Males have very short lifespans and so a high degree of allochronic isolation can result even with no underlying change in the genetic component of breeding period. The shift to a new host may be an adaptive response to egg parasitoids whose emergence time is independent of the host plant phenology.

In *Rhagoletis*, different species and host-associated populations within species have diverged in their seasonal emergence patterns, apparently in response to selection for correspondence with the fruiting times of their host plants (Smith, 1988). Thus allochronic isolation enhanced the barrier to gene exchange that already existed due to differences in mating location.

An extreme example of temporal isolation is provided by the periodical cicadas in which there is synchronous adult emergence every 13 or 17 years in several species. Each species has several 'broods' which emerge in different years, and gene flow between the broods must be severely restricted due to these periodicities. Indeed, hybridization may not occur on the rare occasions when different broods emerge in the same year (Martin & Simon, 1988), which suggests that some broods may in fact be distinct species. The selection which acts on mating period in this case is complex. It probably arises from predators or parasites whose shorter life cycles cannot be synchronized with the pattern of adult cicada emergence in prime numbers of years (Simon, 1979).

More obvious sources of direct selection on mating period can be inferred when signalling is restricted to times of day when predators are inactive or when environmental conditions are most suitable for the effective propagation of signals. In the guppy, *Poecilia reticulata*, courtship signals are primarily visual, and the major predators use vision to hunt. Guppies restrict courtship activity to times of day when the light conditions favour visibility to conspecifics at close range but not to predators from longer ranges and with different colour sensitivities (Endler, 1992). Since localities vary in water quality, in lighting conditions and in the predator species present, this type of selection can lead to divergence in the timing of mating activity among populations.

3.3 Mating signal systems

Differences in mating signals between species most often form the main barrier to gene exchange. This can be appreciated by placing together receptive males and females of closely related species in the laboratory. For example, *Drosophila melanogaster* and *D. simulans* are very closely related and hard to distinguish. However, female *D. simulans* will almost always reject male *D. melanogaster* despite their persistent and vigorous courtship. When male *D. simulans* and female *D. melanogaster* are placed together the result is even more striking. The flies usually ignore one another, even though they will mate immediately if placed with a partner of their own species. The block to mating between these two species must be due to mating signals, in this case probably a combination of acoustic signals produced by vibrating the wings and contact pheromones (Bennet-Clark & Ewing, 1969; Kyriacou & Hall, 1988; Welbergen, 1992).

Figure 3.1 gives just three examples of the range of different signals observed among closely related species. Many more could be given, including those using other signal channels such as the highly species-specific chemical attractants (pheromones) used by many moth species (Roelofs & Brown, 1982). However, it is very important to note that signal divergence alone is insufficient to produce isolation; there must be corresponding divergences in the sensitivity of the signal receiver. For this reason we will discuss 'mating signal systems', so including both the signal and receiver components, rather than simply signals. In the *Drosophila* example given above, it seems that *D. simulans* males and *D. melanogaster* females have non-overlapping signal *and* response systems, whereas the receiver systems of a *D. melanogaster* male are sensitive to signals from a *D. simulans* female.

In fact it is usually wrong to think of sexual communication systems consisting of only a single signal and response. In perhaps the majority of cases, the progression towards mating is more complex in that there is likely to be an interaction, with each signal from one sex eliciting another signal from the other. The courtship of the Queen butterfly (Fig. 3.3) provides an example. Often the whole sequence must be intact for mating to occur, and so the set of signals, receivers and responses must evolve together in a co-ordinated way.

Although most sexual communication systems are complex, it may not be true that all components of the system must precede every mating. Receptive females may mate after only very brief interactions with males, while prolonged courtship may be necessary in other cases. From the point of view of speciation, it is the subset of essential signals that is most interesting and much effort has gone into identifying these. As mentioned above, the well-studied grasshopper genus *Chorthippus* was the subject of a classic demonstration of the role of acoustic signals in assortative mating (Perdeck, 1958). The 'calling songs' of males in this genus vary enormously, even among morphologically very similar species (Fig. 3.1). They elicit a response song from females, and this has been used in play-back experiments to determine the characteristics of their receivers. These experiments suggest that the ratio of sound to pause in the individual syllables that make up the song is the most important feature in making

Female behaviour   **Male behaviour**

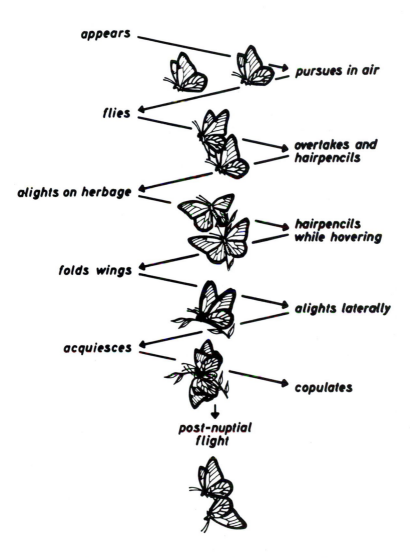

appears ——— pursues in air

flies ←———

overtakes and
hairpencils

alights on herbage ←———

hairpencils
while hovering

folds wings ←———

alights laterally

acquiesces ←———

copulates

↓
post-nuptial
flight

Fig. 3.3. Courtship of the Queen butterfly involves a series of reciprocal signals exchanged between the two sexes, eventually resulting in copulation and a post-nuptial flight. The sequence involves at least two signal channels, visual and chemical. Interruption at any point in the sequence would prevent mating. (Reproduced, with permission, from Brower, Brower & Cranston, 1965.)

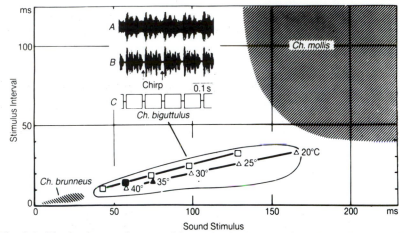

Fig. 3.4. Playback experiments with synthesized song of the form shown in the inset (C) produce female responses only in the shaded areas. This demonstrates that the ratio of chirp duration to stimulus interval is the most important feature of *Chorthippus biguttulus* song (inset B) in eliciting a female response, and that the females of three related species have non-overlapping requirements for this character. (Reproduced by permission of Kluwer Academic Publishers, from Elsner & Popov, 1978.)

the female respond. Interestingly, this characteristic is much less temperature sensitive than many other features of the song. The ranges of signal/pause ratios that elicit responses from different species are largely non-overlapping (Fig. 3.4). Although the natural songs differ in many characteristics, synthetic songs which vary in only this feature will elicit species-specific responses.

The role of signal characteristics as barriers to gene exchange has been demonstrated in a rather different way in the planthopper, *Nilaparvata lugens*. This small insect signals by vibrating the rice plant stem on which it lives in a stereotyped pattern (Fig. 3.5a). Geographically separate populations differ in the pulse repetition frequency (PRF) of the males' signals and they also mate assortatively in the laboratory. The importance of the PRF in this partial isolation was demonstrated by comparing the distribution of PRFs from the few males that did fertilize females from other populations with the distribution of PRFs from a random sample of males (Fig. 3.5b; Claridge, Den Hollander & Morgan, 1984). The signals were significantly displaced towards the PRFs typical of the female's own population.

Thus differences in mating signal systems among species make a major contribution to their genetic isolation. The process of speciation must

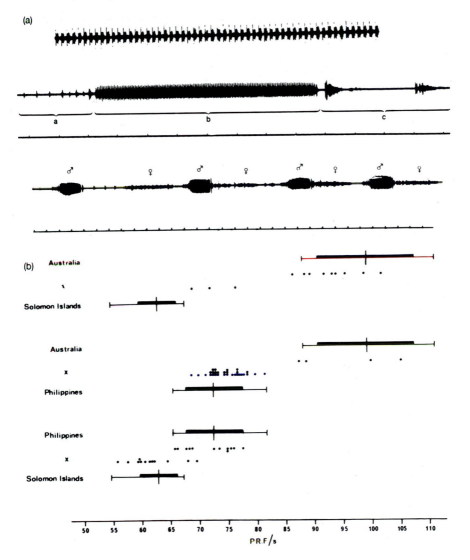

Fig. 3.5. (a) The substrate-transmitted signals of female (top) and male (middle) brown planthopper, *Nilaparvata lugens*, and a section of an alternation between male and female (bottom). Time marks are at 0.25-s intervals. (b) The pulse repetition rate in section (b) of the male song above is plotted for three populations. The bars give the mean (vertical line), ±1 standard deviation (heavy bar), and range of random samples from the population. The points are PRFs of males that mated successfully with females from the other population in each pair. In each case successful males have PRFs displaced, on average, towards the mean of the males from the female's own population. (Reproduced, with permission, from Claridge, Den Hollander & Morgan, 1984, 1985.)

frequently involve the evolution of novel signals and preferences and the problem is to understand how this occurs.

3.3.1 *The problem of new variants*

New genetic mutations influencing the mating signal system are unlikely to spread for two reasons: minority disadvantage, and co-ordination of signal and receiver. Imagine a single locus that influences the mating pattern such that matings occur more frequently between individuals of the same genotype than between individuals of different genotypes. A new allele at such a locus is initially rare and selection acts against it because individuals possessing it are unlikely to find a mate (Moore, 1981). There is thus positive frequency-dependent selection which opposes the introduction of any new variants into the population.

A locus with such an effect may seem improbable, but one example does fit this model closely. Some populations of the land snail *Partula suturalis* from the Pacific island of Moorea are polymorphic for the direction of coil of the shell, and this is controlled by two alleles at a single locus. It is physically difficult for snails with left and right coiling shells to mate with each other and so there is assortative mating. As expected, this leads to positive frequency-dependent selection and polymorphic populations are only maintained in a narrow contact zone between a 'pure left' and a 'pure right' region (Johnson, 1982).

The situation is even worse if the signal and receiver characteristics are inherited independently. This is the second difficulty: if a new signal variant appears and is not accompanied by a corresponding receiver, animals showing it will fail to find a mate and it will be removed by selection. The same argument clearly applies also to new receiver variants.

The influential theory of 'genetic coupling' was proposed as one way in which co-ordination between signal and receiver could be maintained (Alexander, 1962b). It supposes that signal production and reception share a common genetic basis. It may be, for example, that a gene controlling the production of an acoustic signal could also influence a 'template' against which the signal is compared in the receiver. One gene that affects signal structure has been studied in *Drosophila melanogaster*. Acoustic signals are produced by male wing vibrations during courtship. Two signals are produced, a 'hum' and a string of pulses (Fig. 3.6). The interval between pulses (interpulse interval, or IPI) has a mean value of about 35 ms that is characteristic of a species but cycles around this value with a period of about 55 s. This rhythm also differs among species, and females

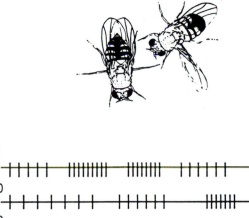

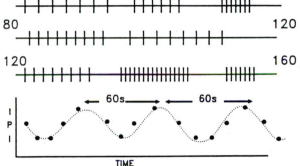

Fig. 3.6. Courting male *D. melanogaster* produces song by wing vibration. The gaps between pulses (interpulse intervals, IPIs) oscillate around a mean of 35 ms, with a period of about 1 minute, and there is evidence that normal females mate fastest when stimulated with song containing appropriate IPIs. (Figure courtesy of C. P. Kyriacou. Reproduced with permission from *Ultradian Rhythms in Life Processes*, ed. D. Lloyd & E. Rossi, Springer-Verlag, 1992.)

mate more rapidly if stimulated with song of the correct rhythm and IPI (Kyriacou & Hall, 1988). Mutations at a locus called *period* (or *per*) have been identified which have a major effect on the rhythm. The *per*[l] allele produces a 80-s period and the *per*[s] allele a 40-s period. However, recent tests by Greenacre *et al.* (1993) have shown that females from stocks differing only in their *per* alleles are not stimulated to mate more by song of their 'own' period. Thus the *per* locus does not have common effects on both signal production and preference in *Drosophila*; therefore there is no 'genetic coupling'.

The same conclusion has been drawn from the small number of interspecific crossing experiments with various animals that have been taken beyond the F1 stage. All have shown independent inheritance of

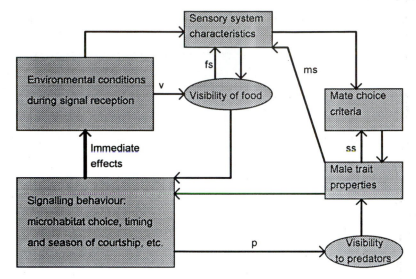

Fig. 3.7. The range of influences on sexual signals and receivers. v, Effects on visibility of prey; p, effects on visibility of predators; fs, feeding success; ms, mating success; ss, sexual selection. (Adapted, by permission of the University of Chicago Press, from Endler, J. A., (1992). *American Naturalist*, 139, S125–53.)

signals and preferences (Butlin & Ritchie, 1989). There is no well-documented example of genetic coupling in animal mating signals.

The alternative to coupling is some form of co-evolution. If signals and receivers are not alternative developmental outcomes of the same genes, then changes in one must precede changes in the other, and the co-ordination between them necessary to achieve successful fertilization must be maintained by natural selection. The forces responsible for initiating divergence in mating signal systems may act on either signal or preference alone, but selection must continually maintain the co-ordination between the two components.

3.4 Divergence in mating signal systems

Both mating signals and receivers are subject to a variety of selection pressures arising not only from the need to achieve successful matings but also from other aspects of the life of the organism (see later sections on 'Adaptation to the environment' and 'Predators, parasites and prey'). The situation is summarized in Figure 3.7 (Endler, 1992). The key point about this diagram is that receptors typically have functions in addition to the reception of mating signals that may influence their evolution, and that signals may be received by predators and parasites as well as by potential mates.

Any one of this complex set of selection pressures may cause divergence between geographically isolated populations. This divergence may be sufficient to provide a complete barrier to gene exchange between the populations. When this happens, speciation is not distinct from adaptive evolutionary change but simply a consequence of independent evolution in geographically or otherwise extrinsically isolated populations. The conditions that favour such speciation are those that favoured subdivision of distributions and directional change within populations.

Alternatively speciation may sometimes be a direct result of selection if divergence in mating signal systems results from selection *for* reproductive isolation. This important and controversial idea will be considered first.

3.4.1 *Reinforcement*

Suppose two populations of a species diverge genetically in allopatry, either because of differential selection or genetic drift, to the point where matings between populations produce offspring of lower fitness than matings within populations. Imagine also that the populations diverge slightly in mating behaviour so that they have a small tendency to mate assortatively (i.e. mating partners would most often be from the same population). What would happen if the ranges of these two populations were to expand and meet? Dobzhansky (1951) argued that individuals of population A with alleles that increased their tendency to mate with other A individuals would leave offspring of higher average fitness than those that mated more often with individuals from population B and thus had hybrid offspring. For this reason the alleles would spread within population A, and alleles of complementary effect would spread within population B, until assortative mating was complete and hybrids were no longer produced.

This model is known as the 'reinforcement' model of speciation because premating barriers to gene flow evolve to strengthen existing isolation. An analogous process can occur in sympatry where, for example, subpopulations evolve specialization on alternative resources and 'hybrids' use both resources inefficiently. Reinforcement is important because it is the only mechanism by which selection directly favours speciation, where barriers to gene exchange are elaborated by selection rather than being incidental consequences of divergence. However, there are substantial theoretical difficulties with the process and unequivocal examples have yet to be found.

There is a flaw in Dobzhansky's argument as it is expressed above. When the two divergent populations meet and mate, hybrids are produced. In the next and subsequent generations more first-generation hybrids are produced and so are backcrosses and 'F2' generations. These hybrid genotypes are at a disadvantage but are continually produced. Repeated generations of crossing result in a 'hybrid zone', an area separating the two populations in which a wide range of intermediate genotypes occurs. This may be stable over long periods as a result of a balance between selection against hybrids and flow of genes in from the parental populations. These hybrid zones are common and provide extremely valuable insights into evolutionary processes (Barton & Hewitt, 1989). From the point of view of reinforcement, they present a very different picture from that which considers only the simple effect of new signal system variants on mating within or between populations. In a hybrid zone a new variant promoting assortative mating may originally appear in an individual of type A, but recombination may break down this association so that it is sometimes found in type B individuals. Indeed, the mixing of genes from the two parental populations within the hybrid zone will probably blur the distinction between types to the extent that no individual within the zone is of pure A or B type, all individuals having a mixture of A and B genes. Reinforcement in these circumstances can only operate by building associations between genes influencing mating behaviour and those directly responsible for selection against hybrid genotypes. It must do this against the randomizing force of recombination. Population genetic models of reinforcement suggest that it can only happen where selection against hybrids is intense, where assortative mating is strong and where the genes for the two characters are closely linked (Felsenstein, 1981). If there is recombination between separate loci influencing the signal and receiver components of the mating signal system, then conditions for reinforcement will be even further restricted.

The effect of recombination is not the only difficulty. Reinforcement may be opposed by other selection pressures, including stabilizing selection resulting from the need for co-ordination between signal and receiver. The selection pressure for divergence only operates where hybrids are produced, but these areas experience a continuous inward flow of genes from populations that are not exposed to hybridization. Finally, reinforcement is self-limiting: as it proceeds it reduces the production of hybrid genotypes

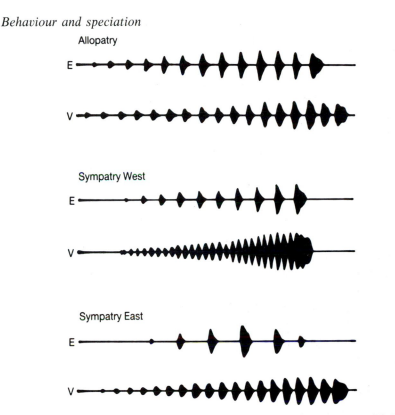

Fig. 3.8. Oscillograms of the calls of *Litoria ewingi* (E) and *L. verreauxi* (V). Note that in allopatry they are very similar, but in sympatry they differ markedly in pulse rate and number. In western sympatry, close to the allopatric range of *ewingi*, *verreauxi* is most divergent, while the reverse is true in eastern sympatry. (Reproduced, with permission, from Littlejohn (1965), *Evolution*, 19, 234–43.)

and thus the selection pressure for divergence. It is very hard to see how it could produce complete isolation.

The search for empirical evidence for reinforcement has concentrated on the prediction that it will lead to greater divergence in mating signals in sympatry than in allopatry. This pattern has been clearly demonstrated in a few cases, although it is difficult to separate it adequately from other possible sources of signal variation. In the Australian frogs *Litoria ewingi* and *L. verreauxi*, divergence in the characteristics of the mating call is closely associated with sympatry (Fig. 3.8; Littlejohn & Watson, 1985). Furthermore, the significance of this variation for female phonotaxis has been demonstrated, as has variation in preference in line with the call variation. However, this type of evidence presents a problem. It could result from any of three evolutionary pathways: reinforcement, reproduc-

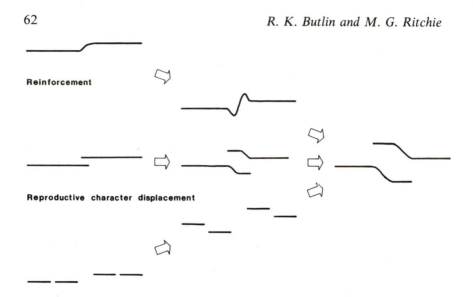

Fig. 3.9. Three possible pathways leading to the pattern of extra divergence of sexual signals in sympatry. Time progresses from left to right and the vertical scale is an arbitrary measure of a sexual signal. See text for explanations of the three processes.

tive character displacement, or divergence in allopatry. There is no easy way of telling them apart after the event (Fig. 3.9).

Reproductive character displacement is defined here as the process of divergence in mating signal systems between reproductively isolated species. This usage differs from that of some other authors who either use the term synonymously with reinforcement or simply to describe the pattern of increased divergence in sympatry (Butlin, 1987). As defined here, reproductive character displacement is distinct from reinforcement since there is no mixing of genes from the two interacting populations. It provides an alternative explanation for the situation in the *Litoria ewingi/verreauxi* overlap zone that is hard to discount on the evidence available and which does not suffer from all of the theoretical difficulties of reinforcement. It will be discussed further below.

The third option is that allopatric populations diverge in mating signal and receiver characteristics to the extent that individuals from some pairs of populations are unable to complete the mating sequence. Following range expansion, these populations can become sympatric (provided they are ecologically compatible), whereas other pairs of

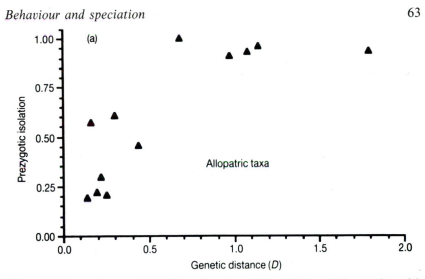

Fig. 3.10. Comparisons of isolation between pairs of *Drosophila* species with genetic distance. Prezygotic isolation is derived from laboratory studies of assortative mating and is plotted separately for species pairs whose ranges do not overlap (allopatric, A) and pairs whose ranges do overlap (sympatric, B). The measure of postzygotic isolation (C) combines F1 viability and fertility. See text for explanation. (Reproduced, with permission, from Coyne & Orr (1989), *Evolution*, 43, 362–81.)   (*continued*)

populations which show less divergence in signalling behaviour interbreed and either become indistinguishable or remain separated by hybrid zones. The resulting pattern is, once again, one of greater difference in sympatry.

An important comparative study of *Drosophila* species (Coyne & Orr, 1989) provides support for reinforcement with allowance for these altern- ative possibilities. Laboratory mating experiments and hybridizations have been attempted between many pairs of *Drosophila* species. Coyne and Orr collected data on assortative mating (premating barrier to gene exchange), on viability and fertility of hybrids (postmating barriers to gene exchange), and on genetic distances from allozyme studies as a measure of the time since the species diverged. Both pre- and postmating barriers increase with time but, interestingly, there are some species groups with genetic distances of less than 0.5 (indicating divergence over less than about 2.5 million years) that have low levels of postmating isolation but high levels of assortative mating. If the distributions of these recently diverged species are considered, there is a clear difference between groups of taxa which do or do not have overlapping geographic ranges (Fig. 3.10). This cannot be explained by reproductive character displacement since postmating isolation is clearly incomplete. Neither can it be explained by

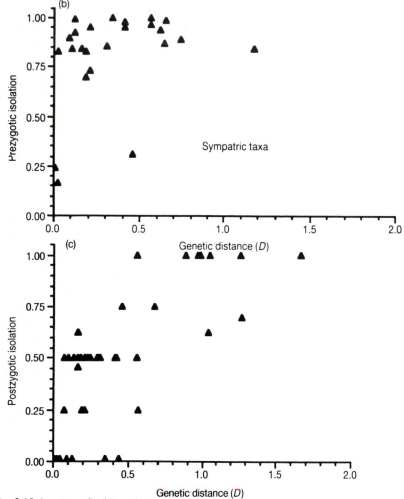

Fig. 3.10 (*continued*) (b) and (c).

the 'divergence in allopatry' hypothesis since this predicts that premating isolation in sympatric pairs should be a subset of the levels of
isolation observed among allopatric pairs and this is not the case (Fig.
3.10). Nevertheless, this study does not provide conclusive evidence of
reinforcement.

Reinforcement is most likely where there is strong selection against
hybrids, whereas in these recently diverged species pairs postmating
reproductive isolation was weak. Coyne and Orr suggest that the
laboratory estimates of postmating isolation may be too low because they

exclude ecological factors and mating success. Alternatively, the data may be explained if selection against hybrids in sympatry triggers sexual selection through female choice (see Chapter 6), incidentally creating premating isolation.

Perhaps the best opportunity to demonstrate reinforcement directly is to study hybrid zones. In many of these there is no doubt that selection operates against hybrids but that there is also gene exchange. In these circumstances, the reinforcement model predicts greater divergence in signal system components close to the zone. As yet there is no fully convincing example of such a pattern. In the meadow grasshopper, *Chorthippus parallelus*, there is a hybrid zone in the Pyrenees. There is selection against cross-mating because male hybrids have reduced fertility, but there is clear evidence of gene flow as there is a smooth change in morphological characters. Features of the male songs that differ between the two hybridizing subspecies change in a simple clinal pattern across the hybrid zone (Fig. 3.11) without any increased divergence that might be attributed to reinforcement. There is a more complex pattern in female mate preferences (Fig. 3.11) but statistically a simple cline provides as good a fit to the data as a more complex cline involving increased divergence in the zone (Butlin & Ritchie, 1991). Ideally, a demonstration of reinforcement needs to include not only an appropriate pattern of change in the mating signal system but also a reduction in the production of unfit hybrid genotypes resulting from the altered pattern of mating. Ritchie, Butlin & Hewitt (1992) failed to find evidence that assortative mating within the hybrid zone in *Chorthippus parallelus* produced fitter offspring. Until such results are found for cases where deleterious gene flow is also demonstrable, reinforcement must be considered an important, but so far unsubstantiated, idea.

3.4.2 Reproductive character displacement

The characteristics of mating signal systems may be influenced by other species signalling in the same environment. Two types of selection pressure may be involved: wastage of reproductive effort, and signal interference.

Where closely related species signal in the same environment and their signal characteristics are similar, there is a risk of cross-matings that produce no viable or fertile offspring. Clearly this will impose a strong selection pressure for signal divergence. However, where the risk of cross-matings is high, this may prevent co-existence and, further,

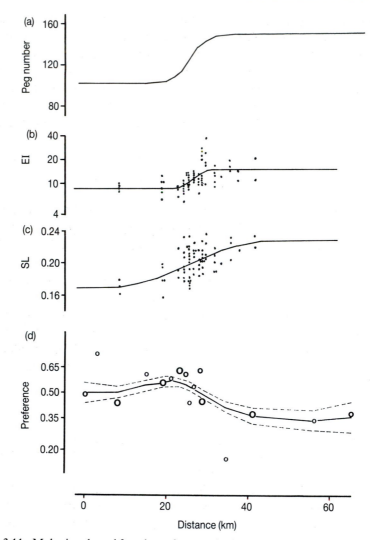

Fig. 3.11. Male signals and female preferences in the *Chorthippus parallelus* hybrid zone. (a) The cline in stridulatory peg number – a morphological character used as a reference. (b and c) Clines in two features of the male 'calling song', the chirp interval (EI) and the syllable length (SL). Both change in a simple clinal pattern, although with very different widths. (d) The pattern of female preferences, expressed as the proportion of females choosing to mate with a *C. p. parallelus* male (the northern subspecies, to the left in these figures). Note some evidence of stronger preference close to the zone.

as the risk decreases due to assortative mating, the selection pressure favouring further assortment will also lessen and the process is unlikely to go to completion, as in the case of reinforcement. By contrast, signal interference may occur between more distantly related taxa and does not necessarily imply great similarity between the signals, only that they use similar parts of a signalling channel (e.g. overlapping frequencies in acoustic or visual signals). This can produce complete separation of signal characteristics.

Reproductive character displacement may explain the example of *Litoria ewingi* and *L. verreauxi* discussed above (see Fig. 3.8). It is also one possible explanation for the enormous range of songs amongst Hawaiian crickets of the genus *Laupala* (Fig. 3.12; Otte, 1989). A striking feature of these songs is that they overlap less within localities for the important parameter of pulse rate than would be expected from random assemblages of the species in the genus.

Reinforcement involves selection for divergence in the mating signal system because of the disadvantages of gene exchange, whereas reproductive character displacement involves selection for divergence *in the absence of* gene exchange. The remaining mechanisms to be discussed do not involve selection for divergence as such. They operate within populations and may, incidentally, lead to changes that cause populations not to interbreed when they subsequently meet.

3.4.3 Adaptation to the environment

Mating signals are both attenuated and degraded by the physical properties of the environment. Attenuation is the loss of signal intensity with distance, while degradation is the corruption of information characteristics of the signal. An acoustic signal, for example, is attenuated by distance in a way that depends on the signal frequency. Low frequencies attenuate less with distance than high frequencies, especially in a spatially complex environment like a forest (Morton, 1975). A potentially informative character like syllable repetition rate is degraded by reflection or echoes off structures, and again this is more of a problem in a spatially complex habitat (for example, an acoustic signal propagating through leaves).

These conditions suggest that a signal that functions efficiently in one environment may not be ideal in another environment. Therefore populations of a species inhabiting different habitats may diverge in important properties of their mating signals. There is some evidence that this does

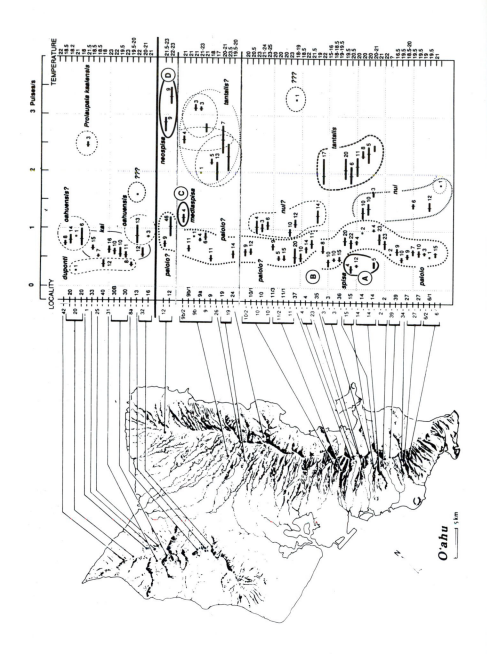

happen. The frog species *Ranidella ripara* and *R. signifera* occupy different habitats and there is only a small overlap in their distributions. *R. signifera* occurs in heavily vegetated creeks and has a loud call with slow pulse repetition rate; by contrast, *R. riparia* occupies open rocky streams and has a quiet call with fast pulse repetition rate. The *R. riparia* call is degraded more rapidly with distance in the creek habitat than is the *R. signifera* call (Odendaal, Bull & Telford, 1986). Variation in the same call *within* a species of frog, *Acris crepitans*, has also been attributed to selection for efficient propagation (Ryan & Wilczynski, 1991).

Comparable habitat associations may occur with other signalling channels. Endler (1992) has shown how the spectrum of light reaching a forest stream varies dramatically with the size of overhead gaps in the canopy as well as with time of day and weather conditions. The conspicuousness of a visual pattern is dependent on the spectrum of the incident light and so different colours and colour combinations may be favoured as mating signals in different parts of a stream. Thus Endler has suggested that the purplish light in the morning when male guppies, *Poecilia reticulata*, normally court females has favoured blue and orange spots rather than yellow spots in the males' sexual coloration. However, in some streams the water is coloured by tannins and this may favour orange spots over blue spots. It is likely that the female mating preferences will have diverged in concert with the spotting patterns in these circumstances, and there is some evidence for this in the guppies.

3.4.4 Predators, parasites and prey

Conspicuousness may be advantageous in mate attraction but it may also have costs. In the case of the guppies, predatory fish also use male coloration and males have more, larger spots in areas of low predation than in areas of high predation.

Predators and parasites are also known to home in on non-visual mating signals. A parasitic tachinid fly, *Euphasiopteryx ochracea*, uses the acoustic signal of the cricket, *Gryllus integer*, to find its host (Cade, 1979),

Fig. 3.12. Song characteristics of crickets (*Laupala*) from the eastern part of O'ahu, Hawaiian Islands. Collecting localities are shown on the left, and the distribution of pulse rates of songs from each locality on the right. Note the lack of overlap in this important signal character among species and the apparent changes where species become sympatric, e.g. *palolo* where its range overlaps *spisa*, or *tantalis* where its range overlaps *nui*. (Reproduced by permission of Sinaeur Associates, from Otte and Endler (1989), *Speciation and its Consequences*.)

and frog-eating bats home in on the calls of the male tungara frog, *Physalaemus pustulosus* (Ryan, 1991). In the fireflies this is taken even further, with the predatory species in the genus *Photuris* mimicking the flash signals of female *Photinus* species to attract males to their deaths (Lloyd, 1984; Fig. 3.1).

Clearly, all of these examples involve direct selection on signal characteristics and, if the suite of predators or parasites varies among localities, there is the potential for signal divergence to result.

The process may also operate in reverse. Female receptors must be tuned not only to receive male signals but also to detect potential predators or prey. The resultant sensitivities of the receptors may give rise to selection on male behaviour. For example, consider the courtship behaviour of the water mite *Neumania papillator*. Courting males tremble towards females in a manner which seems to mimic copepod prey, inducing feeding attacks from the female during which males deposit a spermatophore (Fig. 3.13; Proctor, 1991).

In these cases, adaptive changes in signal sensitivities, for reasons independent of sexual signalling, are followed by an evolutionary response in the signals. This type of signal evolution has recently been called 'sensory exploitation' (Ryan *et al.*, 1990). This theory predicts that the evolution of preferences in the females of a species may precede the evolution of corresponding signal characters in males. This prediction has been tested in comparative studies of the sexual behaviour of closely

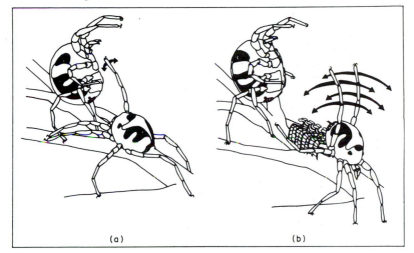

Fig. 3.13. Courtship in *Neumania papillator*. Males move their forelegs in ways which resemble prey, inducing attention from the larger female. Spermatophores deposited by the male can be seen in (b). (Reproduced, with permission, from Proctor (1991), *Animal Behaviour*, 42, 589–98.)

related species of frog (*Physalaemus*), and is supported by the earlier origin of a preference for a particular call component in the evolutionary history of the group than of the call character itself (Fig. 3.14). In this case it is more difficult to understand the origin of the preference than in the water mites. Perhaps sensory biases which arise as chance side-effects of the neurophysiology of female perception can exert significant selective forces on signals.

3.4.5 Sexual selection

Many of the processes described above could properly be considered as modes of sexual selection since they operate through differential mating success. Other forms of sexual selection may also influence signal characters in ways that could, potentially, create barriers to gene exchange. These processes of inter- and intrasexual selection are considered in Chapter 6. In the context of speciation, competition of the latter sort is more important since reproductive isolation must involve incompatibilities between the sexes. In particular, Fisher's theory of sexual selection has received much attention due to its potential to produce rapid evolutionary changes in arbitrary directions (Fisher, 1930; Partridge & Halliday, 1984).

Fisher's process requires a combination of genetic drift, to initiate

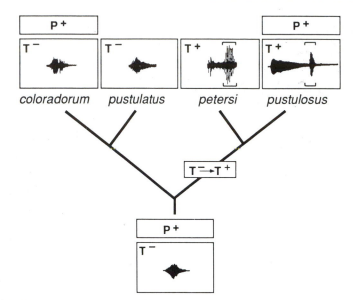

Fig. 3.14. Male song and female preferences in *Physalaemus* frogs. Only the closely related *P. pustulosus* and *P. petersi* add 'chucks' to their calls (in square brackets), and so this probably evolved in their immediate ancestor. Females of both *P. pustolusus* and *P. coloradorum* prefer calls with chucks, from which it is inferred that the common ancestor of all four species (bottom) had the preference. The preference for the chuck therefore evolved before the chuck itself. (T⁻, no male chuck; T⁺, male chuck; P⁺, female preference for chuck.) The oscillogram at the bottom is typical of other calls in the genus. (From Kirkpatrick & Ryan (1991), *Nature*, 350, 33–8. Reproduced with permission from Macmillan Magazines Ltd.)

movement away from a specific equilibrium, and a large enough population to maintain a strong association between genes influencing preference and those influencing the signal. If these conditions are met, then it is likely that allopatric populations will evolve in different directions since the process has no inherent directionality. Thus Fisherian selection is an attractive explanation for the elaborate and often bizarre secondary sexual traits that characterise some groups and that tend to be species specific (Carson, 1982; Lande, 1982). The effectiveness of these characters in preventing interspecific mating is largly unknown and may not be as great as appearances suggest. For example, hybrids are found where the ranges of the pheasant species shown in Figure 3.15 overlap in China.

The likelihood of appropriate conditions for Fisherian selection and its importance relative to other forms of sexual selection are matters for

Fig. 3.15. Males of (a) Lady Amherst's Pheasant (*Chrysolophus amherstiae*) and (b) the Golden Pheasant (*Chrysolophus pictus*). (Reproduced, with permission, from Futuyma (1986), *Evolutionary Biology*, 2nd edn. Sunderland, Mass.: Sinauer Associates.)

debate. However, instability of signals may not be dependent only on this specific mechanism. West-Eberhard (1983) has argued that competition among members of the same species in a wide range of contexts can lead to rapid, indeterminate evolution. This is because fitness depends on the interactions among individuals rather than their relationship to the environment, so that there is no defined 'optimum'. Competition for mates, either by male–male aggression or through female preferences, is a special case of what West-Eberhard calls 'social selection', and the Fisherian process is one class of competition for mates. This view suggests that divergence of isolated populations in socially selected characters will be the rule rather than the exception. In many cases it will incidentally result in novel barriers to gene exchange.

3.4.6 Population processes

In small populations, weakly selected characters will diverge by genetic drift alone. Most attention has been given to this possibility in the context of extreme bottle-necks in population size, especially so-called 'founder effects', and much of the impetus has come from studies of the many species of *Drosophila* found on Hawaii.

Out of a worldwide total of about 2000 species of *Drosophila*, some 800 are believed to be endemic to the Hawaiian islands. They are descended from one, or possibly two, original species of colonist. The oldest island that now supports *Drosophila* populations is 5.1 million years old and the youngest island, Hawaii itself, is only about 0.5 million years old. Nevertheless, Hawaii has 26 endemic species in the well-studied 'picture-winged' group alone. Thus speciation has been both frequent and rapid in these islands.

The Hawaiian *Drosophila*, especially the 'picture-winged' group, exhibit an enormous diversity of mating behaviour and of secondary sexual characteristics (Kaneshiro & Boake, 1987). A particularly striking example is the wide head of male *D. heteroneura* relative to its sister species *D. silvestris* (Fig. 3.16). Since both species occur only on Hawaii, this extreme character must have evolved in the last 0.5 million years, and probably in a much shorter time. What is the source of this evolutionary innovation?

The answer may lie in the unusual population structure of the flies that results from the volcanic nature of the islands. Each of the islands in the group has arisen in sequence as the ocean floor has moved over a 'hot spot' in the earth's crust. New islands have been colonized from older islands by rare founder events, possibly involving just single mated females that have been blown there by chance. Similar processes occur within islands. The forest is frequently destroyed by lava flows except for small pockets called 'kipukas'. Isolated populations of flies may persist in the kipukas and later expand to colonize the regenerating vegetation on the lava flow. Thus the populations of flies are characterized by extreme contractions in numbers followed by expansions into new territory.

Carson and Templeton (1984), in particular, have proposed that these 'founder-flush' cycles might promote speciation. They argue that the founder event extracts a small, and potentially atypical, random sample from the original population. The particular set of genes, and gene combinations, present in this sample may bias the evolution of the

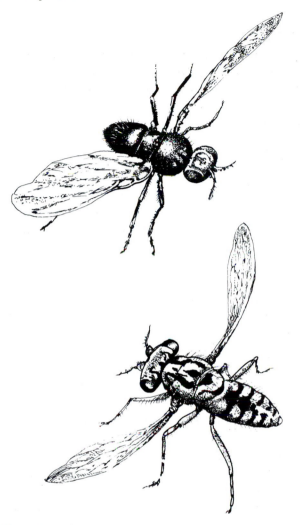

Fig. 3.16. Hawaiian *Drosophila. Top. D. silvestris*, which has a normal head shape. *Bottom D. heteroneura*, which has an elongated head.

population as it expands during the 'flush' phase of the cycle. Since some components of selection may be less intense in an expanding population, novel genetic combinations may arise and spread. As the population stabilizes, it may reach an equilibrium that is distinct from that of the original population, and the two populations may be reproductively isolated. This mechanism is not specific to the mating signal system, but sexually selected traits may be particularly prone to disruption if, for example, their evolution depends on genetic correlations between preferences and signals or between signals and viability genes, or if these systems are characterized by multiple equilibria.

Kaneshiro (1989) has made specific predictions about the evolution of mating behaviour in founder-flush cycles. He argues that founder events are associated with low population density and an absence of related species. Both of these factors will favour females that mate relatively indiscriminately, since there is little risk of cross-matings, and rejecting a male may be very costly if the chances of encountering other males are small. In the newly founded population, rapid, less discriminate mating will thus spread among females. This will reduce the selection for 'courtship persistence' in males and may also relax selection on signal characters, allowing an increase in variability. Eventually, as population densities increase and other species colonize the new habitat, selection for discriminating females will increase and this will impose selection on male signals. However, the population may well not return to the original pattern of signals and preferences.

This idea is supported by the widespread observation of asymmetric behavioural isolation, especially amongst Hawaiian *Drosophila* species. *D. silvestris* and *D. heteroneura* are believed to be recently derived via a founder event from related species on the next oldest island. Their nearest relatives are *D. differens* and *D. plantibia*. In laboratory mate-choice tests involving males of one species and two types of female (the same and another species), a clear pattern emerged in this group. Females of the derived species, *heteroneura* and *silvestris*, mate readily with their own males and with males of the 'ancestral' species, *differens* and *plantibia*. But females of the ancestral species discriminate strongly against males of the derived species, preferring their own males. This can be explained if the derived females are less discriminating and the derived males have less intense or persistent courtship, or more variable signals.

Founder-flush cycles have also been studied experimentally. Powell (1978) induced repeated founder events followed by population expansion in laboratory lines of *Drosophila pseudoobscura*. He found that after 5 years and four founder-flush cycles, there was detectable asymmetrical premating isolation between some lines and the base population. There was no significant postmating isolation, and no isolation of either type between the base population and control lines that lacked bottle-necks or inbred lines. The mechanisms underlying these evolutionary changes are not yet fully clear, but the results do provide support for a role of population processes in behavioural divergence.

3.5 Recognition, isolation and the nature of species

This chapter has followed Mayr's definition of species based on two properties: the sharing of a common gene pool within species, and barriers to gene flow between them. For most practical purposes, such a definition presents no difficulties, but the process of speciation leads to situations in which the boundaries of species are unclear. Because evolution is a continuous process, there will be cases in which pairs of populations are neither completely isolated nor freely exchanging genes. Hybrid zones, such as the contact between subspecies of the grasshopper *Chorthippus parallelus* described above, illustrate this point clearly. The two subspecies do interbreed and some of their hybrid offspring are viable and fertile. Thus they can exchange genes, and yet gene flow is clearly impeded since the characters of one subspecies only spread a few kilometres into the other subspecies' range.

Geographically isolated populations present a further problem. They do not exchange genes because of extrinsic barriers and this makes it difficult, and of questionable evolutionary significance, to determine whether they *could* exchange genes if the extrinsic barriers were removed. In practice, the specific status of such populations is usually judged by the degree of morphological similarity between them. While this may often correlate with the degree of genetic similarity, the degree of behavioural variation may often be more relevant.

Two things are clear from the analysis of behavioural involvement in speciation: that differences in behavioural characters, especially mating signal systems, are important as barriers to gene exchange, and that these barriers are most likely to be incidental consequences of evolution within extrinsically isolated populations. Only reinforcement involves selection for reproductive isolation and this mechanism has not yet been clearly demonstrated in any case study. For this reason, the common use of the phrase 'reproductive isolating mechanism' for a barrier to gene exchange has been avoided here. This phrase implies a device selected for the function of reproductive isolation whereas, as we have seen, this is usually an incidental effect of mating behaviours (Paterson, 1982).

Considerations such as these have led some workers to search for species concepts that match more closely the evolution of reproductive isolation. The recognition concept proposed by Paterson (1985) is of particularly interest because of the central role it gives to behaviour. Paterson argues that for sexually reproducing organisms to share a gene pool, the key requirement is for them to have a 'fertilization system' in common. The fertilization system is the set of characters required to ensure the meeting and fusion of gametes. It must involve a process of 'recognition' consisting of the exchange of signals between mating partners and also between gametes. Paterson calls the signals, preferences and responses involved in the exchange between potential mates the specific-mate recognition systems (SMRS). Within populations, selection will favour the evolution of the SMRS to achieve efficient fertilization, given the conditions of the environment, and the SMRS will be subject to strong stabilizing selection because variants will disrupt the signal–response chain. Since organisms tend to stay within their favoured habitat, the SMRS is unlikely to vary among populations. Given that recognition is essential for fertilization and fertilization defines the common gene pool the SMRS defines the limits of species.

The principal advantage of the recognition concept is that it shifts emphasis away from reproductive isolation between species towards gene exchange within species. Looked at from some perspectives, these may appear to be two sides of the same coin but, when considering speciation, the selection pressures acting on fertilization success within populations are probably much more important than those related to interactions between species. On the other hand, the recognition concept may go too far in denying that heterospecific interactions may exert any significant selection on mating signals or that processess occurring after fertilization may also be important. The emphasis placed on the stability of the SMRS contrasts sharply with the expectation of rapid, undirected change predicted by some models of sexual selection. Examples of the arguments on both sides of this issue are the papers by Coyne, Orr and Futuyma (1988) and Masters and Spencer (1989).

3.6 Conclusions

Speciation is one of the most important processes in the evolution of sexual organisms. Observations of the barriers to gene exchange between closely related species suggest that behavioural changes are frequently the key events initiating speciation in animals. Yet it is likely that selection only very rarely, if ever, acts directly to create these barriers. The behaviour patterns contributing to speciation appear to evolve primarily as a result of processes occurring within species such as sexual selection or host plant choice. The resulting barriers to gene exchange are incidental consequences, rather than functions, of the characters involved. Thus, contrary to common usage, mating signal systems rarely function as 'reproductive isolating mechanisms'. This does not mean that behaviour is unimportant in speciation. On the contrary, the evolution of behaviour is central to understanding processes of speciation in animals, and also what evolutionary biologists mean by the term 'species'.

4

The phylogeny of behaviour

J. L. GITTLEMAN and D. M. DECKER

4.1 Introduction

A primary strength of behavioural study is direct observation. Virtually every major problem or approach in animal behaviour, whether it concerns development, communication, social behaviour or aggression, is open to direct data collection: except one, phylogeny. Phylogeny, defined as the evolutionary history of an organism or lineage, is explicitly historical and thus closed to observational study. Unfortunately, our only direct window into history, the fossil record, '. . . is practically useless as a means of understanding the evolution of behaviour' (Tinbergen, 1951, p. 186). Animal morphology, and occasionally the physical structures animals create (e.g. nests or burrows), are more often fossilized than are examples of animal behaviour. Sometimes fossils do provide indirect clues about behaviour. For example, the wormlike tracks left by ancient invertebrates that lived on the ocean floor in the Paleozoic indicate certain types of locomotion and foraging (Seilacher, 1967). But these are only broad kinds of information that do not reveal detail or individual variation, and cannot provide us with any real historical record of behavioural evolution.

Phylogenetic questions about behaviour are not only hampered by hidden data. A fundamental assumption behind phylogenetic approaches is often questioned. Behaviour patterns reflect individual variability, plasticity, and responses to environmental change. Therefore, if behaviour changes with environment, then lineage-specific ties with behaviour (i.e. homology) should be rather weak (Atz, 1970). Consider, for example, different forms of parental care in fishes. In some species only the male protects the eggs, whereas in other species both the male and female display parental care. Such differences in parental care are generally associated with differences in predation pressure or variation in life-

history patterns such as egg size or number of eggs. Intuitively, then, it was thought that species differences in parental care would not relate to phylogeny (Breder & Rosen, 1966; Eibl-Eibesfeldt, 1975). We now realize that parental care in fishes (Gittleman, 1981), as with many other quite malleable patterns of behaviour, such as mating systems, is *constrained by phylogeny* and shows some *phylogenetic inertia* (Wilson, 1975). The causal reasons why closely related species typically behave in the same way are that they: (1) tend to evolve into similar niches, (2) have similar genetic variance for selection to act upon, and/or (3) have similar phenotypes and so tend to respond to environmental change in like fashion (Harvey & Pagel, 1991; see also Futuyma, 1992).

Phylogenetic studies of behaviour involve two kinds of questions. First, we may be interested in how an aspect of behaviour evolved. For example, we can ask whether mating systems in mammals evolved in a certain sequence, such as from a more common polygynous type in ancestral groups to monogamy and then perhaps, more recently, to polyandry. A similar problem may be referred to as 'behavioural systematics'. It is very much in the tradition of Lorenz (1939, 1941a, b) who, with his strong background in comparative anatomy, asserted that behaviour should not be neglected as a potentially valuable taxonomic character. That is, again assuming that behaviour follows a phylogenetic path, perhaps the mating system or pattern of parental care can be used in the same way as DNA sequences or morphology for taxonomic purposes. This is currently a very active area from both behavioural and systematic perspectives (Brooks & McLennan, 1991).

The second type of question considers the relative degree to which a given behaviour pattern is influenced by phylogeny or by the animal's environment. In the recent debate about 'adaptation' (see Gould & Vrba, 1982), this is parallel to arguments between 'historical genesis' and 'current utility'. In the parental care example, we might want to know if the frequent occurrence of male parental care is related more to the ancestry of different species as opposed to some common ecological condition such as predation or food resources. Ideally, information from the fossil record would again be instructive. However, as mentioned above, the fossil record will not be useful for this type of detailed, causal question. Our most accessible and robust technique is through the comparative method. The comparative method rests on the Linnean hierarchy that exists of species, genera, families, etc., which should represent historical changes through evolutionary time. The structure and assumptions upon which classifications are based must be carefully evaluated before using them in comparative studies of behaviour. In a sense, the systematic blueprint of how taxa are related to one another will dictate observed patterns for evolutionary changes in behaviour. That is, whether species are in the same or a different genus will influence whether behaviours in these taxa should be considered independent evolutionary events. In this chapter we will present a primer on the assumptions and techniques of modern systematics which are important for comparative studies of animal behaviour.

Given that our classification for a group of animals is solid, new comparative methods are available which are quite robust statistically and may be applied to an assortment of problems in the evolution and phylogeny of behaviour. Behavioural data are typically either categorical (e.g. mating system: monogamous or polygynous) or continuous (e.g. number of displays; frequency of fighting). Likewise, comparative methods for studying the evolution of behaviour are structured differently depending upon whether categorical or continuous data are being analyzed; both types of analysis, however, depend on using data that represent independent evolutionary events. In the case of categorical data, we will discuss comparative methods for: (1) counting the number of independent evolutionary origins of a behavioural trait; (2) determining the evolutionary direction of a trait; and (3) examining the amount of evolutionary time a trait in one taxon has taken to evolve relative to other taxa. Comparative methods for examining continuous data address similar problems but, as with analyzing quantitative data in general, they involve a greater range of statistical techniques. Some assume particular models for evolutionary change (e.g. gradual, punctuational or random change) and some make no assumptions about evolution at all. In present examples of comparative phylogenetic studies, we must emphasize the need to understand the assumptions behind particular techniques and encourage the simultaneous use of different techniques to assess how conclusions are affected by the comparative method chosen rather than by the data being analyzed.

In essence, modern studies of the phylogeny of behaviour are new and exciting because of advances in methodology: ways of peering into the past. Our chapter therefore focuses on these new methodologies, using particular behaviour patterns to illustrate various methods. Our depiction of this field may sometimes appear more statistical than substantive, especially in delivering conclusions about behavioural phylogeny. This is the current state of the field, however. There is every hope that the recent building of more rigorous comparative methods will rapidly give rise to more substantive empirical studies in the near future.

4.2 Early studies: forerunners of the computer age

Computers have dramatically changed all facets of comparative biology, from initial data collection to systematics to tests of evolutionary hypotheses about behaviour. Prior to the advent of computers handling large data sets, early studies of animal behaviour only occasionally included phylogeny as a worthy subject. Darwin (1959) advanced a way of looking at natural variation which systematically linked organisms to one another. Thus, in explaining differences in the hunting behaviour of wolves or in the development of deer, it was natural that Darwin would repeatedly refer to the evolutionary history of these taxa (see also Gould, 1986). Similarly, Whitman (1898) described ancestral patterns of parental care in amphibia, reptiles, birds and mammals. Heinroth (1911) carefully detailed movement and reproductive patterns in waterfowl in order to argue that systematics, as a study of whole-animal biology, must include behavioural traits in systematic classifications. Later work benefited greatly from more complete data bases on taxonomic classification and behaviour. For example, Emerson (1938) was able to use a sizeable literature to study the phylogeny of nesting behaviour in termites. Interestingly, important features of his work were the following assumptions about comparative behavioural study: that there is a hereditary component, that evolutionary sequences are known, and that adaptive modifications may be demonstrated. He showed that the phylogeny of termite nests, from simple ground excavations, to mound nests, to arboreal dirt nests, to more specialized types, followed a clear phylogenetic pattern. Further, presaging Lorenz, Emerson strongly encouraged the use of behaviour as an equally viable taxonomic character to morphology and, indeed, warned that classification schemes will be inconsistent and incorrect until behaviour is incorporated into systematics.

Lorenz (1939, 1941) was most overt about the importance of systematics to all behavioural studies. 'Even in the psychological field, living organisms are phylogenetically derived entities whose specific origin and form can only be interpreted in the light of their phylogenetic history' (p. 16, 1941). Lorenz's most convincing empirical work was on locomotor patterns in waterfowl. He assembled comparative data on 18 species for behavioural 'characters' ranging from preening to mating displays and vocalizations (Fig. 4.1). Based on general reasoning that trait evolution moves from a more common to a less common condition, he mapped out a phylogeny of these behaviour patterns. Without a computer, Lorenz resorted to 'using a bundle of stiff wires for arranging the species, united into sub-groups with thin wires representing "common characters"' (p. 111). He concluded that some characters (e.g. display drinking) were indeed more primitive than others (e.g. some contact calls), a pattern also detected with 'good' morphological characters (e.g. bill type, pelage coloration). It is strong testimony to Lorenz's insight that his analysis of these patterns stands up to new sophisticated computer analyses and a variety of assumptions about evolutionary rate and systematics (Burghardt & Gittleman, 1990). In parallel, early studies of the importance of behaviour in systematics were developed by Tinbergen (1951, 1959) on display behaviour in gulls, by Andrew (1956) on tail flicking in passerines, by Mayr (1958) on mating behaviour in finches, and by Alexander (1962a) on cricket songs (see also Cullen, 1959).

We stated above that phylogeny was studied only occasionally in early behavioural research. This raises two important general points about the historical development of phylogenetic studies of behaviour. First, to pursue phylogenetic questions, a fairly complete comparative data base across taxa is necessary (Gittleman, 1989). Only in the last few decades have we had this luxury. Second, during the early days of behavioural

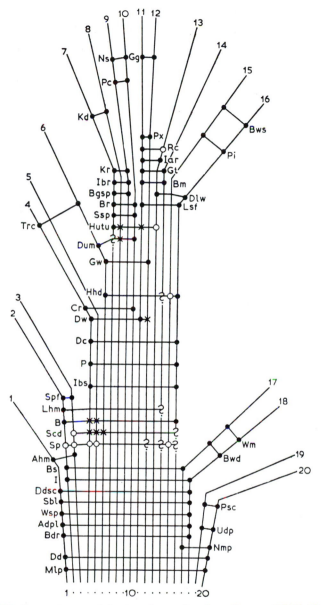

Fig. 4.1. Phylogenetic tree of the Anatinae taken from Lorenz (1941). The *vertical* lines represent species; the *horizontal* lines characters common among them. A *cross* indicates the absence of a character in a species. A *circle* indicates special emphasis and differentiation of the character. A *question-mark* indicates the author's uncertainty. The species are as follows: 1 = *Cairina moschata*, Muscovy duck; 2 = *Lampronessa sponsa*, Carolina wood-duck; 3 = *Aix galericulata*, mandarin duck; 4 = *Mareca sibilatrix*, Chiloe wigeon; 5 = *Mareca penelope*, wigeon;

study, the kind of data that behavioural variables represented were questioned as legitimate material for scientific study – speculating on unknown history and unobservable behaviour patterns was not very convincing to the sceptic. Now, with behavioural data on more solid ground, simultaneous development of rigour in systematics has combined with advances in evolutionary theory and the accumulation of large behavioural data bases to present the phylogeny of behaviour as a vibrant field of study.

4.3 A primer of systematic zoology

As the evolutionary biologist G. G. Simpson (1958) pointed out, 'similarity of behaviours tends, like structural similarity, to be proportional to

Caption for fig. 4.1 (*cont.*)
6 = *Chaulelasmus streptera*, gadwall; 7 = *Nettion crecca*, teal; 8 = *Nettion flavirostre*, South American teal; 9 = *Virago castanea*, chestnut-breasted teal; 10 = *Anas* as genus, including mallard, spot-billed duck, Meller's duck, etc.; 11 = *Dafila spinicauda*, South American pintail; 12 = *Dafila acuta*, pintail; 13 = *Poecilonetta erythrorhyncha*, red-billed duck; 14 = *Poecilonetta* (?) *erythrorhyncha*, red-billed duck; 15 = *Querquedula querquedula*, garganey; 16 = *Spatula clypeata*, shoveller; 17 = *Tadorna tadorna*, shelduck; 18 = *Casarca ferruginea*, ruddy shelduck; 19 = *Anser* as genus; 20 = *Branta* as genus. The characters are as follows: **Mlp** = monosyllabic 'lost-piping'; **Dd** = display drinking; **Bdr** = bony drum on the drake's trachea; **Adpl** = Anatinae duckling plumage; **Wsp** = wing speculum; **Sbl** = sieve bill with horny lamellae; **Ddsc** = disyllabic duckling social contact call; **I** = incitement by the female; **Bs** = body-shaking as a courtship or demonstrative gesture; **Ahm** = aiming head-movements as a mating prelude; **Sp** = sham-preening of the drake, performed behind the wings; **Scd** = Social courtship of the drakes; **B** = 'burping'; **Lhm** = lateral head movement of the inciting female; **Spf** = specific feather specializations serving sham-preening; **Ibs** = introductory body shaking; **P** = pumping as prelude to mating; **Dc** = Decrescendo call of the female; **Br** = bridling; **Cr** = chin-raising; **Hhd** = hind-head display of the drake; **Gw** = grunt-whistle; **Dum** = down-up movement; **Hutu** = head-up-tail-up; **Ssp** = speculum same in both sexes; **Wm** = black-and-white and red-brown wing marking of Casarcinae; **Bgsp** = black-gold-green teal speculum; **Trc** = chin-raising reminiscent of the triumph ceremony; **Ibr** = isolated bridling not coupled to head-up-tail-up; **Kr** = 'Krick-whistle'; **Kd** = 'Koo-dick' of the true teals; **Pc** = post-copulatory play with bridling and nod-swimming; **Ns** = nod-swimming by the female; **Gg** = Geeeeegeeeee-call of the true pintail drakes; **Px** = Pintail-like extension of the median tail-feathers; **Rc** = R-calls of the female in incitement and as social contact call; **Iar** = incitement with anterior of body raised; **Gt** = graduated tail; **Bm** = bill markings with spot and light-coloured sides; **Dlw** = drake lacks whistle; **Lsf** = lancet-shaped shoulder feathers; **Bws** = blue wing secondaries; **Pi** = pumping as incitement; **Dw** = drake whistle; **Bwd** = black-and-white duckling plumage; **Psc** = polysyllabic gosling social contact call of Anserinae; **Udp** = uniform duckling plumage; **Nmp** = neck-dipping as mating prelude. (Reproduced by permission of the publishers from *Studies in Animal and Human Behaviour*, Vol. II, by Konrad Lorenz, Cambridge, Mass.: Harvard University Press, Copyright © 1971 by Konrad Lorenz.)

phylogenetic affinity' (p. 54). To examine the relationship between taxonomy and behaviour we must understand the assumptions and methodologies of systematic zoology.

To arrange organisms in some system it is first necessary to arrive at some agreement about taxon, characters, evolutionary rate, etc. Taxonomic names, such as whether a species is placed in a genus with other species or placed in a monotypic genus, or indeed placed as a sole member of a family, are largely arbitrary. Further, systematic classifications may be based on morphology, molecules, behaviour, ecology, or any other measurable character. These issues are obviously very important (see Brooks & McLennan, 1991), but we are mainly concerned here with how this information is turned into systematic classification and what impact it will have on a phylogenetic (comparative) study of behaviour. There are three competing schools of systematics: evolutionary systematics (or phyletics), phenetics, and cladistics (or phylogenetic systematics). Differences among them centre around whether evolution can or should be the basis of biological classification (Ridley, 1986; Hull, 1988; Funk & Brooks, 1990). The question, 'What is a classification and what type of information should it store?' is controversial. A classification is an artificial system, used by humans, to represent an order or arrangement of a group of animals (e.g. the lion is a felid, felids are carnivores, carnivores are mammals, etc.). To be effective, classifications should be stable. That is, they should not be different each time a new analysis is performed with a different set of characters or traits. Additionally, for classifications to be of general use, particularly in animal behaviour, the information used to form a classification should be measurable and retrievable (i.e. there should be a full data listing, preferably on computer). In this section, we will briefly describe the logic and relative merits of the three schools of systematics. We then give some simple examples of phylogenetic pictures generated by each method, which are sometimes the same but sometimes different.

Evolutionary systematics is a compromise between phenetics and cladistics (see below). Phylogenies are based on both genealogy and overall similarity. Evolutionary systematists assume that ancestors can be identified and that evolutionary rates are constant. The major problem with evolutionary systematics is that it completely ignores convergent evolution; that is, similar traits (e.g. wings in birds and bats) which evolve independently in unrelated taxa. Although evolutionary taxonomists claim that they do not classify polyphyletic groups (i.e. unnatural groups of multiple evolutionary origin; see Ashlock, 1971) together, by ignoring convergent evolution they are recognizing unnatural groups. Another problem is that the ónly method of distinguishing homologous from analogous characters is based on imprecise, subjective methods. By excluding convergent evolution in a subjective manner, what is left is a classification of very similar organisms (Ashlock, 1979). There is no need to discuss evolutionary systematics any further because the remaining methods take elements of evolutionary systematics but improve them and make them more objective.

Phenetic taxonomy is the arrangement of taxa by overall similarity based on all available characters without any preference (i.e. weighting) given to one character over another. Phenetic taxonomy was a child of computers. In the sixties, with the first computers, many characters could be objectively and rapidly analyzed. Phenetics assumes that the evolution of organisms cannot be known. Thus emphasis is given to what is (or may be) known, namely the measurement of as many character values as possible; a phenetic hierarchy is then formed from the *similarity* of all the characters that are measured. In phenetic taxonomy, therefore, classification is by similarity not evolutionary history. Classifications based on phenetics should not be used to represent phylogenies.

Cladistics (or *phylogenetic systematics*) determines the genealogical relationships of evolutionary descent of a group; groups are formed to include animals with recent common ancestry. Thus primitive and derived characters are identified, and are then used to cluster taxa on the basis of shared characters. Classifications based on cladistics are therefore explicitly phylogenetic, with the provision that phylogenetic inference is a working hypothesis about history which should be continually updated and revised. These working hypotheses may come from many sources: character similarities; ontogeny; the fossil record; character sequences; or correlation of a suite of characters (see Stevens, 1980, for a review of the relative merits of different sorts of information for phylogenetic construction).

What kind of differences may arise from the approaches and methods of these last two schools? Figure 4.2 shows different phenetic and phylogenetic inferences for three data sets. Both classifications give the same phylogenetic answers when the rate of evolution is approximately constant and its direction is clearly divergent, as when comparing the human, chimp and rabbit (Fig. 4.2a). The two schools of classification differ when there is convergence or different rates of evolution. In Figure 4.2b, the barnacle and limpet are phenetically closer due to convergence because the barnacle looks more like the limpet.

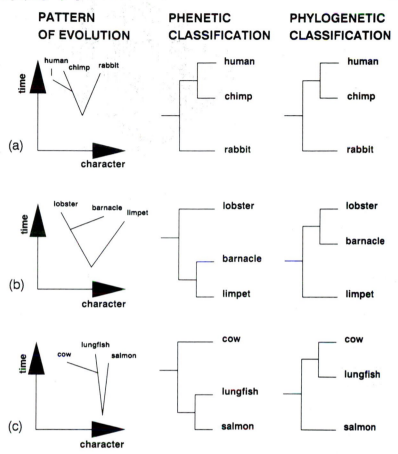

Fig. 4.2. Various phylogenetic inferences depicting how patterns of evolution are handled by phenetic classification and phylogenetic classification. See text for details. (From Ridley, 1986.)

Figure 4.2c also illustrates the last type of disagreement, as the lungfish clearly looks more like a salmon, although it shares a closer ancestry with the cow.

To summarize, even from these simple illustrations it is apparent that different schools will produce vastly different classifications. One form of classification (phenetic) designates similarity of appearance as important, whereas another (phylogenetic) views common descent as the critical criterion. Therefore, one must be aware of which perspective was adopted in a given classification. If a classification is purely phenetic (based on similarity), it cannot be used for studies of the phylogeny of behaviour.

4.4 Phylogenetic approaches to behaviour

As should now be clear, the comparative method using variation between taxa is our most effective technique for analyzing the phylogeny of behaviour. At a general level, the comparative method can be used to detect how behaviour has changed through evolutionary time relative to other traits, typically morphological ones. This involves overlying behavioural traits onto systematic classifications which are formed from independent characters. Such 'behavioural systematics' usually involves categorical (or discrete) traits. We begin this section with examples of comparative techniques for handling this type of trait, first with simple approaches and then moving on to more difficult statistical methods. A further form of phylogenetic comparative work involves measuring the extent to which behaviour evolves (or is correlated) with phylogenetic history and then removing this correlation so that comparative data are statistically independent. The majority of recent comparative research falls in this category and involves primarily continuous (or quantitative) traits; again, using examples we will review various comparative methods motivated by this problem.

4.4.1 Categorical traits

Behavioural systematics

Although Lorenz (1941) recognized the value of behavioural traits in systematics, few earlier studies used them. However, with larger data bases including a more complete representation of taxa, behaviour patterns are now included among the systematic characters used (e.g. Greene & Burghardt, 1978; McLennan, Brooks & McPhail, 1988; Brooks & McLennan, 1991). Furthermore, the criticism that behaviour patterns are too flexible to reveal evolutionary history for systematic purposes, is now seen to be unjustified. Many behavioural characteristics, such as mating systems, display behaviour and other aspects of reproduction, show strong relations with phylogeny. Indeed, a comparative study of the consistency indices – a statistic for the minimum number of trait changes on a given phylogeny divided by the total number of possible trait changes (Swofford, 1991) – between morphological (e.g. categories of colour; presence or absence of fins) and behavioural (e.g. presence or absence of nest building) traits from 22 data sets show no significant differences between the two types of data (de Queiroz & Wimberger, 1993). This is not without exception though. Other traits, such as home range size or social behaviour, tend to be more susceptible to environmental change. The important point, however, is that behaviour may be useful in systematics and that this can be examined empirically. Comparative data sets including both morphological and behavioural characters can be used at the outset to describe the relationship with phylogeny, either with some form of a consistency index or with various descriptive statistics (Nested Analysis of Variance or Moran's *I*: see Gittleman & Luh, 1992).

A recent example of using behaviour for phylogenetic analysis is the study of display patterns in Neotropical manakins by Prum (1990). Display behaviour varies widely among species and genera of manakins. These birds may display on different surfaces, vary in territory utilization, and show complex fighting and display movements (Fig. 4.3). Prum categorized 21 species according to 44 display characters and performed three types of phylogenetic analyses: (1) using only display behaviour; (2) using display behaviour and syringeal characters (anatomical features of the vocal apparatus – a common morphological trait in avian systematics) in a single phylogenetic analysis; (3) using display behaviours overlapped onto a phylogeny based on syringeal characters. The first two analyses showed that display elements follow phylogenetic lines similar to syringeal morphology. Further, 37 of the 44 display characters revealed significant congruence with independent phylogenies which had been based on morphology, with particular display movements representing a gradation of change from primitive to derived taxa. Several caveats must be applied to this type of phylogenetic analysis. First, the data typically come from disparate sources, which may introduce some error in analyses, and, second, detailed behavioural information is usually missing for many taxa. These problems, however, are more concerned with data collection than with the method itself. The third problem is more difficult and concerns us in the next two sections: how should comparative data on behaviour and morphology be analyzed, when either testing for phylogenetic or adaptive hypotheses, if the two types of data are *not statistically independent*?

Detecting evolutionary independence, direction and duration

Comparative analyses across species frequently cannot be taken to involve statistically independent data if one species shares evolutionary history

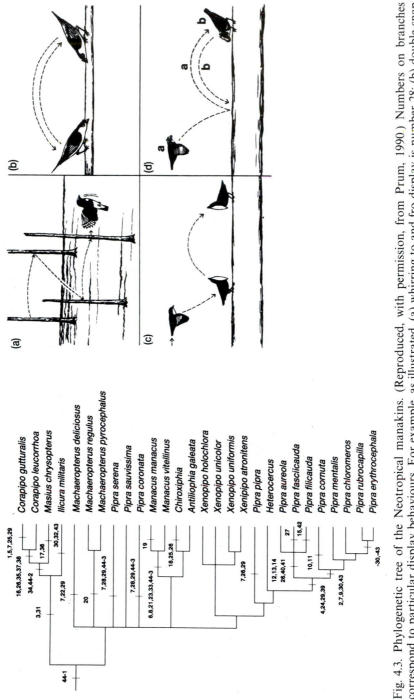

Fig. 4.3. Phylogenetic tree of the Neotropical manakins. (Reproduced, with permission, from Prum, 1990.) Numbers on branches correspond to particular display behaviours. For example, as illustrated, (a) whirring to and fro display is number 28; (b) double-snap jump display is number 32; (c) log-approach display in bill-pointing posture is number 34; and (d) co-ordinated log-approach display in chin-down posture is number 36.

with another species (Felsenstein, 1985; Ridley, 1989). This makes intuitive sense. Organisms are products of their history and, given that classifications are dependent on evolutionary history to a certain extent, traits shared by closely related taxa are not independent. Only after taxa split from their most recent common ancestor (i.e. after a branching or node in a phylogenetic tree) do they represent an independent evolutionary event. Three methods are available that establish statistical independence in categorical behavioural data and, in so doing, provide different kinds of phylogenetic information: (1) counting the number of (independent) evolutionary events by reconstruction from terminal taxa (Ridley, 1983); (2) inferring directionality (Maddison, 1990; Maddison & Maddison, 1992); and (3) determining the amount of evolutionary change (Pagel & Harvey, 1989). Results from these methods are of course subject to the problems of any comparative analysis (e.g. spurious data, measurement error of variables: Jarman, 1982; Clutton-Brock & Harvey, 1984; Gittleman, 1989) and all of them rely upon a well-established (or at least acceptable) phylogeny being available for the taxa under study.

To illustrate these methods, we take an example from patterns of parental care and diet in the family Canidae (Fig. 4.4). Species vary in parental duties, which typically include either non-communal rearing of offspring (i.e. by the parents alone, as in the grey fox) or communal rearing

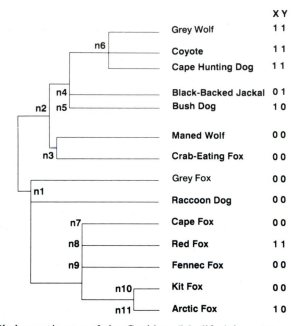

Fig. 4.4. Phylogenetic tree of the Canidae. (Modified from Wayne *et al.*, 1989, and Harvey & Pagel, 1991.) Numbers represent nodes and are referred to in the text. Character X relates to diet, where state 0 is omnivory and state 1 is carnivory. Character Y denotes types of parental care, in which state 0 is non-communal care and state 1 is communal care.

(i.e. by the parents and helpers, as in wild hunting dogs or grey wolves). If we search for correlations between these types of parental care and ecology, we find that across canid species, diet is influential; carnivorous canids tend to be communal more than omnivorous canids (Gittleman, 1985). Given the relatively complete phylogenetic picture we have for canids (Wayne *et al.*, 1989), how do modern comparative methods assess this result in relation to phylogeny?

Ridley's (1983) technique simply counts the number of taxa at the end of a phylogenetic tree (terminal or end-points) which have a given combination for a pair of traits. Thus the numbers of a particular combination are tallied. When combinations of characters are overlayed onto the phylogeny, there are certain *transitions* that are presumed to have occurred; only transitions from an immediate common ancestor are considered to be *independent* data points. In the canid example, we tally seven taxa that exhibit omnivory and non-communal care (0,0 in Fig. 4.4) and then work through the entire branching transitions to erect a

Table 4.1. *Observed values of parental care and diet in canid species. A Fisher exact test revealed a non-significant association between diet and parental care ($p > 0.05$)*

	Omnivore	Carnivore	Total
Non-communal care	7	2	9
Communal care	1	4	5
Total	8	6	14

two-by-two contingency table showing all the transitions. A Fisher exact test (Table 4.1) on the relationship between diet and parental care reveals a non-significant association, but does suggest that some combinations are more (or less) common than expected by chance. This example illustrates a general problem in the statistical analyses of behavioural phylogeny: that the techniques used require large data sets (large numbers of species) which may not always be available or, indeed, appropriate in a particular context. In essence, Ridley's method only uses data points representing independent evolutionary transitions. This approach has been applied to lekking and sexual dimorphism in birds (Hoglund, 1989) and mating frequencies in insects (Ridley, 1990). It is important to recognize that a counting technique assumes equal weighting to each transition, equal branch lengths, and data calculations only when a transition has occurred (Harvey & Pagel, 1991).

Maddison (1990; see also Maddison & Maddison, 1992) devised a test for *directionality* in comparative data (called the 'correlated-changes test'). That is, in the canid example, we might ask whether omnivory evolved only after non-communal care. The correlated-changes test calculates the probability that one trait evolved in certain regions (clades) of a phylogeny given the overall distribution of trains in a clade and given the null model that trait change is random. To determine whether omnivory is associated with non-communal care in the canid example, the total number of nodes at which omnivory occurs regardless of the occurrence of non-communal care is counted (it does so seven times: n1, n2, n3, n4, n7, n9 and n10), and the number of times omnivory occurs in the presence of non-communal care is tallied (six times: n1, n2, n3, n7, n9 and n10). The probability that omnivory evolved only in the presence of non-communal care is 6/7 or 0.86. This is an extremely simplified version of the correlated-changes test: the new MACCLADE computer package discusses the assumptions and applications of the method more fully (Maddison & Maddison, 1992) as well as those of other types of phylo-genetic analysis. A nice example of this approach is in Maddison's (1990) re-analysis of data (from Sillen-Tullberg, 1988) on the association between gregariousness in butterflies and warning coloration of their larvae.

Finally, the method of Pagel and Harvey (1989) and Pagel (1992) considers variable branch lengths and consequently different amounts or *duration* of evolutionary change. They assume that character changes are more likely to occur on longer branches than on shorter ones, simply because more evolutionary time will have elapsed (see Maddison & Maddison, 1992, for discussion). Thus, the expected values for characters and variances of character change are determined for each branch in a phylogeny. The expected value is based on the probability of a character changing from one state to another, and its expected variance depends on the length of the branch. These expected values can be standardized by subtracting from each its mean and dividing by the square root of its variance. This will result in all sets of observations having a mean of zero and a standard deviation of one. The standardized observed values are positive if there was a transition from 1 to 0 or 0 to 1, and negative if there was no change. However, Pagel and Harvey arbitrarily assign a positive sign to any branch ending in 1, and a negative sign to any branch ending in 0. Therefore, two characters evolving from 0 to 1 and two retaining state 1 are treated the same (for a review of this method, see Harvey & Pagel, 1991). Following these procedures through the formal equations, with the canid data we find that diet and parental care are significantly correlated (Spearman rank correlation: $r_s = 0.55$, $P = 0.01$). However, as might be expected, when the statistical test is restricted to only the branches where a transition occurs, the correlation is not significant ($r_s = 0.50$, $P = 0.32$). Other examples of this technique are found in Harvey and Pagel (1991) and Packer, Lewis & Pusey (1992). Computer programs for easy application of the method are available from Pagel (1991) or Purvis (1991).

To summarize this section for analyzing the phylogeny of categorical traits, various methods are available depending on the type of question being addressed. One technique simply examines statistical independence of traits across taxa given that evolutionary histories, to a certain extent, are associated with trait variation (Ridley). Another technique addresses directionality in phylogeny, primarily by considering the chances of one trait evolving from another given the overall number of evolutionary transitions (Maddison). Last, there is a method to establish phylogenetic independence in the data but also to detect the duration of change in trait evolution (Pagel & Harvey, 1989). Clearly, these techniques are very statistical. Yet, they offer the first opportunity to address questions previously intractable to Lorenz and other comparative ethologists and to examine specific hypotheses about phylogenetic independence, and the direction and duration of behavioural change.

4.4.2 Continuous traits

Most behavioural traits are measurable in a quantitative or continuous fashion. Even variables such as amount of arboreality in a primate or intensity of territorial calls in birds can be described by a numerical value. This difference in the kind of comparative data compared to those considered in the last section opens up a variety of new techniques for analyzing the phylogeny of behaviour. Again, studies are based on testing a hypothesis about behavioural phylogeny against some blueprint of phylogenetic structure. But in using quantitative traits, assumptions about evolutionary rate and an appropriate statistical null model may be selected by the researcher. At present there are at least seven different comparative methods for analyzing quantitative traits in a phylogenetic manner (for reviews, see Harvey & Pagel, 1991, and Gittleman & Luh, 1992). Here, we will describe the basic assumptions and the techniques involved in two which probably represent the extremes of comparative analysis: independent contrasts and autoregression.

To describe these methodologies we will revert to the canid example (see Fig. 4.4), but assume that we now have quantitative information rather than discrete data: communal care and diet are measured in the form of the number of adults assisting in care and the percentage of meat in the diet. To simplify matters, we use hypothetical data for four canid species; we take diet as the independent variable (X) and parental care as the dependent variable (Y). Species values for these variables are then overlaid onto the phylogeny (relationships as in Fig. 4.4), which already gives us branching diagrams for these taxa. The independent contrasts and autoregressive approaches can work through these quantitative behavioural data using the following assumptions and procedures.

Independent contrasts

Felsenstein (1985) first suggested the idea of independent contrasts in order to get around the problem of phylogenetic correlation in comparative

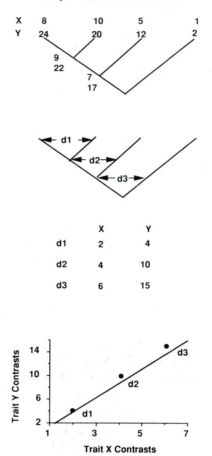

Fig. 4.5. Simplified illustration of the procedure used in the 'independent contrast method' to account for phylogeny in comparative analyses. (Taken from Harvey and Pagel, 1991.) X and Y correspond to the characters in Fig. 4.4. Hypothetical phylogeny includes the grey wolf (8, 24), coyote (10, 20), black-backed jackal (5, 12), and grey fox (1, 2).

data. He suggested a procedure which calculates 'contrasts' (i.e. weighted differences) between taxa at each split or node in a known phylogeny (Fig. 4.5). This procedure results in (N − 1) contrasts from N original tip species. If the ancestral nodes for the phylogeny are known, then each of the contrasts between taxa which share a similar ancestry is not confounded by phylogeny between them and therefore reflects an *independent* evolutionary event. (In most cases, a true phylogeny is unavailable, in which case various mathematical algorithms can reconstruct ancestral nodes from species tip values, as shown in Fig. 4.5.) If two variables are related (e.g. diet and communal care), then a large contrast with one variable at a specific node should be associated with a large contrast in the other. Notice that, in the hypothetical canid example, larger contrasts are evident in more distantly related taxa (e.g. grey fox and coyote), thus giving greater contrast values.

This comparative technique has helped towards understanding the evolutionary function and phylogeny of a number of quantitative traits (see Harvey & Pagel, 1991; Harvey & Purvis, 1991; as for the analysis of discrete traits, computer programs are available from Pagel and Purvis for executing contrast methods). A nice illustration is Møller's (1991) study of testis size in birds. Cross-species variation in testis size may relate to sperm competition, as males with relatively larger testes will compete better with other males, or to sperm depletion, which may lead males that copulate often to have larger testes. Thus Møller calculated independent contrasts across 247 bird species for relative testis size. These were correlated with contrasts of copulation frequency per female (reflecting intensity of sperm competition) and copulation frequency per male (reflecting intensity of sperm depletion). A significant positive correlation was observed with copulation frequency per female but not with that per male. Therefore, the sperm competition hypothesis was supported by the comparative evidence after controlling for phylogeny.

Other comparative analyses using the independent contrasts approach have considered locomotion behaviour in lizards (Losos, 1990), fecundity and longevity in Hymenoptera (Blackburn, 1991), and brain size evolution in carnivores (Gittleman, 1991). The contrast methods used are of different types, depending on evolutionary assumptions, measures of phylogenetic branch length, and algorithms for standardizing contrasts (see Grafen, 1989; Harvey & Pagel, 1991; Martins & Garland, 1991). All of them do, however, share some general features that should be recognized. First, as stated by Felsenstein (1985), the independent contrast approach is built around some model of evolution; this is typically Brownian motion, because it is a testable null model of random evolutionary change. That is, contrasts are calculated at each node with an expectation that traits have evolved to some degree. Second, the method calculates the number of evolutionary events (at each node) as statistically independent data; for example, ten species in one genus yield one contrast value, just as do five species in a single genus. Lastly, the strength of a contrast approach, especially in comparison to the method we will describe next, is that it uses all of the variations in a trait to test a comparative result with a phylogenetic framework rather than partitioning traits into phylogenetic and non-phylogenetic (adaptive) components. Further, when contrasts are calculated for the nodes at all hierarchical levels in a phylogeny, interesting comparative results can be detected at deeper phylogenetic levels (e.g. the original splitting of canids from ancestral carnivores).

Autoregression

The autoregressive method was originally developed by Cheverud, Dow and Leutenegger (1985). Essentially, trait values are partitioned into a phylogenetic component and a component due to independent (adaptive) evolution. Such partitioning is analogous to dividing phenotypic values into genetic and environmental components. In a similar way to linear regression, an autoregressive approach takes the form (Fig. 4.6):

$$y = \rho \mathbf{W} y + \varepsilon$$

with y (the standardized trait value) equal to $\rho \mathbf{W} y$ (a phylogenetic component) plus ε (the residual). Analytically, $\mathbf{W}$ is an n × n weighting matrix, ρ measures the correlation between the trait (y) and phylogenetic values ($\mathbf{W} y$), and ε depicts the independent evolution of each species. A simple illustration of the model using the canid data is given in Figure 4.6. In general, it should be recognized that the phylogenetic distance among taxa (represented by the weighting matrix) can be set in relation to various phylogenies; also, the value of ρ is a useful descriptive statistic to gauge overall correlation between a given trait and phylogeny.

An example of this method can be taken from a comparative analysis of olfactory bulb size in carnivores. Differences in brain size across mammals are typically related to overall body size, phylogeny and ecology (Harvey & Krebs, 1990). For example, in carnivores around 85% of the variation in brain size is accounted for by differences in body size (ranging from the weasels up to the bears) and by phylogenetic (family) associations. Ecological variables appear to be unimportant, at least as far as total brain size is concerned (Gittleman, 1986). Nevertheless, perhaps functional differences in brain size are more attributable to specific kinds of information processing which occur in particular areas of the brain. Thus, in carnivores, the olfactory bulbs may reveal functional trends as these animals have a relatively good sense of smell. After analyzing data on body size and olfactory bulb size for 146 carnivore species with the

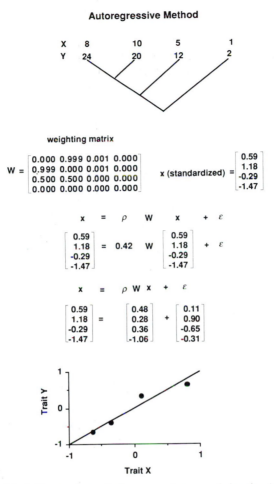

Fig. 4.6. Simplified illustration of the procedure used in the 'autoregressive method' to produce residual values that are used in comparative tests to account for phylogeny. (Updated and modified after Cheverud *et al.*, 1985, Gittleman & Kot, 1990, and Gittleman & Luh, 1992.) Characters and species are the same as those illustrated in Fig. 4.5.

autoregressive model, clear ecological patterns emerged: residual values (the specific components) for the olfactory bulbs of carnivores that live in aquatic environments (e.g otters and the fishing cat) are significantly smaller than those for species residing in other habitats (Fig. 4.7). A possible functional explanation for this finding is that aquatic carnivores communicate less through scent-marking and other olfactory cues than do more terrestrial species (Gorman & Trowbridge, 1989), thus having less-developed olfactory

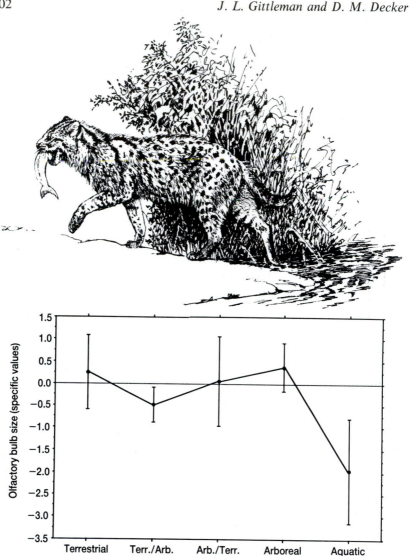

Fig. 4.7. Illustration of the average zonal categories for carnivores following application of the autoregressive method to species values. This example shows that functional trends are apparent after phylogenetic correlation is removed in the comparative data. See text for details. (Taken from Gittleman, 1991.)

senses. Similar examples have been described for studies of life-history patterns in carnivores (Gittleman, 1993), clutch size in birds (Gittleman & Kot, 1990), body size dimorphism in primates (Cheverud *et al.*, 1985), co-operative breeding in perching birds (Edwards & Naeem, 1993), and ornament dimorphism and mating systems in birds (Bjorklund, 1990).

As with the independent contrast approach, there are a number of key features in the autoregressive method. First, residual values are calculated for each taxonomic unit (usually species) under study. Second, at the heart of autoregression is the connectivity matrix W. This weighting matrix matches the relative phylogenetic distance of all pairs of species and then generates anticipated trait similarities given these distances. Third, this weighting matrix may be set up to represent different proposed phylogenies and, equally important, to incorporate some measure of distance. Gittleman and Kot (1990) applied a variable weighting index (using a maximum likelihood procedure) to provide greater flexibility to deal with both conservative and plastic traits.

4.4.3 Selecting a method

Both the independent contrast and autoregressive methods are relatively new. It is difficult to make simple suggestions which are appropriate for different sorts of behavioural study. Nevertheless, there are some obvious criteria that will, at least initially, aid in selecting an appropriate method (see Gittleman & Luh, 1992, 1993). The independent contrast approach has a model of trait evolution built into it, whereas autoregression is a purely statistical approach that removes phylogenetic correlation in traits. When a given model of evolution is necessary or desired, then the independent contrast method is more appropriate. The analytical procedures used clearly differ between the two methods. The independent contrast approach calculates the result of a (presumed) independent evolutionary event at each node in a tree. Therefore it assumes that phylogenetic correlation is apparent in trait differences. Autoregression generates a correlation matrix which will indicate whether any correlation between traits and phylogeny occurs and, in some cases, may suggest that no data transformation be made (with *any* comparative method) when phylogeny is uncorrelated with comparative traits. Using a variable weighting matrix in the autoregressive model may, however, overestimate the contribution of phylogeny to trait variation (Gittleman & Kot, 1990; Gittleman & Luh, 1992). Last, if sample sizes are small (i.e. there are few evolutionary events and thus nodes), the small number of calculated independent contrasts may make it impossible to use the technique.

As a final word of advice, because these techniques are new and rely heavily on phylogenetic pictures which often are questionable, it is wise to apply different types of analysis to the same data (see, for example, Garland, Huey & Bennett, 1991; Gittleman, 1991). This will evaluate whether a behavioural conclusion depends on the comparative method that was adopted. Another approach is to run computer simulations using different data sets (e.g. with varying sample sizes; showing phylogenetic correlation in traits; using phylogenetic versus taxonomic information) to find out the statistical likelihood of arriving at similar conclusions under different conditions (see Martins & Garland, 1991; Gittleman & Luh, 1992).

4.5 Conclusions and future problems

Many readers will have reached the end of this chapter with the nagging question, 'Where is the phylogeny of behaviour?'. We have presented few studies that show proposed historical pathways for the evolution of behaviour. This is because only recently have we had the necessary data (i.e. comparative, cross-taxa data bases and associated phylogenies) and the comparative methods that will permit us to pry into the behavioural past *accurately*. We have reviewed both the historical approaches to the phylogeny of behaviour and the new comparative methods that should give rise to exciting studies. We will finish by pointing to a few problems that seem eminently fruitful areas for future study.

Behavioural systematics

Are behavioural traits as useful for systematic studies as morphological and genetic characteristics? Undoubtedly, there will be some behaviour patterns and some taxa that are more suited to such studies than others. An interesting problem will be to assess if behavioural traits are more diagnostic for taxa that have had brief evolutionary lifespans. Relations between behavioural transitions in evolutionary terms, should relate to phylogenetic time intervals.

Phylogenetic constraints

Various reasons were given at the beginning of this chapter for the correlation of behaviour and phylogeny. This correlation is supported on general theoretical grounds, but specific hypotheses are lacking. For example, what is the likelihood that two related taxa responding to similar environmental stimuli really do evolve in a like manner as far as their behaviour is concerned?

Rate of evolution

Others (e.g. Bateson, 1988a; Wcislo, 1989) have suggested that behaviour is a most influential factor in subsequent evolution. Qualities such as learning to exploit novel environments, behavioural flexibility to compensate for morphological or environmental change, and mechanisms of behavioural ontogeny are all areas that demand phylogenetic study.

Tinbergen (1951), in describing the evolution of behaviour, warned that comparative studies could not be 'haphazard' if they are to be of value, and therefore should be treated in the same careful manner as designing experiments. The comparative methodologies described in this chapter should bring us closer to that vision.

Acknowledgements

We thank Peter Slater, Tim Halliday and Gordon Burghardt for helpful comment and criticism on the manuscript.

5
Strategies of behaviour

I. F. HARVEY

5.1 Introduction

A behavioural strategy is an act, or sequence of acts, that an animal performs in a given situation. Strategies can be divided into two broad classes, for which different techniques of analysis are needed. These are strategies for which fitness (defined in the next section) is independent of the frequencies of other strategies in the population, and those for which fitness is frequency dependent. In this chapter, techniques for analysing strategies will be given, and the use of such techniques will be explained, taking the evolution of mating patterns as the main example.

Much of the recent interest in behavioural strategies has come from the functional approach to behaviour: this aims to understand why some strategies are maintained by natural selection within a population, why other strategies are eliminated, and in particular how variation in behaviour, both within and between species, can be maintained. In answering such questions, the principle that natural selection acts as an optimizer is relied upon. Why should natural selection act in this way? Natural selection favours the fittest genotypes in a population and so will tend to increase the mean fitness of a population, less fit genotypes being eliminated. Therefore, if a population has been subjected to the same selection pressures for a relatively long period of time, then only those individuals with the fittest genotypes will remain in the population, and hence selection has acted as an optimizing agent. This is a simplified view of the way in which selection acts (indeed, it is not true if fitnesses are frequency dependent, a situation which, as shown later in this chapter, is very common), but it provides the main rationale for assuming that an animal's behaviour serves to maximize its Darwinian fitness.

The idea that selection acts as an optimizer has been used as the basis of a series of theoretical models that have been of great value in our understanding of the evolution of behavioural strategies. Some of these models form the basis of this chapter. One of the advantages of using models is that they force the explicit statement of assumptions and generate testable predictions. Whilst it may be true that some technical developments (such as DNA fingerprinting) have made a contribution, most of the observations and experiments discussed in this chapter could have been carried out a century ago. That they were not is due to the fact that, in many cases, the theoretical basis that led to questions being asked was not available.

5.2 Fitness and reproductive success

Much confusion exists over the use in biology of the concept of fitness. Indeed, Dawkins (1982) paraphrased Lewis Carroll and entitled a chapter 'An agony in five fits', suggesting that at least five definitions of fitness have been used. Here, the population geneticists' 'zygote-to-zygote' definition of fitness is adopted. This term reflects the fact that the counting of genotypes is done at the zygotic stage. The model outlined below is applicable to organisms with discrete generations, such as many insects. If generations overlap, then defining fitness becomes harder (see Box 3.2 in Maynard Smith, 1989).

Fitness (as defined by population geneticists) is a property of a genotype – individuals do not possess fitness, nor do genes; rather, the fitness of a genotype is an average of the survival and reproductive success of individuals of that genotype. A hypothetical example may help to explain how the fitness of a genotype is calculated. Consider one locus with two alleles (A_1 and A_2), giving three genotypes, which differ in egg-to-adult survival and in the number of offspring produced:

Genotype	A_1A_1	A_1A_2	A_2A_2
Survivorship (egg–adult) (S)	0.9	0.8	0.7
Offspring produced (reproductive success) (R)	7	9	11
Absolute fitness (R × S)	6.3	7.2	7.7
Relative fitness	0.82	0.94	1.00

Relative fitness is calculated by dividing the fitness of each genotype by that of the fittest genotype (which therefore has a fitness of 1). As this shows, the reproductive success of an individual is only one component of the fitness of a genotype.

Fitness is a property of a genotype in a particular environment; if the environment changes, then fitness is likely to change. The rapid increase and then gradual decrease in the frequency of the melanic form of the peppered moth *Biston betularia* in Britain, in response to an increase and then, following the introduction of the Clean Air Act (1964), a decrease in industrial pollution, illustrates this point. It must also be remembered that fitness as defined above is valid for one generation only: if selection is taking place (i.e. there are differences in fitness among genotypes), then gene frequencies, and hence relative fitness, will change. A final point is that there is no such thing as the fitness of a population because populations do not reproduce themselves (Maynard Smith, 1989). You may come across a term called the mean fitness of a population (often symbolized as $\bar{W}$, pronounced bar w), but this is just a mathematical abstraction used in calculating changes in gene frequencies under selection. It is true that selection will tend to increase $\bar{W}$, but not under all circumstances and especially not if fitnesses are frequency dependent (see below).

Why is fitness important? When considering the evolution of

behavioural strategies, they are assumed to be genotypes that will spread *via* natural selection. In principle, this means that the fitness of a strategy must be measured, but in practice this is extremely difficult. Therefore, the assumption must be made that something that *can* be measured is a good correlate of fitness. A variety of such *currency assumptions* has been used. Early models of foraging behaviour and their experimental tests relied on the maximization of rate of energy intake as the currency assumption. More sophisticated measures, such as lifetime copulatory or reproductive success, have been made possible by studying relatively short-lived animals (e.g. dragonflies and damselflies) or by detailed long-term population studies of animals such as red deer (*Cervus elaphus*) and a wide variety of bird species (see Clutton-Brock, 1988; Newton, 1989). However, it must be stressed that, even if techniques such as DNA fingerprinting (see Chapter 7, this volume) are used to determine parentage accurately, lifetime reproductive success will still not correlate precisely with fitness, and may in some cases be misleading. In the case of dragonflies and damselflies (Odonata), for example, the number of mates a male obtains will undoubtedly influence his fitness, but the part of the season in which he mates is also likely to be important. For example, individuals that emerge and mate early will give their larvae an advantage in lowering their exposure to the risk of cannibalism.

5.3 Frequency-independent versus frequency-dependent fitness

For many behavioural decisions the outcome, in terms of fitness gain, is not influenced by the decisions of other individuals in the population. For example, the energy gained from eating a particular prey item will not generally depend on the behaviour of other individuals. Frequency-independent fitness consequences of behavioural decisions probably occur most commonly among feeding and foraging decisions. Such behaviour patterns can be studied using *optimality theory*.

In contrast, the fitness consequences of many other behavioural decisions are frequency dependent: what an animal receives, in terms of fitness gain or loss, depends on strategies adopted by other individuals in the population. For example, the success of a strategy for all-out aggression in a population of submissive individuals may be high, but in a population consisting of a large number of other aggressive individuals, the risk of serious injury during escalated fights may be so high that the aggressive strategy has a rather low fitness. The framework for the formal analysis situations in which fitnesses are frequency dependent was set out in 1973 by John Maynard Smith who, together with George Price, defined the term *evolutionarily stable strategy* (ESS) and used the theory of games (a technique devised by economists for analysing co-operation and conflict in humans, Von Neumann & Morgenstern, 1944) in the search for ESSs. The importance of frequency-dependent fitness has been realized long before 1973, going back to the discussions of the sex ratio by Darwin (1871) and Fisher (1930). Hamilton's (1967) search for an unbeatable sex ratio under local mate competition continued these ideas and foreshadowed many of the developments of ESS theory. Parker's fieldwork on competitive mate searching by male dung flies, *Scathophaga stercoraria* (e.g. Parker, 1970a), implicitly assumed the idea of the ESS.

Much of the early ESS theory was aimed at analysing the behaviour of animals during aggressive interactions (Maynard Smith & Price, 1973; Maynard Smith, 1974; Parker, 1974; Maynard Smith & Parker, 1976). It was concerned with pairwise interactions which have been termed *contests*. Another common situation arises when an animal is competing not against another individual, but against some average property of the population as a whole. For example, the ability to attract mates may depend on the amount of resources an animal puts into sexual advertisement (for instance by producing an extravagant train like that of a peacock, *Pavo cristatus*). In such cases it might not be absolute size that is important, but size of the advertisement relative to that of other males in the population. Such situations are termed *playing the field* (Maynard Smith, 1982b) or *scrambles* (Parker, 1984).

In addition to the distinction between frequency-dependent and frequency-independent fitnesses, one can also divide strategies into two classes: discrete and continuous. The distinction between these two types of strategies is clear. In discrete strategy sets the animals adopt one of a limited range of possible types of behaviour: for example, in some species where males defend territories to obtain mates, other males may adopt a satellite or sneaky mating strategy. In contrast, with a continuous strategy set, the choice is in the value of a particular strategy (e.g. the time to spend feeding in a patch). The adoption of a particular behaviour pattern may be under genetic control, or an animal may make a decision based on its assessment of various aspects of the environment, such as the distribution and availability of resources, or the behaviour of competitors.

5.4 Optimality theory – with reference to foraging behaviour

The most efficient animal would encounter food at an infinitely high rate and would handle (pursue, subdue, consume and digest) food items in a negligible period of time. Clearly no animal could behave like this. In practice, animals act under a variety of constraints. For example, physical limitations set by skeleton and musculature mean that animals cannot move infinitely fast, and prey items take time to subdue and consume. Given a defined series of constraints (*constraint assumptions*), making assumptions about the appropriate currency (*currency assumptions*) and about the behavioural options open to the animals (*decision assumptions*), it is possible to show mathematically how an animal should behave if it is to maximize its fitness. Two examples of this type of model will begin below: the prey model and the patch model. Further details of these models can be found in Stephens and Krebs (1986), on which this account is largely based.

5.4.1 The prey model

Consider an animal presented with an environment containing two prey types, for which it can search simultaneously and which differ in energy content (E) and handling time (h). The currency assumption that the animal's goal[1] is maximization of long-term rate of energy will be used. Should the animal concentrate on one of the two prey types, or should it take one type only occasionally, or perhaps should it take both prey types every time it encounters them? The prey model answers these questions. The model assumes that prey items are encountered at random and at a constant rate, and that energy values and handling times are fixed constraints. 'Omniscience' on the part of the predator is also assumed, i.e. the animal is continuously aware of all of the parameters in the model.

Let the two prey types be labelled 1 and 2, so their energy contents are E_1 and E_2 and their handling times h_1 and h_2. The rate at which the two prey types are encountered is given by λ_1 and λ_2. The *profitability* of a prey item is defined as the ratio $E\!:\!h$, that is, the rate of energy intake achieved while the animal is actually feeding on a prey item. In our example, it is assumed that prey type 1 is more profitable than prey type 2, i.e.:

$$\frac{E_1}{h_1} > \frac{E_2}{h_2} \tag{1}$$

Consider a predator that forages for a time T_s and takes both prey types whenever encountered. It will obtain the following amount of food (E):

$$E = T_s(\lambda_1 E_1 + \lambda_2 E_2)$$

This energy will be gained in time T, which is the search time T_s plus the handling time:

$$T = T_s + T_s(\lambda_1 h_1 + \lambda_2 h_2)$$

The rate of energy intake (E/T) is therefore given by:

$$\frac{E}{T} = \frac{T_s(\lambda_1 E_1 + \lambda_2 E_2)}{T_s + T_s(\lambda_1 h_1 + \lambda_2 h_2)}$$

which, by dividing through by T_s, can be simplified to:

$$\frac{E}{T} = \frac{\lambda_1 E_1 + \lambda_2 E_2}{1 + \lambda_1 h_1 + \lambda_2 h_2}$$

Note that the search time has thus cancelled out. This gives us the rate of energy intake when both prey items are taken. By a similar argument, one can show that the rate of energy intake when only type 1 is taken is given by:

$$\frac{E}{T} = \frac{\lambda_1 E_1}{1 + \lambda_1 h_1}$$

Clearly, if the animal receives a higher rate of energy intake by concentrating on prey type 1 only, then it should do so. (In fact it is possible to show mathematically that a strategy of taking a prey type on a proportion of encounters cannot give a higher payoff; see Stephens & Krebs, 1986.) Therefore, the animal should specialize when:

$$\frac{\lambda_1 E_1}{1 + \lambda_1 h_1} > \frac{\lambda_1 E_1 + \lambda_2 E_2}{1 + \lambda_1 h_1 + \lambda_2 h_2}$$

that is, when the rate of energy gain from specializing on the more profitable prey type is greater than when taking both together. This condition can be rearranged as follows:

$$\lambda_1 > \frac{E_2}{E_1 h_2 - E_2 h_1} \tag{2}$$

This inequality gives the rate of encounter with the more profitable prey type, above which the animal should specialize on just the more profitable prey type. Note that it is independent of λ_2, the rate of encounter with the less profitable prey type.

This model makes three basic predictions:

1. Ranking by profitability. Prey are ranked according to E/T.
2. The zero-one rule. A prey item is either always taken or always ignored.
3. The most profitable prey type is always included in the diet. The decision to include less profitable prey types depends on inequality (2).

There have been various tests of this basic optimal diet model (often with slight modifications to reflect more accurately the biology of the animal under study; see Stephens & Krebs, 1986). The two shown in Fig. 5.1 are both on damselfly larvae of the same family (Odonata: Coenagrionidae), which are generalist aquatic predators. The behaviour of *Enallagma cyathigerum* (Charpentier) fits the model well, especially when compared with the dotted line which shows the expected pattern of prey selection if the animal foraged randomly for the two prey types (instars of the yellow fever mosquito *Aedes aegypti* (L.)). Results for *Pyrrhosoma nymphula* (Sulzer) are rather different. In this case the random predation model (the two-prey version of Rogers' (1972) random predator equation) provides a much better fit to the animal's behaviour. Does this mean that the behaviour of *E. cyathigerum* is optimal but that of *P. nymphula* is not? Some caution needs to be exercised. The interpretation of departures from optimality models in general and optimal foraging models in particular generated a heated argument in the late 1970s and 1980s (see, for example, Pierce & Ollason, 1987; and the reply by Stearns & Schmid-Hempel, 1987), part of a wider debate about the use of adaptive explanations in biology (Gould & Lewontin, 1979; Maynard Smith, 1978a; Mayr, 1983). What is clear from this generally cloudy debate is that the idea of optimality is an assumption and is not under direct test. When an optimality model is tested, the constraint assumptions of the model, along with the model's predictions, are tested, and the appropriateness of the model for describing an animal's behaviour in a particular set of circumstances is determined. If a model fits an animal's behaviour, then there are good grounds for thinking that the model encapsulates important aspects of the constraints under which the animal acts. If, on the other hand, a poor match is found, then this might be for one of four reasons (Krebs & Davies, 1981): there may be errors in the constraint assumptions; the currency assumption may be wrong; costs and benefits of particular behaviour patterns may not have been properly identified; and finally the basic assumption of optimality may be incorrect. The first three of these reasons can be

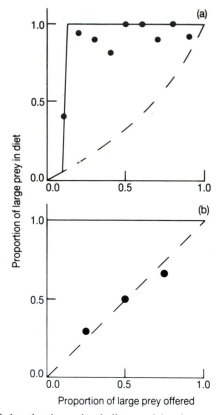

Fig. 5.1. Tests of the classic optimal diet model using two species of damselfly larvae, which were offered a choice between large and small larvae of the mosquito, *Aedes aegypti*. Solid lines show the predictions of the optimal diet model, dashed lines show predictions of the two-prey version of Rogers' (1972) random predator equation which assumes that the predators show no preference. (a) *Enallagma cyathigerum*; (b) *Pyrrhosoma nymphula*. (Redrawn from Chowdhury *et al.* (1989) and Harvey & White (1990).)

investigated by modifying the existing model, or by trying new models. However, it is extremely difficult to suggest ways of investigating the fact that the animal may not be optimal.

The usefulness of testing simple models is illustrated by the studies on the two damselfly species mentioned above. The fact that the prey model provides a good description of the behaviour of *E. cyathigerum*, but not of that of *P. nymphula*, may well lead to the development of further hypotheses about other constraints (such as the need to avoid being eaten) acting on the foraging behaviour of the two species.

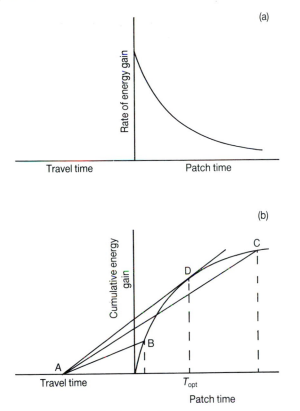

Fig. 5.2. The patch model: (a) relationship between time spent in a patch and rate of energy gain; (b) relationship between time spent in a patch and cumulative energy gain – the optimal patch time is given where the line A–D forms a tangent to the gain curve. (*Continued*)

5.4.2 *The patch model*

The next example of an optimality model is one of more general use because it can be applied in a wide variety of situations other than that for which it was first produced, animals feeding in patches of prey. Consider an environment consisting of a series of discrete patches containing prey items. Assume that an animal has no limit to the amount of time it can forage in a patch (patch time, T_p), but that the act of moving between patches takes a certain amount of time (travel time, T_t). When an animal first enters a patch, it encounters prey items at a high rate, but as it eats them, so depleting the patch, the rate of encounter with prey items inevitably declines (Fig. 5.2a). After a period of exploitation, it may therefore pay the animal to move to a new patch of prey. When should this

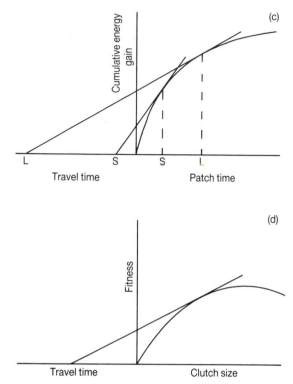

Fig. 5.2 (*cont.*). (c) the effect of changing travel time on optimal patch time – S and L refer to short and long travel times; (d) using the model to predict clutch size for a gregarious insect parasitoid. For further details, see text.

decision be made, assuming again that the animal wishes to maximize its long-term rate of energy intake? If an animal leaves a patch very soon after its arrival, it will spend most of its time moving between patches (where it cannot encounter prey items) and so will have a low rate of energy gain. Similarly, if it spends a very long time in a patch of food, then it will be foraging for long periods in patches of low prey density (because it has already eaten most of the food items present), and so again will have a low rate of energy gain. Clearly, an intermediate value of T_p must be optimal. It is easy to determine the optimal value of T_p (referred to as T_{opt}) graphically (for mathematical derivation, see Stephens & Krebs, 1986). Figure 5.2b shows the same data as Figure 5.2a, but this time expressed as a cumulative gain curve, rising steeply at first as the rate of energy intake is high on first entering a patch, but gradually approaching an asymptote as the patch is exploited. Now consider line

A–B: the slope of this line gives the average rate of energy intake [energy gained/$(T_t + T_p)$] for a predator with a relatively short T_p. If T_p is increased, then the rate of energy intake will increase. However, the slope of the line A–C gives the rate of energy intake for a long T_p: if the value of T_p is decreased, the rate of energy intake still increases. The maximum average rate of energy intake occurs when the line just forms a tangent to the curve (A–D). This gives us T_{opt}. Such a diagram is known as a *rooted-tangent diagram*. Figure 5.2c shows that increasing travel time increases the T_{opt}.

Two examples of experimental tests of this model will be given. The first also serves to illustrate how the model can be used to predict 'resource-emigration thresholds' (Parker & Stuart, 1976) in any situation where an animal gains fitness on a curve of diminishing returns. Rather than thinking in terms of travelling time and patch time, it is possible to recast the model in terms of search cost (equivalent to travel time) and investment cost (equivalent to patch time). Consider the case of a species of gregarious insect parasitoid (an insect whose larvae develop in or on the immature stages of other insects and which kills its host at eclosion). As clutch size increases, fitness will increase, but at a decreasing rate because small adults will result from large clutches. Thus it is possible to plot a graph of clutch size against fitness, which will take the form shown in Figure 5.2d (the eventual decline occurs because of the decrease in survivorship and/or fecundity as adult size becomes very small in the largest clutches). The x-axis, clutch size, represents the investment cost. Search cost might be the mean time to locate a host. Drawing the rooted-tangent diagram, it is possible to predict optimal clutch sizes for different rates of encounter with hosts (Charnov & Skinner, 1985).

Our first example of the use of the patch model concerns the copulatory behaviour of male dung flies (*S. stercoraria*). Males of this species aggregate at dung pats where they intercept and mate with females that come to lay eggs. Females mate many times and so will usually contain the sperm of previous males. Parker (1970b) demonstrated that, by extending copulation time, a male will fertilize a greater proportion of a female's eggs. but that this relationship is a curve of diminishing returns (Fig. 5.3a). Through careful fieldwork, he also discovered that the average time for a male to find and guard a female is 156.5 minutes. Using these data, he predicted an optimal copulation time of 41 minutes, which was close to the observed copulation time of 36 minutes (Fig. 5.3a; Parker & Stuart, 1976). For a discussion of the 5-minute discrepancy, see Kitcher (1985), Parker (1992) and Parker *et al.* (1993).

The other example is a study of the feeding behaviour of the greater

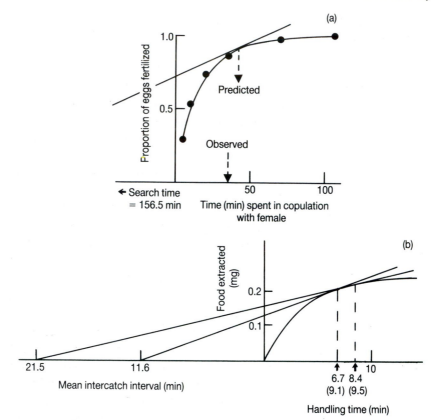

Fig. 5.3. Two tests of the patch use model: (a) predicting optimal copulation duration for the dungfly, *Scathophaga stercoraria* – the gain curve in this case shows proportion of eggs fertilised with increasing copulation time (redrawn from Parker & Stuart, 1976); (b) predicting optimal handling time for the water boatman, *Notonecta glauca* – in this case the gain curve shows the amount of food extracted whilst feeding on a single prey item (observed times are shown in brackets). (Data from Cook & Cockrell, 1978.)

water boatman (*Notonecta glauca*) by Cook and Cockrell (1978). Like other notonectids, this species is a predator that feeds by sucking body fluids from its prey. The rate at which food is extracted from the prey declines with time spent feeding (Fig. 5.3b). Therefore, handling time would be predicted to increase with mean interval between capture of food items. Figure 5.3b shows, for two mean intercatch intervals, the predicted optimal feeding times together with feeding times that were observed. There is a good qualitative fit: feeding time rises with mean intercatch interval. In the middle of the range of intercatch intervals, the model also provides a good quantitative fit, but the fit is less good for short and long intercatch intervals.

5.5 Extensions to the basic optimality models

Much criticism of optimality models in general and optimal foraging models in particular stems from their unrealistic simplifying assumptions. For example, the two models outlined above both assume omniscience – clearly an unrealistic assumption. That animals do approach the solutions predicted for an omniscient animal must show that they have efficient mechanisms for obtaining information about the environment and for using such information in making foraging decisions. This has led to the development of models of foraging behaviour which use learning as an essential component (e.g. Ollason, 1980; McNamara & Houston, 1985). Another drawback of the models above is that they use maximization of rate of energy intake as the currency assumption. This may not be the appropriate assumption for many animals. Consider shrews, which have very high metabolic rates. If a shrew selected prey according to the classical optimal diet model, it could find itself in a situation where the most profitable prey type is sufficiently abundant that it pays the animal to specialize, but sufficiently rare that the shrew is likely to starve to death before one is encountered! Clearly the animal should take a less profitable prey type if the alternative is death. This rather exaggerated example shows that maximization of rate of energy intake may not always be the appropriate currency assumption.

Another criticism of optimal foraging models is that the state of the animal in motivational and physiological terms is ignored. This objection has been overcome by the application of the technique of dynamic programming (Bellman, 1957) to 'state-variable models' (see Mangel & Clark, 1988). Such models view an animal's behaviour as a series of steps at which decisions are made. These decisions may be contingent on the animal's physiological state (e.g. energy reserves) and will in turn influence the animal's state at the next decision point. Dynamic programming models are proving a very useful means of studying animal behaviour.

5.6 Frequency-dependent fitness and evolutionarily stable strategies

Situations in which the payoff (fitness gain) to a strategy depends on the behaviour of other individuals in a population will now be considered.

Here, evolutionarily stable strategies, defined as '*a strategy which, if adopted by most individuals of a population, cannot be invaded by another strategy that is initially rare*' (Maynard Smith, 1982b), must be found. In this section, ESSs are found for a simple model, the 'Hawk-Dove' model. This model was produced by Maynard Smith and Price (1973) in order to explain the existence of ritualised fighting behaviour, rather than all-out aggression, during intraspecific conflicts. Prior to 1973, explanations for the absence of all-out aggression had often been couched in group-selectionist terms: it was for the 'good-of-the-species' if individuals avoided inflicting serious damage on conspecifics (Lorenz, 1966). However, a correct evolutionary explanation for the existence of ritualised behaviour must be in terms of benefit to the individual, not to the group or species. Maynard Smith and Price used the mechanics of games theory to provide such an explanation.

The first stage in the production of a games theory model is to specify the strategy set, that is a statement of the behavioural options that are open to the animals. In the Hawk–Dove model it is assumed that the animal can behave in three possible ways: 'escalate' – attack the opponent as aggressively as possible; 'display' – do not escalate but do not retreat; and 'retreat' – leave the contest. An animal that displays will not injure its opponent; one that escalates may succeed in causing injury. These three behaviour patterns are combined into two strategies:

Hawk (*H*): immediately escalate, retreat only if injured;
Dove (*D*): retreat immediately if opponent escalates, otherwise display.

The model assumes that the animals are fighting over an indivisible resource and that the fitness of the individual that wins a contest (so gaining access to the resource) is enhanced by V fitness units. If an animal is injured in a fight, it pays a cost, C. The population consists of a proportion p Hawks and $(1 - p)$ Doves. Individuals interact in pairs at random and each individual is involved in only one interaction per generation. To simplify matters, the model assumes that the species concerned is haploid and asexual, thus H individuals give rise only to H offspring.

The outcomes, in terms of fitness, when two individuals fight are:

H v. *H*: each contestant has a 50% chance of winning the resource (thus gaining V), and a 50% chance of being injured (thus losing $-C$), hence, on average, H gains $(V - C)/2$;
H v. *D*: H escalates immediately, causing D to retreat. H therefore gains V; D gains nothing, but pays no cost of injury.
D v. *D*: each D has a 50% chance of winning the resource, thus D gains $V/2$. At this point the model assumes that there is no cost to display: this assumption will be relaxed below.

$(V - C)/2$ is the *payoff* to H when played against H. Note that payoffs refer to *changes* in fitness, not fitness itself. Payoffs can be arranged in a *payoff matrix*, which shows the change in fitness when each strategy is played against itself and against all other strategies in the model. The payoff matrix for the Hawk–Dove model is:

		Played against:	
		Hawk	Dove
Payoff to:	Hawk	$\dfrac{V - C}{2}$	V
	Dove	0	$\dfrac{V}{2}$

A shorthand for the elements of the payoff matrix is $E(D, H)$, the payoff to D when played against H. This payoff matrix can be used to simulate evolution within a population that consists of a mixture of H and D. Let $W(H)$ and $W(D)$ be the fitness of the H and D strategies. If the individuals playing the two strategies interact at random, then H will meet H in a proportion p fights and D in $(1 - p)$ fights. Equations for $W(H)$ and $W(D)$ are:

$$W(H) = W_0 + pE(H, H) + (1 - p)E(H, D)$$

$$W(D) = W_0 + pE(D, H) + (1 - p)E(D, D)$$

Here W_0 is the basic fitness of the animal, that is fitness arising from activities other than the game under consideration. It is assumed to be the same for all strategies.

Mean fitness $\bar{W}$ is calculated at

$$\bar{W} = pW(H) + (1 - p)W(D)$$

p', the proportion of H in the next generation, is given by:

$$p' = \frac{W(H)}{\bar{W}}$$

Using this technique, it is possible to simulate evolution within a population and hence look for ESSs. However, a simpler way of finding ESSs is to use the 'standard conditions' of Maynard Smith and Price (1973). These are based on the criterion of non-invadability, which is the basis of the formal definition of an ESS. Consider a population consisting entirely of individuals playing strategy I in pairwise encounters. The payoff to strategy I is therefore $E(I, I)$. Now consider a mutant strategy J which, because it is a mutant, will be rare. It will therefore play only against individuals of strategy I and hence have an average payoff of $E(J, I)$. It will increase in frequency if $E(J, I) > E(I, I)$. But consider the case if $E(I, I) = E(J, I)$. There will be no selection for J to change in frequency, but J may increase through random genetic drift. Strategy J is then likely to play against itself as well as against strategy I. Selection for strategy J to increase further will occur if $E(J, J) > E(I, J)$. Therefore the following conditions must be satisfied for I to be an ESS:

Either: $E(I, I) > E(J, I)$

Or: if $E(I, I) = E(J, I)$, then $E(I, J) > E(J, J)$

These are the ESS standard conditions. Their use can be demonstrated with reference to the Hawk–Dove model. Two cases can be considered: $V > C$ and $V < C$.

Case (i) $V > C$. Let $V = 4$ and $C = 2$. The payoff matrix is:

		Played against:	
		Hawk	Dove
Payoff to:	Hawk	1	4
	Dove	0	2

$E(H, H) = 1$, which is greater than $E(D, H)(=0)$, so H is an ESS. D is not an ESS, since $E(H, D) = 4$ is greater than $E(D, D)(=2)$. If the value of the resource is greater than the cost of serious injury during fights, then animals would be expected to adopt a hawkish strategy and fight aggressively. The ESS conditions begin to make intuitive sense if one thinks of the fitness of a rare mutant in a population. Imagine a population of Doves. They will fight only other Doves, and thus their average payoff will be 2. Now consider a mutant Hawk. Because it is a mutant, and therefore rare, it is unlikely to meet another Hawk, and thus it will fight Doves. Its payoff will therefore be 4, and so the mutant Hawk will have a higher fitness than the Doves in the population and hence will spread via natural selection. So Dove is not an ESS. If one considers a population of Hawks, each Hawk will receive an average payoff of 1. A mutant Dove, which will only meet Hawks, receives an average payoff of 0; Dove cannot invade the population, so Hawk is an ESS.

This example also shows that natural selection does not always maximize the mean fitness of a population. A population of Doves has a mean fitness of $W_0 + 2$, which is higher than the mean fitness of a population of Hawks ($W_0 + 1$), but Hawk is the ESS.

One example of a situation where the value of the resource is greater than the cost of fighting is provided by the larvae of solitary parasitoid wasps: a host can only support the development of one larva and if more than one larva occurs in a host through superparasitism, a fight inevitably occurs. Early instars are equipped with large mandibles for fighting and one opponent is invariably killed.

Case (ii) $V < C$. Let $V = 2$ and $C = 4$. The payoff matrix is:

		Played against:	
		Hawk	Dove
Payoff to:	Hawk	−1	2
	Dove	0	1

In this case $E(D, H) > E(H, H)(0 > -1)$ and $E(H, D) > E(D, D)(2 > 1)$ so neither pure H nor pure D can be an ESS. Therefore if there is an ESS, it must consist of a mixture of H and D – a 'mixed ESS'. In a mixed ESS both strategies must have equal fitness. The stable proportion of H in the population can be calculated by setting:

$W(H) = W(D)$

Cancelling W_0 on each side gives:

$pE(H, H) + (1 - p)E(H, D) = pE(D, H) + (1 - p)E(D, D)$

and substituting for the payoffs gives:

$$p\left(\frac{V - C}{2}\right) + (1 - p)V = p \cdot 0 + (1 - p)\frac{V}{2}$$

which reduces to:

$$p = \frac{V}{C}$$

For the case of $V = 2$ and $C = 4$, a population which consists of 50% H and 50% D would be stable against invasion. This population composition can come about in one of two ways: either half the population are H individuals and half the population are D, or all individuals play the same pure strategy 'play H with a probability 0.5'.

The Hawk–Dove model helps to explain the evolution of ritualised display behaviour: if the cost of injury exceeds the value of the contested resource, then a mixture of H and D in the population would be expected. If C is very much greater than V, then hawkish behaviour would be very rare and most contests would be settled by display.

5.6.1 More than two strategies

Consider a strategy called Retaliator (R): 'display, but if the opponent escalates, escalate as well'. How will R fare in a population of H and D? The payoff matrix is as follows:

	H	D	R
H	$\dfrac{(V-C)}{2}$	V	$\dfrac{(V-C)}{2}$
D	0	$\dfrac{V}{2}$	$\dfrac{V}{2} \times 0.9$
R	$\dfrac{(V-C)}{2}$	$\dfrac{V}{2} \times 1.1$	$\dfrac{V}{2}$

The additional assumption has been made that, on 10% of occasions that R and D meet, the Retaliator discovers that the Dove is not prepared to escalate and so escalates, winning the resource at no cost. This increases $E(R, D)$ and decreases $E(D, R)$ slightly. Inspection of the matrix reveals that, if $V > C$, then R is an ESS ($E(R, R) > E(D, R) > E(H, R)$). R is also an ESS if $V < C$. But what of a strategy I, which plays H with a probability $p = V/C$ and D on the remainder of occasions? If $V = 2$, $C = 4$ and $p = 0.5$, $E(I, I)$ is 0.5 (the average of $E(H, H) + E(D, H) + E(H, D) + E(D, D)$). $E(R, I) = 0.05$ so that I is an ESS. This illustrates an important point: a game with more than two strategies may have more than one ESS. The eventual composition of the population will depend on the starting conditions: a population consisting of a large number of H and D and few R will go to the $H - D$ mixed ESS: one in which R is common will evolve towards R.

5.6.2 Continuous strategies

The Hawk–Dove game is an example of a discrete-strategy game: animals are able only to use one of a series of discrete strategies. Often, however, animals can vary their investment in a situation on a continuous scale. Examples might include: allocation of time between competing activities, such as foraging and vigilance; optimal body size when reproductive success is determined partly by relative body size; and sex ratio of offspring to produce. The question then arises: what is the ESS level of investment? The answer to this question will be given with reference to a model termed 'the war of attrition' (Maynard Smith & Price, 1973), which arose as an extension of the Hawk–Dove model. It was assumed above that when two Doves interacted, the contest was settled quickly and without escalation and that each competitor had a 50% chance of winning the encounter. How might a contest be settled under such circumstances? One possibility is that each contestant may decide on a cost that it is prepared to incur during a contest, and not modify that investment during the contest: the animal that is prepared to accept the highest cost wins the encounter. If it is assumed that cost is a linear function of time, then one can imagine each animal choosing a persistence time at the start of a contest: the one which is prepared to persist for the longest period wins. What is the ESS persistence time? This explanation is based on Parker (1984). Imagine a population in which all individuals play the same persistence time T. Cost is accumulated at a rate c when competing for a resource worth V. In such a population the average payoff to an individual is:

$$E(T, T) = \frac{V}{2} - cT$$

Imagine a mutant individual playing a time T^* slightly greater than T. It will always win the resource, but will pay the cost cT determined by the shorter persistence time. The average payoff is, therefore:

$$E(T^*, T) = V - cT$$

and so the mutant will have a higher fitness. However, this does not mean that the population will come to consist of individuals playing the longest possible persistence time because if $cT > V$, then any mutant playing $T = 0$ will spread. Hence no single choice of persistence time can be an ESS. The ESS in this case is to choose persistence time from a probability distribution $p(t)$. This distribution must have the property that any particular time played against it must have the same payoff (remember that, in a mixed strategy, payoff must be equal to the different components of the strategy). If I is the ESS, then the expected payoff to an individual playing time t' is:

$$E(t', I) = \int_0^{t'} (V - ct)p(t) \, dt - \int_{t'}^{\infty} ct'p(t) \, dt = \text{constant A}$$

Contests: won   lost

From this it can be shown that the ESS distribution of persistence times is:

$$p(t) = \frac{c}{V} \exp\left(-\frac{ct}{V}\right)$$

A strategy in which animals choose their persistence time from this

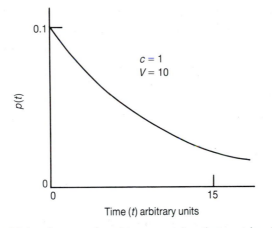

Fig. 5.4. The ESS for the war of attrition, assuming that cost is a linear function of the display time.

distribution (Fig. 5.4) cannot be invaded by individuals playing different persistence times, or distributions of persistence times.

In continuous strategies games, the ESS (if one exists) may be either a pure strategy (play one particular value), or a mixed strategy such as the probability distribution outlined above for the war of attrition. In the case of a pure strategy, no mutant strategist that plays a value close to the ESS could invade. In the case of a mixed strategy, all values of the strategy give the same average payoff.

5.6.3 *Asymmetric games*

In the models outlined so far it has been assumed that the animals playing the games are equal in all respects other than the strategy that they adopt during the game. However, animals often differ in other ways: some individuals may be larger due to a better food supply during development; some may be older; some may have more to gain from access to a particular resource; or individuals may differ in some arbitrary way that is not related to fitness, such as being the resident rather than the intruder in a fight. Parker (1984) classified asymmetries into three categories, which are largely based on models of fighting behaviour, but are also more generally applicable (Table 5.1).

Asymmetric games may be considered by adding a third strategy to the Hawk–Dove model. This strategy uses ownership as a convention for settling disputes over resources in which one individual (the owner) is

Table 5.1. *A classification of asymmetries that can occur during conflicts between animals (Parker, 1984)*

(i) *Asymmetries in resource value.* One opponent may have more to gain from winning than its opponent.

(ii) *Asymmetries in 'resource-holding power' (RHP).* Opponents may differ in their fighting ability or in some factor which affects their ability to defend a resource.

(iii) *Uncorrelated asymmetries.* These are arbitrary asymmetries that are uncorrelated with any effect on payoff. An example would be the owner–intruder asymmetry described in the text.

---

present at the start of the contest. The third strategy, termed bourgeois (B), behaves as follows: 'if owner play H, if intruder play D'. Assuming that individuals have a 50% chance of being an owner and that ownership is uncorrelated with strategy, then the payoff matrix is as follows:

	H	D	B
H	$\dfrac{V-C}{2}$	V	$\dfrac{3V-C}{4}$
D	0	$\dfrac{V}{2}$	$\dfrac{V}{4}$
B	$\dfrac{V-C}{4}$	$\dfrac{3V}{4}$	$\dfrac{V}{2}$

If $V < C$, then B is the ESS. This shows that an arbitrary asymmetry may be used as a cue to settling contests quickly, and without escalation. Indeed, in nature it is relatively common to find owners winning contests, as illustrated by Davies' (1978) study of the speckled wood butterfly (*Pararge aegeria*). It should also be noted that the strategy anti-bourgeois ('play D if occupant, H if intruder') is also an ESS if added to the basic Hawk–Dove model. Evidence for such paradoxical (as opposed to commonsense) ESSs is rather rare. Maynard Smith (1982b) gives an example of the colonial spider *Oecibus civitas* (Burgess, 1976). If an intruder is introduced on one side of a colonial web, it will displace the nearest animal, which will then oust another individual and so on, giving a wave of replacements across the web.

The bourgeois strategy is an example of how an asymmetry uncorrelated with payoff can be used to settle contests. A similar case can be made to show that strategies based on asymmetries that are correlated with fitness can be ESSs. A strategy 'In role A play I, in role B play J' can be an ESS where role A and B may be 'bigger:smaller' or 'more to win:less to win'. Such strategies are termed conditional strategies. Because conditional strategies are often based on phenotypic differences such as size, they are also termed 'phenotype-limited strategies'. Normally it would be expected that individuals in role *A* (larger, more to win) would use the strategy that gives the higher payoff. Paradoxical ESSs are possible for correlated asymmetries, but only if the asymmetry is small (Maynard Smith & Parker, 1976).

The important difference between mixed ESSs and conditional ESSs of the type mentioned above is that, for mixed ESSs the payoffs to each component of the ESS must be equal, but this is not the case for conditional ESSs: here, there is no reason to expect the strategies that make it up to have equal fitness. In the next section, examples will be given to show how it is possible, in practice, to distinguish between mixed and conditional ESSs.

5.7 Alternative strategies – determining payoffs

5.7.1 Salmon life-histories

Coho salmon (*Oncorhynchus kisutch*) lay eggs in the upper reaches of rivers on the North-West Pacific coast of America. After hatching, young fish swim to sea where they grow, returning to rivers for a single breeding attempt, after which they die. Females take 3 years to reach maturity. Males show variation in life-history: some individuals return after 2 years at sea; others after 2 (Fig. 5.5a). Those which return after 2 years, termed 'jacks', are small; individuals that return after 3 years are larger and are termed 'hooknoses', because their mouths are specially adapted for fighting. As well as being different in size and morphology, hooknoses and jacks differ in behaviour. Gross (1985) studied this variation in detail to discover if it represented a mixed or a conditional ESS. Hooknoses fight with each other to gain access to spawning females; jacks sneak close to females by hiding behind rocks. Assuming that the success of males in fertilising females depends on how close they are able to get to females, it is possible to work out the relative mating success of hooknoses and jacks of different sizes – jacks obtain 66% fewer mates than hooknoses (Fig 5.5b). This short-term measure of mating success is not an adequate measure of fitness because it does not take account of the fact that hooknoses and jacks differ in their breeding lifespan (determined by observation of marked individuals) and in their chances of surviving to maturity (determined from fisheries data). Putting these data together, the relative fitnesses of hooknoses and jacks are calculated

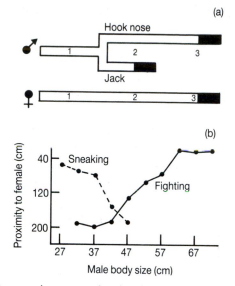

Fig. 5.5. Alternative mating strategies in the coho salmon: (a) life-histories of females and hooknose and jack males – solid areas represent reproduction; (b) effect of male body size and strategy (fighting or sneaking) on the ability of males to get close to spawning females. (Redrawn from Gross, 1985.)

as follows:

$$\frac{W_{\text{Jack}}}{W_{\text{Hooknose}}} = \frac{0.13}{0.06} \text{ (survivorship to maturity)}$$

$$\times \frac{8.4}{12.7} \text{ (breeding lifespan – days)}$$

$$\times \frac{0.66}{1.00} \text{ (relative mating success)}$$

$$= 0.95$$

The fitness of the jack relative to the hooknose is therefore 0.95, which is not significantly different from 1. This is therefore an example of a mixed ESS with hooknoses and jacks having the same fitness. It is also easy to see that the fitnesses of the two strategies are likely to be frequency dependent: if the frequency of hooknoses rises, then there will be more competition among the fighting males and so mating success will decline. Similarly, an increase in the frequency of jacks will cause a decrease in their mating success through increased competition for the limited number

of hiding places. Such frequency dependence is essential for the maintenance of a mixed ESS.

5.7.2 Dragonfly mating strategies

Males of some species of dragonfly are territorial, defending sites that
usually contain an oviposition substrate. Females mate with a territorial
resident before laying eggs in his territory, and are frequently guarded by
him as they oviposit. However, in some species, males who are excluded
from territories adopt alternative strategies for obtaining mates. Often this
takes the form of 'sneak' behaviour in which they sit at the edge of other
males' territories, escaping the notice of the territorial male and trying to
intercept females before they reach the resident. In most species this
appears to be a conditional strategy, based on age and/or size, and it is
often dependent on male density. Males of *Calopteryx maculata* adopt
such a role when young and old (Fig. 5.6). In *Libellula quadrimaculata*,
as male density rises, large males adopt sneak roles: this slightly counter-
intuitive result (one normally expects large individuals to be better
fighters) can be explained by the fact that small males are more aerobatic
and thus better able to defend territories (Convey, 1989). In all these cases,
there is no reason to expect payoffs to the sneak and territorial males to be
equal: sneaky males are probably making the best of a bad job. However,
few studies have looked in detail at the payoffs to such strategies.

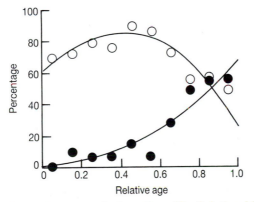

Fig. 5.6. Alternative mating strategies in a damselfly. Relationship between male
age and the adoption of territorial (open circles) and sneaky (closed circles) mating
strategies by males of the damselfly, *Calopteryx maculata*. Data show the
percentage of males engaged in each activity during each tenth of their lifespan.
(Redrawn from Forsyth & Montgomerie, 1987.)

In principle it is easy to distinguish between mixed and conditional ESSs. Notwithstanding the difficulties of measuring payoffs, in practice further complexities can arise. Consider the case of the anthophorid bee, *Centris pallida* (Alcock *et al.*, 1977). Males vary much more in size than females and the mate-finding strategy of a male is dependent on his size: large males dig into the ground for females which are emerging from pupae. When several males are competing for the same female, the largest male usually wins the fight. Small males do not dig for emerging females, but hover above the ground and hunt for females that have been missed by the digging males. At first sight, this seems a clear case of a conditional ESS, 'if large fight, if small hover'. However, the fact that the variance of male size is larger than that of females suggests that the situation is more complicated. Offspring size is determined by the number of eggs a mother lays in a brood chamber and the amount of food provided for the developing larvae. That females can control offspring size quite closely is shown by the uniform size of daughters. This suggests that the wide variation in male size may be adaptive: by producing sons of a range of sizes, females may be playing a mixed strategy, and so forcing a pure conditional strategy of their offspring.

5.8 Mating patterns

In the remainder of this chapter the techniques outlined above will be used to analyse patterns of mating (often known as mating systems, Emlen & Oring, 1977; Davies, 1991). The mating pattern of a species arises as a result of options associated with searching for mates (should the male search? or the female? or both?), and mating and care of the young (which sex gives parental care?). The ESS for such cases consists of a pair of strategies, I_f and I_m for the female and male. These strategies determine the mating pattern. It is important to view mating patterns as the product of selection acting separately to maximize the fitness of each sex. Thus the reproductive interests of the sexes do not always coincide; the outcome of evolution in such cases can prove interesting. There exists considerable variation in mating patterns, both within and between species. Several classifications have been proposed, but most currently accepted schemes are based on that of Emlen and Oring (1977). Table 5.2 shows an example.

One core idea that has influenced much work on mating systems over the past two decades is Trivers' concept of relative parental investment (PI). PI is defined as '*any investment by the parent in an individual offspring*

Table 5.2. *Classification of mating patterns (Alcock, 1979, after Emlen & Oring, 1977)*

Monogamy: an individual mates with one partner per breeding season

Polygyny: individual males may mate with more than one female per breeding season
A. Parental investment polygyny: males vary in attractiveness to females because of great individual variation in male parental investment
   1. Resource defence polygyny: males monopolise resources that are useful to females; several females may mate with a male that controls access to superior resources
   2. Parental care polygyny: males vary in the quality of direct parental care they offer their progeny; several females may choose a superior parent
B. Pure dominance polygyny: males vary in dominance status (and, therefore, genetic quality), with several females selecting a high-ranking male
   1. Harem defence polygyny: males compete for priority of access to groups of females that have formed for non-reproductive reasons (usually for mutual protection against predators)
   2. Lek polygyny: males compete for high dominance ranking within a group, usually at a traditional display arena; females mate with the top males

Polyandry: individual females may mate with more than one male per breeding season
A. Parental investment polyandry: females mate with several males that provide parental investment
   1. 'Prostitution' polyandry: females mate with several males to gain the parental investment benefits offered by their mating partners
   2. Resource defence polyandry: in cases in which males make a GREATER parental investment per offspring than females, a female may control a resource attractive to more than one male
B. Pure dominance polyandry: in cases in which the male parental investment per offspring exceeds that of the female, females may compete for dominance positions in female aggregations; high-ranking females may be chosen by more than one male

that increases the offspring's chance of surviving (and hence reproductive success) at the cost of the parent's ability to invest in other offspring' (Trivers, 1972). Trivers argued that because females, by definition, produce larger gametes than males, their PI in a batch of offspring, even in the absence of parental care, is fundamentally higher than that of males. In some viviparous species, such as mammals, females produce relatively small eggs, but the costs of nurturing the embryo represent considerable PI and usually raise it above that of the male. This asymmetry in relative PI has important consequences: members of the sex with the lower relative PI can increase their reproductive success by mating as many times as possible. In contrast, members of the sex with the higher relative PI are likely to be limited by the resources they can put into reproduction.

Increasing their number of matings is unlikely to increase their reproductive success and so they become a resource for which the other sex competes. Trivers proposed that this generates sexual selection: the sex which has the lower relative PI competes with other members of the same sex for access to members of the opposite sex (intrasexual selection). For the sex with the higher relative PI, mating with an inappropriate partner (a member of the wrong species or a conspecific that will provide little parental care or 'bad' genes) may be extremely costly as the investment in a whole breeding attempt may be lost. It therefore pays members of this sex to be choosy about their mates. This generates intersexual selection. Because of the fundamental differences between males and females in investment in sperm and eggs, sexual selection is usually of the form in which the intrasexual component consists of male–male competition and the intersexual component takes the form of female choice. However, it is important to recognise that this is a generalisation and that in species where males make large investments in offspring it is possible to get sex-role reversal and for females to compete for access to males and males to be the choosy sex.

Trivers' ideas of the relationship between gamete dimorphism, relative PI and sexual selection provide one central theme in the analysis of mating strategies. A second important theme concerns the dispersion of resources, and this can be of two types. First, there are resources that the female requires for laying eggs and perhaps for rearing offspring and, second, the resources can be females themselves. The ability of single males to control access to either of these two types of resource greatly influences mating patterns.

Three topics will be discussed: first, which sex expends effort in searching for partners; second, which sex provides parental care; and third, the evolution of polygamous mating systems. Finally, by discussing the highly variable mating pattern of the dunnock (*Prunella modularis*), it will be shown that mating patterns are emergent properties of decisions made by individuals.

5.8.1 *Which sex searches?*

Given that females generally invest more in offspring and so are a resource for which males compete, one might intuitively expect males to invest energy in searching and females to expend little effort on this activity. Such is certainly the case for the gametes of anisogamous species, where eggs are non-motile and sperm highly motile (Parker *et al.*, 1972). But is such a result an inevitable consequence of relative PI and sexual selection? Hammerstein and Parker (1987) give details of a 'mobility game', an ESS model in which the two evolutionary variables are the mobility levels of each sex. They consider a situation in which individuals search for a mate on a chessboard by moving between squares. When two individuals of the opposite sex meet they leave the board in order to mate and rear offspring and also for a period of recovery which is proportional to the search costs incurred in finding a mate. Thus search costs have the effect of lengthening the reproductive cycle. Individuals with short reproductive cycles are assumed to have higher fitness because they will be able to complete more reproductive cycles in their lifespan. Mobility levels v_f and v_m for females and males vary between 0 and 1 and represent probabilities of moving between squares per unit time. The interesting prediction of this model is the existence of two alternative ESS pairs of strategies, one of which has a female mobility of level 1 and a male mobility level of 0 (i.e. the female makes all the moves and the male sits and waits), and the other vice versa. Hammerstein and Parker (1987) term these alternatives the 'roaming female' and 'roaming male' ESSs.

The existence of alternative ESSs always poses a problem: to which will evolution proceed? Remember from above that this is often dependent critically on the initial frequency of the strategies. Thus, if a population started with v_m near to 1 and a small value of v_f, then evolution would proceed to the roaming male ESS and the opposite if females started as mobile. But what would happen with intermediate levels of movement in each sex? Hammerstein and Parker (1987) suggest that the 'roaming male' ESS is more likely to evolve because the selective force stabilizing this ESS is stronger than that stabilising the roaming female ESS. This arises because of the assumption that females invest more in a batch of offspring than males. Thus, while differences in relative PI do not guarantee that the male will be the sex that searches actively for mates, they do make it more likely.

5.8.2 *The parental care game*

Which sex should care for the offspring? Benefits of parental care vary between species, a variety of factors, such as predation rate on juveniles, increasing the value of parental care, the prime value of such care in increasing the survivorship of offspring. The survivorship of young cared for by 0, 1 and 2 parents is referred to as p_0, p_1 and p_2, respectively. Offspring are likely to survive as well or better if cared for by two parents

rather than by one, and even uniparental care is likely to be better than
no parental care. So it is reasonable to assume that $p_0 \leqslant p_1 \leqslant p_2$.
However, the marginal increase in survivorship between offspring cared
for by two parents as opposed to one may often be far less than the
increase given by a single parent caring as opposed to no parent caring
(i.e. $p_0 < p_1 \cong p_2$). If this situation holds, then there is obviously a conflict
between the sexes in who should give parental care. The parent left caring
for the offspring cannot breed again until they have become independent,
whereas the parent which leaves can potentially find another partner and
so produce more young. Interestingly, there appear to be broad taxonomic
differences in the sex which 'wins' this evolutionary battle (Dawkins &
Carlisle, 1976): in mammals and birds, if uniparental care is given it is
usually by the female, but in fish it is often the male. It is convenient to
think in terms of two strategies that can be played by either sex, 'guard'
(G) and 'desert' (D). If both sexes guard, then biparental care is given, if
one guards and the other deserts, then uniparental care results, and if
both desert then no parental care is given. If ecological conditions favour
uniparental care (i.e. if $p_0 < p_1 \cong p_2$), then there is a conflict between the
sexes in who cares for the offspring. It is clearly better to be the sex that
deserts, as long as the other sex does not desert as well.

An early consideration of this situation was by Trivers (1972). He
argued that the sex that should desert first is the one that has invested
least in the offspring up until the moment of desertion. This analysis was
criticised by Dawkins and Carlisle (1976), who argued that this explanation committed the 'Concorde fallacy'[2] – animals should base their
decisions on future potential gains, rather than past investment. Maynard
Smith (1977) pointed out that Trivers was, in fact, strictly correct
because of his definition of PI in terms of loss of future offspring. However,
Maynard Smith (1977) went on to present an explicitly forward-looking
games theory model for the evolution of parental care, and this will be
used to show how different patterns of parental care can evolve.

Maynard Smith's model assumes that the success of a pair during the
breeding season depends both on parental care given after eggs are laid
and on the investment the female puts into the eggs. Female investment
in eggs depends on whether she guards or not: a female that deserts can
produce E eggs, and one that guards e eggs, where $E > e$. A male that
deserts has a probability q of finding another mate. The payoff matrix for
this game is shown in Fig. 5.7. Note that each cell in the table contains
two entries. These represent payoffs to males (below diagonal) and females
(above diagonal). To show how these payoffs have been calculated,

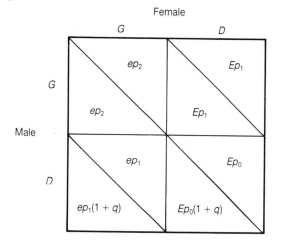

Fig. 5.7. Payoff matrix for the parental care game (Maynard Smith, 1977). See text for details.

consider the case in which both males and females desert. Females will produce E eggs, which will have a survival probability of p_0 because neither sex cares, hence the payoff to females when both sexes desert is Ep_0. Now consider the payoff to males. From their first partner's offspring they will receive Ep_0, but they also have a probability q of mating again. The model assumes that a male's second partner will play the same strategy as his first. The male will therefore receive Ep_0 with a probability q and so, in a population in which both males and females desert, an average male's payoff will be $Ep_0 + qEp_0 = Ep_0(1 + q)$. A similar argument can be used to determine the remaining payoffs for the matrix. What are the ESSs for this game? Using the criterion of non-invadability, it is possible to ask whether populations of various compositions can be invaded by mutant individuals playing an alternative strategy. Consider first the case of a population where all males and females play D. Now consider the fate of a mutant female that guards. Her payoff will be ep_1, and she will have a higher fitness so that the trait should spread through the population provided that $ep_1 > Ep_0$. Now consider the fate of a mutant G male. His payoff will be Ep_1. So if $Ep_1 > Ep_0(1 + q)$, then the mutant trait will spread. This latter condition can be simplified to $p_1 > p_0(1 + q)$. These are conditions under which mutant G males and females can invade a population in which all males and females play D. For such a population to be at an ESS, these conditions obviously must not hold (a population at the ESS is resistant to invasion). So a population in which both sexes desert will be at an ESS if $ep_1 < Ep_0$ (otherwise the female will guard) and $p_1 < p_0(1 + q)$ (otherwise the male will guard). This is the first of four possible ESSs in the parental care game. Conditions for the other ESSs can be determined in a similar manner. Maynard Smith (1977) summarized these conditions as follows:

ESS 1. *D* female and *D* male (as explained above)
    requires $Ep_0 > ep_1$, or the female will guard,
    and $p_0(1 + q) > p_1$, or the male will guard.

ESS 2. *D* female and *G* male (the 'stickleback' ESS),
    requires $Ep_1 > ep_2$, or the female will guard,
    and $p_1 > p_0(1 + q)$, or the male will desert.

ESS 3. *G* female and *D* male (the 'duck' ESS),
    requires $ep_1 > Ep_0$, or the female will desert,
    and $p_1(1 + q) > p_2$, or the male will guard.

ESS 4. *G* male and *G* female
    requires $ep_2 > Ep_1$, or the female will desert,
    and $p_2 > p_1(1 + q)$, or the male will desert.

Now consider the following arbitrary values: $E = 10$, $e = 6$, $p_2 = 0.8$, $p_1 = 0.3$, $p_0 = 0.2$ and $q = 0.7$. Substitution of these numbers into the conditions above will reveal that the requirements for both ESS 1 and ESS 4 are satisfied, and thus there are two alternative ESSs, with both partners guarding or with both partners deserting. Similarly, values can be chosen for which ESS 2 and ESS 3, the stickleback and duck ESSs, are alternatives. Alternative ESSs appear to be a common feature of games between the sexes (Hammerstein & Parker, 1987).

What does this model tell us in biological terms? The first thing is that uniparental care is likely to evolve if $p_1 \gg p_0$ and p_2 is not much larger than p_1 (Maynard Smith, 1977). The stickleback ESS will be favoured if $E > e$ and the duck ESS if q is sufficiently large. Of course, it is perfectly possible for both the stickleback and the duck ESSs to exist. Imagine an ancestral condition with biparental care: selection would favour desertion by one of the two sexes, but it is not clear from the model which sex would win the evolutionary battle. Several suggestions have been made to explain why, in most species, it is the male that deserts. One suggestion, made by Trivers (1972), is that the sex which has the first opportunity to desert should take advantage of this. In a species with internal fertilisation, the male is free to go after he has deposited his sperm, but the female still must lay eggs and so will be 'left with the zygotes' (Hammerstein & Parker, 1987). On the other hand, in species with external fertilisation, females shed their eggs before males release their gametes. This gives females the opportunity to desert first. Support for this hypothesis comes from the correlation between mode of fertilisation and the giving of paternal and maternal care in fishes and amphibians (Gross & Shine, 1981), with paternal care occurring predominantly in taxa with external fertilisation. However, other explanations are equally consistent with the data. For example, males may be much more effective at defending young (because they are larger). If so, male defence would be a resource for which females compete, and a single male could allow several females to lay eggs in his nest, thus reducing the cost of being the caring parent. Another possible reason for favouring desertion by a particular sex is the rate at which gametes can be produced (Bayliss, 1981). If one sex can produce gametes at a higher rate than the other, then this makes the benefits of deserting higher for the first (in other words, that sex has a higher potential 'opportunity cost' (Grafen, 1980) if it guards). Because sperm are relatively cheap to produce, they can be produced at a higher rate than eggs and so this tends to favour male desertion.

Maynard Smith's (1977) model limited each sex to behaving in one of two ways: guard or desert. In practice, however, the decision to invest in offspring may not be an all-or-none one. Rather, the question may be how much to invest in offspring, and in particular how to respond to changes in allocation of resources to offspring by a partner. Consider the case of a pair of birds provisioning a brood of young. The number of visits made to a nest with food by each partner could vary considerably, and individuals that put less effort into feeding are likely to have a higher probability of surviving to the next breeding attempt. If one parent reduces the number of visits it is making, how should the other parent respond? Clearly it could pay to increase feeding rate, thus enhancing brood survival, or it could pay the parent to lower its feeding rate. This situation has been modelled by Houston and Davies (1985), whose main conclusions can best be shown graphically. Their model assumes an asymptotic increase in nestling survival with parental effort and a decrease in parental survival (Fig. 5.8). Figure 5.9a shows how the optimal parental effort of one parent decreases as the effort of the other parent increases. The question is, does a pair of parental efforts exist, such that it pays neither sex to deviate from that level of investment? Four situations are possible (Fig. 5.9b–e). In the first case (Fig. 5.9c) the female line lies above the male line and the ESS is for the female to do all the work and the male

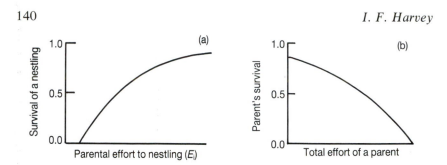

Fig. 5.8. Assumptions of the Houston and Davies model of parental care: (a) nestling survival rises with parental effort; (b) parental survival decreases with the effort they put into chick rearing. (Redrawn from Houston & Davies, 1985.)

to desert. If, on the other hand, the male line lies completely above the female line, then the ESS is for the male to provide all of the investment (Fig. 5.9d). When the male and female investment lines cross, the ESS depends on the slopes of the lines at the point they intersect. Consider the case where the slopes are less than -1 at the point of intersection (Fig. 5.9b), meaning that if one sex increases its effort, the other sex responds by reducing its effort, but not by such an extent that the overall provision to the brood declines (a slope of -1 would result in overall provision remaining constant). If a female invests x, then the male's best investment is shown at point 1. The female's best reply to this is to increase her effort slightly to 2. The male responds by reducing effort slightly to 3, and the process proceeds until a stable equilibrium is reached where the two lines intersect, giving ESS investment levels for each sex from which it does not pay individuals of either sex to deviate. When the two lines intersect with a slope greater than -1, then a decrease in investment by one sex is more than compensated for by the increase in investment of the other sex, thus causing a further decrease investment by the first (Fig. 5.9e). This means that the equilibrium at the intersection point of the lines will be unstable. If one sex is the first to reduce investment, then the other sex will be left doing all of the caring. Note that this could result in all male or all female care, the outcome depending on initial conditions.

In contrast to Maynard Smith's (1977) more general model, which tells us about evolutionary origins of parental care patterns, the model of Houston and Davies (1985) does make some specific predictions about how one partner might respond to changes in investment by the other which can be tested experimentally by manipulating the effort

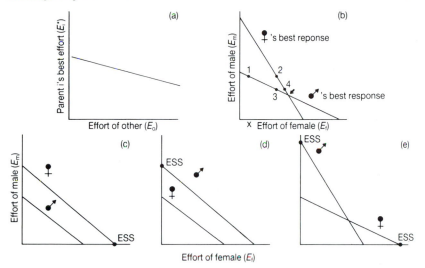

Fig. 5.9. The Houston and Davies model of parental care: (a) optimal parental effort with respect to the effort of the other parent; (b) lines for both parents plotted on the same graph – if the male works harder, then the female responds by reducing her effort, and if the female works harder, the male responds by reducing his effort, and therefore the ESS occurs at the point where the two lines intersect; (c–e) other possible relationships between male effort and female effort resulting in ESSs for female-only care (c), male only care (d), and either female-only or male-only care (e). (Redrawn from Houston & Davies, 1985.)

provided by each parent. One such test has been carried out by Wright and Cuthill (1989), who manipulated the effort that one member of a pair of starlings (*Sturnus vulgaris*) was able to give to a brood by attaching a weight to the bird's tail, thus reducing the number of visits the weighted bird made to the nest. Both males and females responded to the reduction in effort by their partner by increasing their own effort, but not by enough to compensate fully for the loss of care (Fig. 5.10). Various other manipulations have been carried out, usually by removing one member of a pair, and results ranging from reduced effort through no change to increased effort have been found (see Clutton-Brock & Godfray, 1991).

5.9 Polygamous mating patterns

Polygamy can take several forms. First, it is possible to distinguish between those species in which one male mates with several females,

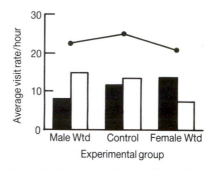

Fig. 5.10. The effect of manipulating feeding nestlings by each sex in starlings (bars) on the overall provisioning rate by each sex. Solid bars represent males, open bars females, and circles show the total rate of provisioning. Each sex responds to a reduction in the provisioning rate of their partner by increasing their own rate of provisioning, but not by enough to compensate the nestlings completely. Wtd = weighted. (Redrawn from Wright & Cuthill, 1989.)

polygyny, and those species in which one female mates with several males, polyandry. Both polygyny and polyandry can result from one sex controlling access to resources required by the other sex: if males control access to resources required by females, then this results in resource-defence polygyny; if, on the other hand, females control the resource, then the outcome is resource-defence polyandry. It is also possible for an individual to monopolise access to a group of individuals of the opposite sex: this is termed female-defence polygyny if a male controls access to a group of females, and male-defence polyandry if a female controls access to a group of males.

5.9.1 *Resource-defence polygyny – the polygyny threshold model*

Verner and Willson (1966) and Orians (1969) suggested that polygyny occurred when females chose to mate with an already-mated male who possessed a good territory, rather than with an unmated male who occupied an inferior territory, because the extra food in the territory more than outweighed the fact that less parental care would be received from the male. It is assumed that there is a cost to polygyny: a female would produce more offspring if she was in a monogamous relationship with a male on a good territory. Orians (1969) produced a simple graphical model of the conditions necessary for the evolution of polygyny. This model, referred to as the *polygyny threshold model*, is illustrated in Fig. 5.11. The two curves represent the reproductive success of individual females in pairs and trios. In an environment of given quality, the

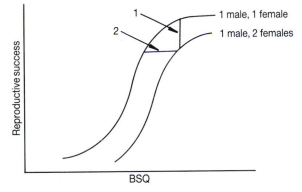

Fig. 5.11. The Orians–Verner–Willson polygyny threshold model. The curves show the reproductive success of females in pairs and trios in environments of different quality (breeding site quality, BSQ). (After Orians, 1969.)

reproductive success of a female in a pair always exceeds that of a female that is part of a trio (distance marked 1) – this is the cost of polygyny. The distance marked 2 is the polygyny threshold, which is the minimum difference between environmental qualities, above which it pays a female to join a pair in a good territory rather than joining an unmated male in a poor territory. This model can easily be extended to account for larger harem sizes. It has also been realised that factors other than the quality of a male's territory may influence the choice a female makes: for example, some males may be prepared to offer more parental care, or some males may be of a higher genetic quality. This can be incorporated into the polygyny threshold model by redefining the x-axis as 'breeding-site quality' rather than just environmental quality (Wittenberger, 1976; Searcy & Yasukawa, 1989).

The polygyny threshold model is deceptively simple and appears to make some straightforward predictions about factors such as female settlement patterns and female reproductive success in different situations. Tests have generally focused on the reproductive success of females in harems of different sizes, and have rarely examined female settlement patterns (Davies, 1989). However, within the basic model it is possible to make various assumptions about the effect of polygyny on females (Davies, 1989): at one extreme one could assume that, as in the case of the pied flycatcher, *Ficedula hypoleuca* (Alatalo *et al.*, 1981), the first female gets all of the male's help; at the other extreme, the male's help could be divided equally among the females in his territory, as sometimes happens in dunnocks, *Prunella modularis* (Davies, 1986). The fact that a variety of different assumptions can be made means that testing the predictions of the polygyny threshold model is complicated, particularly as underlying assumptions are rarely considered (Davies, 1989). Searcy and Yasukawa (1989) suggest that the appropriate approach is to test an exhaustive set of hypotheses, which they list, for the evolution of polygyny. This does not address the central problem of the polygyny threshold model, pointed out by Davies (1989), which is that it assumes males always benefit from polygyny and females are free to make choices about where they settle. This differs from the models of mate searching and parental care, outlined above, which make it clear that a conflict of interest between the sexes almost always arises and that the resolution of this conflict is central to understanding how mating strategies evolve. At its simplest, if paternal care is given, then it is easy to see that, while males must favour polygyny, females should favour polyandry. How such conflicts are resolved is best studied by looking at a species with a variable mating pattern: the final section of this chapter shows how the conflict of interest beween male and female dunnocks has resulted in an extraordinarily varied mating pattern.

5.9.2 Female-defence and lek polygyny

Resource-defence polygyny can evolve when a single male controls access to resources required by several females for their reproduction. However, if resources are highly dispersed, then it may be too costly for males to control access to a resource. Alternatively, male density may be so high that defence of a territory would be very costly due to continual intrusions by neighbouring males. Under such circumstances, the evolution of one of two different forms of polygynous mating system occurs. The critical factor in determining which of these two evolves appears to be the social structure of female groups. If females form stable social groups, then it is possible for males to defend these groups, resulting in female-defence polygyny. This may either take the form of one male controlling exclusive access to a harem of females, such as occurs in red deer, *C. elaphus* (Clutton-Brock *et al.*, 1982), or several males may join in a social group with the females, with the reproductive success of a male being determined by his position in the male dominance hierarchy.

If females do not form social groups, then males may aggregate at a traditional site for the purposes of display and courtship. At such areas, termed *leks*, males often defend small territories. Females visit the lek and choose among the males. In the case of the white-bearded manakin

(*Manacus manacus trinitatis*), females choose the male at a preferred site within the lek: removal of the top-ranking male results in a subordinate male occupying his site and mating with visiting females (Lill, 1974). Why should males aggregate at leks? Three hypotheses have been proposed to explain this (Bradbury & Gibson, 1983; Davies, 1991).

1. Males aggregate at 'hot spots' (Bradbury *et al.*, 1986). Here, males are assumed to gather at sites where females are likely to occur at higher density, for example where two or more female territories overlap. White-bearded manakin leks often occur within a few metres of a stream (Snow, 1962), the junction of two streams being a favoured location. Females visit streams regularly to drink and often use them as flight ways. Thus males appear to have formed leks at a place where there is good chance of intercepting passing females.
2. Males aggregate around 'hot shots'. If some males are particularly good at displaying (hot shots), then it could benefit poorer quality males to aggregate around them and intercept incoming females (Beehler & Foster, 1988).
3. Female choice for sites or male aggregations. It is possible that lekking originated because females favoured certain areas for encountering males (perhaps because the use of those sites minimized search costs), causing males to aggregate at such areas (Parker, 1978).

5.10 Co-operation and conflict in the evolution of mating systems

The dunnock (*Prunella modularis*), often called the hedge sparrow, is a common bird in large parts of Europe. Its familiarity and unobtrusive coloration belie a rich pattern of sexual relations, which have been revealed by N. B. Davies and his co-workers over the last 10 years (Burke *et al.* 1989; Davies, 1983, 1985, 1986; Davies & Houston, 1986; Davies & Lundberg, 1984). These studies reveal how the mating pattern of a species arises as a result of decisions made by individuals trying to maximise their reproductive success. By studying the dunnock in a highly varied habitat (Cambridge University Botanic Garden), a range of mating systems could be looked at in a small area. These mating systems included monogamy (one male, one female), polygyny (one male, two or more females), co-operative polyandry (one female, two or more males), and polygynandry (two or more males and two or more females). In no case were individuals in trios or larger groups related, thus kin selection, a common explanation of co-operative breeding in birds (see Chapter 7), can be ruled out as an explanation for the evolution of these mating systems.

The central determinant of the dunnock's mating pattern is the size of a female's territory, which is in turn a function of the distribution of good foraging patches. Dunnocks are ground gleaners, feeding on small seeds and invertebrates, preferred areas for feeding being flower beds and dense bushes. Territory size was determined by the abundance of these resources, territories being larger in parts of the garden with fewer bushes and flower beds (Davies & Lundberg, 1984). That food availability was important in determining territory size was demonstrated by supplying additional food in some territories: this resulted in a reduction in territory size.

Male territories, which were defended by both singing from song posts and chasing off intruding males, were formed independently of female territories. In contrast to female territories, male territories were not always exclusive, but sometimes shared with other males. This occurred despite the attempts of one male to drive off the other. This was unsuccessful on many occasions for two reasons. First, the generally secretive habits of the bird coupled with dense vegetation in parts of the territory made it hard to locate intruders. Second, females encouraged intruding males to remain in a territory by copulating with them. When two males shared a territory, a clear dominance hierarchy resulted with one male, termed the alpha male, driving the beta male away from food

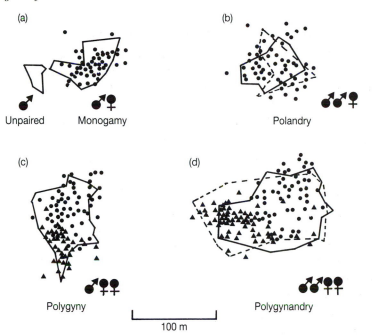

Fig. 5.12. How the overlap of male and female territories results in different mating combinations. Lines (————, – – – –) show male song polygons, circles and triangles represent the first sightings of females on transects of the study site. Female territories are largely exclusive, but male song polygons can overlap when they share one or two females. (Redrawn from Davies & Lundberg, 1984.)

sources and from the female. Alpha males were usually older than beta males. If the beta male succeeded in copulating with the female, which only happened in about half of cases (Davies, 1985) because the alpha male guarded the female very closely during the egg-laying period, then he would assist in provisioning the nestlings. Paternity, which was initially determined by observing copulations (Davies, 1985) and subsequently confirmed using DNA fingerprinting (Burke *et al.*, 1989), was divided on average 0.6 to the alpha male and 0.4 to the beta male.

Figure 5.12 shows how different mating systems arise from female territories of different sizes. In Figure 5.12a, the territories of one male and one female overlap, resulting in monogamy. Polyandry results when the ranges of two males overlap that of a single female (Fig. 5.12b). If female territories are small (Fig. 5.12c), then it is possible for a male to defend a territory that includes two female territories, resulting in polygyny. Finally, it is possible for two males to share a

Table 5.3. *Calculations of seasonal reproductive success of male and female dunnocks,* Prunella modularis, *in different mating combinations. In polyandrous situations, paternity is shared between the alpha and beta males 0.6:0.4.* (*From Davies & Houston, 1986*)

Mating pattern	Nestlings fledged	
	Per female	Per male
Monogamy	5.04	5.04
Polyandry		
Only alpha male feeds	4.41	4.41
Both males feed	6.75	4.05:2.70
Polygyny	3.82	7.64
Polygynandry	3.65	5.26:2.04

territory that contains two female territories, resulting in polygynandry (Fig. 5.12d).

The different mating combinations result in a different number of young fledged per breeding attempt (Davies, 1986), but fledging success per breeding attempt cannot be used as a direct measure of reproductive success because failure rates varied among mating combinations. In polyandrous situations, when the beta male failed to copulate he would attempt to interfere with the breeding attempt by breaking up the nest. In polygynous and polygynandrous systems, interference between females increased the failure rate. Because dunnocks have a long breeding season, several attempts at nesting can be made. Failed attempts are shorter than successful attempts and so this must be taken into account when extrapolating from fledging success in a single breeding attempt to seasonal reproductive output. Such factors were taken into account in estimating the seasonal reproductive successes of males and females in different mating systems shown in Table 5.3. Females clearly have the highest reproductive success under co-operative polyandry; males do best under polygyny. These results show clearly that there is a conflict between the sexes. Male success increases from co-operative polyandry through polyandry, monogamy and polygynandry to reach the male's best option of polygyny. Females do best under co-operative polyandry and worst under polygyny. The outcome of this battle seems to depend on local environmental conditions and particularly the relationship between food supply and female territory size as outlined above.

Most small passerines show monogamy, perhaps with a mild degree of polygyny. However, conflicts of interest such as those outlined above for the dunnock are likely to be common amongst many others: why then is such a rich mating system as this not more common? It may be that the dunnock is unusual in the fact that the structure of its preferred habitat (dense bushes) makes it hard to detect and effectively drive away intruders. If a female cannot drive off another female, the result is polygynandry. If a male cannot drive off another male, and the female encourages the second male to stay by offering copulations, and thus a share in paternity, then polyandry results.

This study of dunnocks clearly shows that mating systems are the outcome of a conflict between individuals striving to maximize their reproductive success. This contrasts with the ideas of the polygyny threshold model which assumed that polygyny simply resulted from the choice of a female to settle on a territory with a bird which already had a partner. Further investigations of animal mating systems will benefit from the insight given by the study of dunnock.

5.11 Conclusions – on the usefulness of models

The development of ESS theory is probably the single most important contribution to the study of behaviour, ecology and evolution in the last 20 years and, along with kin selection (Hamilton, 1964a, b), has provided a stimulus to a generation of theoretical, experimental and field biologists. One particularly pleasing aspect of this has been the degree to which theory has promoted observational and experimental work, while empirical studies in turn have kept theoreticians busy. Probably only in population genetics have other biologists exploited this link between theory and observation so fruitfully.

Notes

1. The use of the word 'goal' is convenient shorthand for the more precise statement that natural selection has favoured individuals with the highest long-term rate of energy intake.
2. Dawkins and Carlisle published at the time when government decisions were made to invest more money in the development of the supersonic airliner Concorde, in spite of the fact that the programme was massively overspent, simply because so much money had already been invested. The term has perhaps lost some of its force in the intervening years.

6

Sex and evolution

T. R. HALLIDAY

6.1 Introduction

During the course of their lives, animals and plants typically show a period
of growth followed by one of reproduction, during which they produce a
number of progeny. Some animals and many plants can reproduce by
dividing into two or more new individuals, so that the distinction between
growth and reproduction becomes blurred. It is a moot point, for example,
whether a *Hydra* that forms a new individual by budding is reproducing
or growing, since the new individual is genetically identical to its parent.
Groups, or clones, of genetically identical individuals can be regarded as
single units in evolutionary terms, since they represent a single genotype.
In sexual reproduction, in contrast, reproduction is clearly distinct from
growth; the essential features of sex is that it produces individuals with
novel and unique genotypes.

Many species of animals and plants reproduce both sexually and
asexually, usually at different times in their lives. The co-existence of
different modes of reproduction among organisms, and sometimes within
the same organism, poses major questions for evolutionary biologists. Do
both kinds of reproduction confer equal fitness? If so, why have both? If
not, why has one pattern not replaced the other during the course of
evolution? The first part of this chapter seeks answers to these questions
by considering the evolutionary forces that maintain sexual reproduction
in natural populations. The question of the evolutionary origins of sex is
not addressed; though an interesting question, it is largely a matter of
speculation, and theories on the subject are not readily amenable to
empirical testing. The second part of the chapter examines the evolutionary
consequences of sexual reproduction in animals, amongst which it is the
predominant mode of reproduction. It looks at why sex leads to the
evolution of two types, male and female, which in turn typically differ in
their behaviour and appearance.

150

6.2 The maintenance of sex

To understand why sexual reproduction is maintained in nature when the alternative of asexual reproduction exists, and *vice versa*, it is necessary to identify and to quantify, as far as possible, the relative costs and benefits of the two modes of reproduction. Much of the challenge involved in trying to understand the maintenance of sex is that asexual females seem to have a two-fold advantage over sexual females. Consider a female in a sexual species that produces several offspring. When she reproduces, she allocates 50% of her genes to each of her progeny, the other 50% being provided by one or more males. If she abandoned sex and produced young by parthenogenesis (meaning 'virgin birth'), in which eggs develop without being fertilised, each of her young would contain 100% of her genes. Thus, an asexual mutant in a sexual population has a two-fold reproductive advantage over other females; she passes her genotype on to the next generation twice as effectively. This leads to the conclusion that, among individuals, asexual reproduction should be favoured over sexual reproduction. This fundamental cost of sex is often referred to as the cost of producing sons (Lewis, 1987; Maynard Smith, 1984).

It is important to note the words 'among individuals' in the argument above. The theory of natural selection is concerned with differential survival and reproduction of individuals and it is generally regarded as incorrect, if not heretical, to envisage selection acting among groups of organisms or among species (Williams, 1966; Maynard Smith, 1976a; Wilson, 1983). In the context of sexual reproduction, however, selection acting at the level of the species is regarded as important, as explained below (Williams, 1975; Maynard Smith, 1978b).

6.2.1 Sex and the species

Sexual reproduction has two essential features that distinguish it from asexual reproduction: meiosis and syngamy. Meiosis is the process of cell division that is intrinsic to the production of haploid gametes; syngamy is the fusion of two gametes to produce a diploid zygote. During meiosis, there is crossing-over between chromosomes and independent assortment of alleles; these processes, together called recombination, cause the genotype of the parent to be 'shuffled' such that each gamete, while containing half the parental genotype, carries a unique combination of parental genes. In asexual reproduction, in contrast, the parent produces diploid progeny that are exact replicas of itself. There is no intermediate

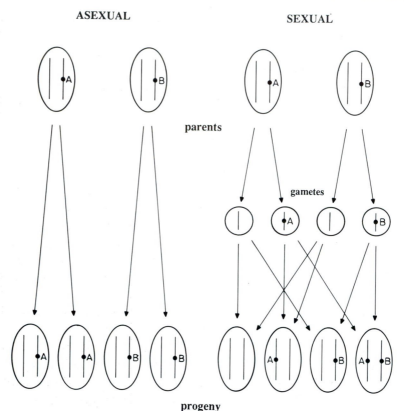

Fig. 6.1. The genetic consequences of asexual (left) and sexual reproduction (right). Asexual reproduction produces progeny that are exact replicas of the parents, carrying the same genes (A, B). Sexual reproduction produces genetically variable progeny that carry different combinations of parental genes. (Modified from Maynard Smith, 1978b.)

gamete stage and thus no recombination or syngamy. As a result, asexual reproduction does not produce the variety of new genotypes that is produced by sexual reproduction (Fig. 6.1).

Suppose that the two alleles A and B, in Fig. 6.1, are deleterious. Whereas the asexual parents inevitably pass them on to their respective progeny, the sexual parents produce some progeny that lack both of them. Sexual reproduction thus has the capacity to eliminate deleterious alleles. Now suppose that alleles A and B are advantageous. Again, they are conserved in their separate lines in asexual progeny, but some of the sexual progeny carry both alleles. Sexual reproduction thus has the capacity to create new combinations of favourable alleles. In an asexual lineage, A and B can only occur in the same individual by mutation; for example, B may arise in a lineage that already carries A.

From this argument it follows that a sexual species can evolve faster than an asexual one (Maynard Smith, 1978b; Cockburn, 1991). Because deleterious alleles are eliminated, and new combinations of favourable alleles are created more rapidly in sexual species, adaptation to environmental changes will be more rapid in sexual than in asexual species. Reproduction has been likened to a lottery in which tickets represent progeny in the next generation (Williams, 1975). To win a lottery (leave surviving offspring), an individual should buy as many tickets as possible (maximise reproductive success). In asexual reproduction, all the 'tickets' have the same number (same genotype); in sexual reproduction, each 'ticket' has a different number. Thus, asexual reproduction will be favoured when the environment changes little between generations. When the environments faced by parents and progeny differ, however, the capacity of sexual reproduction to produce a range of different, novel genotypes will be favoured (Roughgarden, 1991).

As described above, sexual reproduction has the capacity to eliminate deleterious alleles; a corollary of this is that asexual reproduction does not have this capacity. Because asexual species cannot eliminate harmful mutations, they tend to accumulate them over many generations, a process that represents a major cost of asexual reproduction (Muller, 1964; Maynard Smith, 1984). Turning to sex, while a major advantage of sexual reproduction is its capacity to produce new combinations of genes through recombination, the same process may also represent a cost. Recombination tends to break up favourable combinations of genes, by the very same process that brings them together. This is referred to as the cost of recombination (Shields, 1984; Lewis, 1987).

Some support for these general arguments about the relative costs and benefits of sexual and asexual reproduction in different kinds of environment comes from consideration of species that show alternation between the two modes of reproduction during their lives. For example, aphids reproduce asexually during the spring and summer months, producing only daughters, when conditions are favourable. Asexual reproduction enables them to increase in numbers very rapidly. Come the autumn, however, both sons and daughters are produced. These mate to produce eggs that overwinter, hatching the following spring when environmental conditions may be very different from those experienced by the parental generation (Dunn, 1959; Thornhill & Alcock, 1983).

Supporting evidence of another kind comes from comparative surveys of animals. Asexual species are found in all animal groups, except birds and mammals, but, with very few exceptions, they do not predominate in any taxonomic group. This suggests that asexual species arise quite frequently from sexual species during evolution, that they enjoy a relatively brief period of success, but that they eventually become extinct in the face of competition from more adaptable sexual species (Maynard Smith, 1984).

It has been suggested that a major cost of sex at the individual level, that of producing sons (see above), may be adaptive for species. Atmar (1991) argues that the metabolic stress incurred by males during courtship and sex-related aggression subjects them to an arduous physiological 'test' not experienced by females. This test provides a mechanism for screening out genetically defective males prior to mating, and males are, in effect, a 'caste' that is culled at less cost to the species than if females were subjected to a comparable test.

6.2.2 Sex and the individual

Williams (1975) proposed a model for the maintenance of sexual reproduction, in a population containing both sexual and asexual forms, based on the hypothesis that sexual individuals will be at a selective advantage because they produce progeny with more varied genotypes. The model came to be called the sib-competition model, as an essential feature of it is that each individual produces a large number of offspring which must compete with one another for limited resources (Williams, 1975; Maynard Smith, 1976b).

Consider an annual plant that sheds its seeds in large numbers into its immediate environment. The habitat consists of a limited number of patches suitable for seed germination, none of which is large enough to support more than a limited number of mature plants; for simplicity, assume that this number is one. If several seeds land on a particular patch, there will be density-dependent mortality among sibling seedlings. Thus, as only one seedling can survive, the proportion that die increases with the number that germinate (Fig. 6.2a).

Suppose that the patches are not uniform but differ in various ways such that there is variation among patches in the probability that a seedling with a particular genotype will survive to reproduce. The offspring of sexual parents will have varied genotypes and, as a result, progeny from a single parent will be able to survive and reproduce in several kinds of patch (Fig. 6.2b). Those of asexual parents, in contrast,

(a)

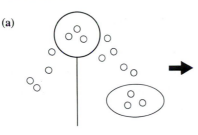

SEED PRODUCTION NEXT GENERATION

(b)

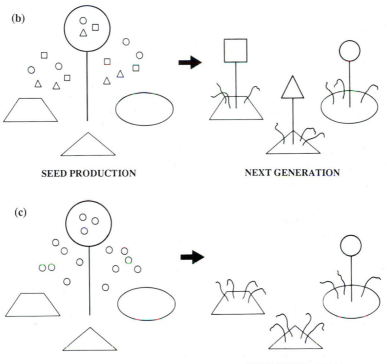

SEED PRODUCTION NEXT GENERATION

(c)

SEED PRODUCTION NEXT GENERATION

Fig. 6.2. The sib-competition model for the evolution of sex. (a) An annual plant sheds it seeds into a patch. However many seeds enter the patch, only one can survive and develop into a mature plant. (b) An annual plant reproduces asexually, shedding genetically variable progeny (three types are shown) into patches that vary in their characteristics. Only one of each seed type survives in each patch, the survivor being the one that is genetically suited to that patch. (c) An annual plant reproduces asexually, shedding genetically identical progeny into variable patches. Only those seeds that are adapted to the same patch type as the parent survive. (Modified from Halliday in Skelton (ed.), 1993.)

having genotypes identical to their parents, will survive and reproduce only in patches that are similar to the parental patch (Fig. 6.2c). Thus, although asexual plants may produce more seeds than sexual plants, they may leave fewer offspring that survive to reproduce. Whether or not sexual individuals are more successful at reproducing in this hypothetical system, depends critically on the heterogeneity of the habitat: the more variable are the patches, the more likely it is that sexual reproduction will confer higher fitness than asexual reproduction. In essence, the sib-competition model is a formal expression of the lottery effect.

Suppose that each patch has a number and that offspring have corresponding numbers; the closer a progeny's number to a patch number, the more likely it is to survive in that patch. Consider three sibling progeny from each of two parents, one sexual, one asexual, landing on a patch with the number 56. The numbers of the three sexual progeny are 23, 54 and 89, those of the three asexual progeny are 47, 47 and 47. In this case, one of the sexual progeny (54) will survive, but it would have been one of the asexual progeny if the patch number had been 47. In this situation, the sexual progeny have three times the chance of 'winning the lottery' than the asexual progeny because they have three times as many numbers among them. Maynard Smith (1976b) has modelled this situation and has identified the conditions under which sex is and is not advantageous. Sex is not favoured if progeny from only one parent enter a patch, however many of them there are. More importantly, sex is not favoured if only one progeny per parent enters a patch; it is critical that there should be competition between the progeny of a given parent, hence the name of the model. Indeed, the advantage of sex increases with the number of progeny produced per parent. Extending the numerical example, if five siblings per parent enter a patch, sexual progeny have a five-fold advantage over asexual progeny.

The sib-competition model thus suggests that sexual genotypes can have higher fitness than asexual genotypes in certain circumstances. But how realistic are the assumptions of the model? The assumption that there is intense competition between siblings is realistic for high-fecundity organisms, such as many plants, particularly if, as in most plants, progeny typically settle close to their parents. Seeds are, however, usually more mobile than asexual progeny, which commonly arise by being formed directly from their parents, as in the stolons of strawberries. A similar relation between mode of reproduction and mobility occurs among animals. In aphids, for example, progeny produced parthenogenetically are wingless, whereas sexual females and males have wings that enable them to disperse from the summer food plant to their overwintering site. Habitat patches that are remote from the parent are more likely to differ from those colonised by poorly dispersing asexual progeny, again suggesting a link between sexuality and varying or unpredictable environments.

This association can be tested by examining the kinds of organisms that inhabit frequently disturbed, and therefore unpredictable, habitats. Among plants, for example, weed species colonise recently cleared patches of soil, and the sib-competition model would predict that weeds should produce sexually. In fact, weeds are rarely fully sexual, contrary to this prediction, but are more commonly parthenogenetic or self-fertilising. It is now thought that what may be more important for weed species than the physical nature of their environment is their biotic environment, specifically the presence or absence of competitors. Weeds thrive in cleared patches of ground where there are few competitors and selection apparently favours the rapid production of as many progeny as possible rather than of offspring that are genetically diverse.

In plants, sexual reproduction is more characteristic of species living in saturated habitats, that is, those in which there are many other plant species competing for the same resources, such as water, nutrients and light, as well as established populations of animal herbivores and a variety of parasites and pathogens (disease-causing organisms such as viruses and bacteria). Sex is advantageous in saturated environments because it generates genetic diversity, enabling plants continually to evolve adaptations against competitors and other enemies that are themselves constantly evolving more effective means of attacking them (Bell, 1982).

This concept that different species of organisms are engaged in a constant evolutionary struggle with one another is central to the 'Red Queen hypothesis', formulated by Van Valen (1973), primarily to account for the observation that, over evolutionary time, species are continually becoming extinct and being replaced by other species. No matter how well a species is adapted to its environment at any particular time, its environment will always be deteriorating as far as its adaptations are concerned. At any one time, its fitness is determined by a variety of biological and physical pressures, such as predation, competition for resources, parasites, etc., but the relative importance of these factors will vary over time. If a species acquires an adaptation in response to one kind of pressure, this may be deleterious for its adaptations to other pressures which, at that time, are relatively weak. As a result, the species will suffer a decline in fitness when the balance of pressures changes. Consider for example, a plant growing in your garden. One year, a major selective pressure may be drought, favouring the ability to retain water; the next year, water may be abundant but there could be a plague of aphids. The balance of selective pressures to which a species must respond is constantly changing, not least because of the continuous adaptive responses that different species living in the same habitat make to one another. Thus any temporary adaptive advantage gained by one species over others represents a worsening of the environment experienced by the others. As a result, the adaptedness of a species relative to others will fluctuate with time.

Metaphorically, each species within an ecological community is engaged in a permanent race with the other species. Hence the reference to the Red Queen in *Through the Looking Glass*, who remarked to Alice, 'Now here, you see, it takes all the running you can do, to keep in the same place'. Sexual reproduction, because of its capacity to produce genetically varied offspring, may be an adaptation that enables species to keep up with their competitors in this race.

These arguments see temporal heterogeneity of habitats as being a critical factor favouring sexual over asexual reproduction. An alternative view, proposed by Ghiselin (1974) and developed by Bell (1982), who called it the 'tangled bank hypothesis', emphasises spatial heterogeneity. In saturated environments where there are few available niches, sex will be advantageous because it generates a greater diversity of genotypes, some of which may be able to exploit vacant niches. Koella (1988) suggests that ecological conditions favouring sexual reproduction are common in nature.

6.2.3 Sex, parasites and disease

Another argument for the adaptive value of sexual reproduction relates to its capacity to produce genetically varied progeny. Most organisms are subject to infection by pathogens and parasites. These may have a profound effect on the individual fitness of their hosts by causing their death or reduced reproductive success. During evolution, hosts have evolved a variety of defences against pathogens and parasites, notably a complex immune system that recognises and attacks alien organisms. On the other hand, pathogens and parasites evolve mechanisms that circumvent or counteract their hosts' defence mechanisms. There is thus co-evolution between an organism and its enemies (Price, 1980; Anderson, 1982).

In this coevolutionary process, pathogens and parasites hold a distinct advantage. They typically have a very short generation time and a very high reproductive rate. A bacterium that infects a human, for example, will generate several dozen generations within a single day. (The generation time of *E. coli*, for example, is 20 minutes under optimal conditions.) As a result, pathogens and parasites can evolve, thus becoming better adapted, within the lifetime of their host. If the host reproduces asexually and thus produces progeny that are genetically identical to itself, they will encounter populations of pathogens and parasites already adapted to them. Sexual reproduction, because it produces genetically variable progeny, can be seen as an adaptation against pathogens and parasites (Hamilton, 1982; Hamilton *et al.*, 1990). It insures that at least some progeny will be less susceptible to attack by those organisms that have evolved to suit the genotypes of their parents.

Evidence that parasites may play a role in determining the mode of reproduction of their hosts comes from studies of a freshwater snail (*Potamopyrgus antipodarum*) in New Zealand lakes (Lively, 1987). Individuals are either male or female, unlike many snails that are hermaphrodites, but females are capable of asexual reproduction by parthenogenesis (see below). Some populations consist entirely of females and must reproduce parthenogenetically; others contain as many as 40% male individuals, indicating that they have the potential for sexual reproduction. In a comparison of nearly 50 sites, Lively found a strong tendency for males to be more frequent in localities where snails were heavily infested by trematode parasites, supporting the hypothesis that sex is maintained in populations where parasites impose a selection pressure favouring the production of genetically diverse progeny.

6.3 Variations on a theme

Asexual and sexual reproduction, as depicted in Figure 6.1, represent only two of a large variety of modes of reproduction. Each mode has different genetic consequences in terms first, of the degree of genetic variation between parents and progeny and, secondly, the level of heterozygosity present in the population. The following list gives a number of modes of reproduction, indicating for each their genetic consequences.

Mitotic parthenogenesis. Female parent produces diploid eggs mitotically. Progeny are genetically identical to parent. Heterozygosity is preserved.

Sexual parthenogenesis. Female parent produces haploid eggs meiotically that can develop without uniting with a male gamete. Progeny will differ genetically from the parent to a small degree, depending on the level of heterozygosity in the parent.

Self-fertilising hermaphroditism. Parent produces male and female gametes meiotically; these unite to form progeny. Progeny will differ genetically from the parent to a small degree, depending of the level of heterozygosity in the parent.

Sex with polyembryony. Females and males produce gametes that unite to form a single fertilised egg; this divides mitotically to produce offspring that are genetically identical to each other but which differ from both parents.

Inbreeding sex. Females and males that are closely related to each other produces gametes that unite to form fertilised eggs. Progeny differ genetically from their parents, but less so than in outbreeding sex. Heterozygosity will decline over successive generations.

Outbreeding sex. Females and males that are unrelated to each other produce gametes that unit to form fertilised eggs. Progeny differ from their parents to a high degree. Heterozygosity will be maintained over successive generations.

This is a very brief and simplistic overview of just a few of the many modes of reproduction that exist among animals and plants. The details of the mechanisms involved are beyond the scope of this chapter. What is important to stress is that there is much diversity in reproductive patterns among organisms and that, as a result, different species can, during reproduction, produce progeny that differ genetically from themselves to a greater or lesser extent. Each mode of reproduction, and the level of genetic diversity that it generates, is assumed to be an adaptation to a particular set of environmental conditions.

6.3.1 Inbreeding and outbreeding

It is common knowledge that inbreeding has genetically harmful effects; in the course of history, dire consequences have befallen various royal families that practised incest (mating between siblings) or frequent matings among cousins. Inbreeding in humans leads to increased mortality, mental retardation, albinism and other physical abnormalities in the offspring. It is commonly assumed that organisms should avoid mating with close relatives, and animal breeders generally insure that they do, but rather little is known about whether or not they do in nature. Models in population genetics, such as the Hardy–Weinburg model, often assume that, within a population, mating is panmictic; that is, mating is random and any individual of one sex is equally likely to mate with any member of the other in the population. In nature, this will not often be true. Plants are sedentary and the probability that one particular plant will fertilize another must be higher if they are close together than if they are far apart. Many animals show some degree of philopatry, a tendency by an individual to return to breed in the locality where that individual was born (Shields, 1984). Thus, for both plants and animals there is good reason to suppose that, in natural populations, mating will not be random and that inbreeding may occur to some extent.

The most well-documented consequence of inbreeding is lowered off-spring fitness, or inbreeding depression. This is a well-known phenomenon in laboratory and domestic animals and in crop plants, but there is very little evidence for it in natural populations. For example, a field study carried out in Britain found that nestling mortality among great tits (*Parus major*) is up to 70% higher among the offspring of related than of unrelated birds (Greenwood *et al.*, 1978). A comparable study of great tits in the Netherlands did not, however, find such an effect (Noordwijk & Scharloo, 1981).

The explanation for inbreeding depression, in genetic terms, lies in the presence of deleterious recessive mutations in populations. In humans, for example, it has been estimated that each of us, on average, carries the equivalent of three to five lethal recessive genes. These are not normally expressed because they are in the heterozygous condition. Close relatives are more likely than non-relatives to carry the same lethal gene or genes, because of their common ancestry, with the result that inbreeding leads to an increased occurrence of such genes in the homozygous condition, and therefore their phenotypic expression in the offspring. The offspring will thus tend to express lethal recessive genes that are not expressed in their parents.

If inbreeding decreases fitness, we would expect animals and plants to have evolved mechanisms that reduce its occurrence. A common mechanism in animals is differential dispersal by the two sexes. In mammals, males generally disperse further than females from their place of birth to breed; in birds, it is more commonly females that disperse further (Greenwood, 1980).

There is also accumulating evidence that animals can recognise and avoid mating with their close kin (Hepper, 1991). Often this is achieved simply by their not mating with those individuals with which they were reared, animals preferring to mate with unfamiliar rather than familiar partners. There is evidence, however, for a variety of groups including mammals, amphibians and arthropods, that kin recognition can occur without individuals having prior experience of one another (see Chapter 7).

Although the available evidence suggests, first, that inbreeding does cause a reduction in fitness and, secondly, that animals and plants possess adaptations that reduce its occurrence, it cannot be assumed that outbreeding is necessarily a more adaptive mating pattern. As discussed above, mating with unrelated individuals will tend to break up adaptive combinations of genes, whereas mating with relatives will tend to conserve such combinations in the progeny. This consideration has led to the idea that there may be an optimum level of outbreeding (Shields, 1984; Bateson, 1983b). Very close relatives and totally unrelated individuals should be avoided as mates, but intermediate relatives should be preferred. Reduction in fitness due to mating with non-relatives is called outbreeding depression. Evidence that both inbreeding and outbreeding lead to reduced fitness is provided by a number of studies of plants (e.g. Price & Waser, 1979), but comparable data are less easily obtained for animals.

Among animals, indirect evidence for optimal outbreeding comes from laboratory studies of Japanese quail (*Coturnix coturnix japonica*), in which individuals prefer to associate with first cousins rather than with their siblings on the one hand, or with unrelated birds on the other (Bateson, 1982). A field study of great tits has shown that females tend to be paired with males whose songs differ slightly, but not very much, from the songs of the females' own fathers (McGregor & Krebs, 1982).

A particular situation in which a degree of inbreeding will be advantageous is where local populations of organisms, called ecotypes, are specifically adapted to local ecological conditions. Mating with individuals belonging to the same ecotype will tend to insure that the offspring inherit parental adaptations to local conditions. The Canadian three-spined stickleback (*Gasterosteus aculeatus*) exists in two forms, an exclusively freshwater ecotype and an ecotype that is anadromous (breeds in freshwater but otherwise lives in the sea). Hay and McPhail (1975) conducted mate choice tests and found that individual sticklebacks, of both sexes, preferred a partner of the same ecotype. This tendency for mating to occur non-randomly, with individuals pairing with partners of a similar type, is called positive assortative mating. In this case, it insures that progeny will be adapted to one life-history or the other. Progeny resulting from mixed pairings would probably be less well adapted to either the freshwater or the anadromous life-history than pure-bred progeny.

Studies of the degree to which organisms are inbred or outbred, and of the fitness consequences of different degrees of inbreeding, are few at present, but this is an area of research that is developing and expanding rapidly (see Chapter 7). A major development in this area is the technique of DNA fingerprinting, which will eventually make it possible to establish, in a natural population, precisely which individuals mate with one another and which descendants they produce. Such studies will make it possible to determine, first, how closely related mating partners are in nature and, secondly, what are the fitness consequences of inbred and outbred pairings. Only then will it be possible to test the hypothesis that, for a given species, there is indeed an optimal level of outbreeding.

6.3.2 Changing mode of reproduction

As indicated in Section 6.1, asexual reproduction appears to have evolved many times during evolution, with the result that, at the present time, asexual species are known within a diverse array of plant and animal groups. If this is so, why has it not evolved more often, given that it appears to confer an immediate selective advantage in certain circumstances? One possible answer is that such circumstances do not arise very commonly. It may also be the case, however, that the number of asexual species is limited because it is very costly to change from one mode of reproduction to another. Evidence for this comes from asexual species that retain certain aspects of sexual reproduction.

A small fish, the Amazon molly (*Poecilia formosa*) is a parthenogenetic species consisting entirely of females (Schultz, 1971). However, before an egg can develop it must be penetrated by a sperm, which triggers cell division but does not fuse with the cell nucleus. To obtain sperm, female mollies must elicit matings from males of a related, sexual species. In many lizards, females require stimulation from males, in the form of elaborate courtship behaviour, if their eggs are to develop fully. Some species of lizard, such as the whiptail lizard (*Cnemidophorus inornatus*), have become parthenogenetic but, though they are emancipated from the need to obtain sperm, they still require 'male' stimulation. This is provided by each lizard in a population spending time courting others like the males of an ancestral, sexual species as well as being courted (Crews, 1987).

What these examples suggest is that, once a species has evolved a particular mode of reproduction, it acquires a range of adaptations, morphological, physiological and behavioural, that enhance the efficiency of that mode of reproduction. While it might be a relatively simple step to switch from producing eggs meiotically to producing them mitotically, or to producing eggs that do not need to be fertilized, effective asexual reproduction may be prevented by a legacy of adaptations for sexuality. This line of argument raises an interesting possibility. It may be that, for a number of existing sexual species, particularly species with low fecundity, it might actually be more adaptive, at least in the short term, if females could switch to asexual reproduction, but this option may be closed to them because it is not possible to switch from one mode to another without going through a series of steps that would be less adaptive than sexual reproduction.

Some organisms, like the lettuce-leaf aphid, employ both sexual and asexual reproduction, switching between them at different phases in the life cycle (Section 6.2.1). A few, such as *Hydra*, can switch between reproductive modes facultatively. When prevailing conditions are good, *Hydra* reproduces asexually, by growing buds that detach as small replicas of their parent. When its pond habitat begins to deteriorate, however, as it dries up in the summer, *Hydra* produces male and female gametes that fuse to form zygotes that can survive for long periods until conditions again become favourable.

Organisms like aphids and *Hydra* appear to enjoy the best of both worlds, being able to switch between sexual and asexual modes of reproduction as environmental conditions change. Why it should be that there are so many other species that do not enjoy this facility is one of many mysteries about the evolution of reproductive patterns that have yet to be explained.

6.4 Anisogamy and its consequences

An essential feature of sexual reproduction is the fusion of two haploid gametes to form a diploid zygote. In a few organisms, such as some of the algae, gametes are identical in form (isogamy), though they may exist in two mating types with fusion occurring only between two gametes of different type. In the great majority of organisms, however, there is anisogamy, with two kinds of gamete being produced, large ova and tiny spermatozoa. The morphology of gametes has no bearing on the evolutionary basis of sexual reproduction *per se*; the genetic consequences of sex are the same, whether gametes are of the same or of different forms. It is important, however, to understand why anisogamy evolved because, as discussed below, it provides the basis for many differences that exist between males and females and which are of great evolutionary significance.

It is generally assumed that the primitive condition is isogamy and that anisogamy evolved from it (Parker *et al.*, 1972; Hoekstra, 1987). In the isogamous conditions, all gametes are similar, though we can reasonably assume that there is some degree of variation in their size and form. Each gamete contains a certain quantity of nutrients stored in its cytoplasm and possesses some means of locomotion, such as one or more tails. The nutrients contained in a gamete not only sustain it until it fuses with another gamete, but also help to sustain the zygote that results from their fusion in the early stages of its development. To be successful, each gamete must find another with which it can fuse. To achieve this it uses its mobility to find another gamete before its reserves are used up and it dies. There are alternative ways by which individual gametes may achieve this objective. They may increase their mobility by having more powerful tails and smaller bodies, so increasing the amount of space they can search in a unit time. Alternatively, they can sacrifice mobility in favour of larger size, expanding their storage capacity and thus increasing their longevity. What will happen when gametes of these two types meet? Two of the smaller, more mobile type will tend to find one another very quickly but, if they fuse, they will produce small zygotes with very limited reserves to sustain the zygote. A fusion of two of the large, less mobile type will produce a very large zygote, but this will be a rare event that is not likely to occur until the two gametes are old and have used up most of their reserves. Fusions between a small and a large gamete will tend to occur quickly, because of the mobility of the smaller one, and will yield a viable zygote, because of the cytoplasmic reserves of the larger one. Such a zygote will, therefore, be more likely either to arise or to survive than those formed by the fusion of two gametes of the same type.

Once two distinct types of gamete have evolved, initially with a small morphological difference between them, disruptive selection will operate on them. That is to say, selection will favour the evolution of smaller, more mobile gametes, provided there are large ones with which they can fuse, and *vice versa*. Furthermore, selection will favour the capacity of each type of gamete to fuse preferentially with one of the opposite type. Selection operates on gametes in a number of ways. First, since they have a finite life and may be existing in an environment that is only temporarily hospitable to them, there will be selection favouring those that find another gamete as quickly as possible. Secondly, since gametes are produced in very large numbers, selection will favour those that fuse to form a large, viable zygote. Third, once a degree of anisogamy exists, there will be competition among the smaller, more mobile ones, creating selection that favours those that reach the larger, less mobile ones first. The first two of these selection pressures are conflicting and will yield either a single type of gamete that is a compromise between conflicting requirements (isogamy), or two distinct types, each of which can exist only if the other type exists (anisogamy).

Sperm, being small, are metabolically cheap to manufacture and typically are produced in vast numbers. They tend to be short lived, though in some species they may live for some time in a nutritious external medium. In some animals this medium is provided by the female, who stores sperm, sometimes for a year or more, between the time of mating and when she fertilizes her eggs. Eggs are much more expensive to make and tend to be produced in relatively small numbers. The differences between eggs and sperm form the basis for many other differences between males and females. Most important, the reproductive success of females is limited by the number of eggs they can produce and, in some cases, care for until they become self-sufficient. In contrast, the reproductive success of males is not generally limited by the ability to produce gametes, but by the ability to obtain matings.

The relationship between egg and sperm is an interesting one because it embodies, at the cellular level, the essential relationship between male and female organisms. Each is dependent on the other if they are to survive to form a viable zygote; their relationship is thus an example of mutualism, in which each benefits from the particular characteristics of the other. Their contribution to the zygote is, however, unequal because, while each contributes equally to the genotype of the zygote, it is the ovum that contributes all of the cytoplasm; in this respect the sperm can be said to be exploiting or parasitising the egg. As described below, sexual reproduction, especially in animals, involves elements of both co-operation and conflict of interests between males and females. The basis of these aspects of sexual reproduction is the phenomenon of anisogamy.

6.4.1 Gender

Dimorphism in gametes represents the basis of a diverse array of differences between males and females that are referred to collectively as gender. Males, by definition, produce sperm; females produce eggs. Gender is not, however, as clear a concept as might at first appear

(Crews, 1988). In everyday human language it refers to a range of characteristic anatomical and behavioural differences between men and women. For some biologists, it refers to chromosomal or other genetic differences between males and females, for others to gonadal differences. Consider, however, the unisexual lizards mentioned above. In terms of their morphology and their gonads, they are female, but in their behaviour they can show the full repertoire of male as well as of female behaviour. Hermaphrodites function gonadally and behaviourally as both males and females; for them, gender cannot be ascribed to the individual organism but only to certain aspects of its reproductive physiology and behaviour.

These conceptual difficulties arise from the fact that, in biological terms, sex and gender are not equivalent or synonymous concepts. Species which, like the sea lettuce, produce isogametes, have sexual reproduction involving meiosis and syngamy, but they have no gender, all individuals having similar morphology, reproductive organs and sexual behaviour. The existence of such systems indicates that anisogamy and gender are not obligatory features or sexual reproduction; rather, they have evolved secondarily, as a consequence of the existence of sexual reproduction.

Because of the difference in the mobility of the two types of gamete, sperm typically have to be delivered to the eggs, a factor that has a fundamental effect on the nature of the mating act. Even this generalisation is not universally true, however; in seahorses, the male cares for the fertilised eggs in a special pouch on his body and the female delivers eggs to him through a penis-like organ (Vincent, 1990).

6.4.2 Sex differences

Eggs, being large, are produced in relatively small numbers; sperm, in contrast, are very small and are produced in enormous numbers. The number of progeny that a female can produce, her reproductive potential, is limited by the number of eggs she can produce; she cannot increase the number of her progeny by mating with several males. The reproductive potential of males, however, is virtually limitless and their reproductive success can be increased by mating with several females (Bateman, 1948). On the basis of this difference we would expect, therefore, that females will typically devote most of their reproductive effort to the production, care and protection of eggs, but that males will devote most of theirs to finding females and competing with other males for them (Trivers, 1972). Some animals are like this, females mating with one male, males with

several females, but very many are not, so clearly there are other factors to be considered. The most important of these is parental care.

The eggs of many animals will not survive unless they are cared for and protected from predators. The eggs of birds, for example, must be kept at the correct temperature by incubation throughout their development and they must be guarded against a variety of egg predators. For birds, reproductive success is limited by the need for incubation; an individual can only incubate as many eggs as it can fit beneath its body. Consequently, the reproductive success of many birds falls well below their reproductive potential; they can lay many more eggs than they can incubate.

For birds living in cold environments, where eggs will die if they are not incubated more or less continuously, or in environments where the risk from egg predators is high, breeding attempts may fail if a parent is not present at the nest all the time. Constant attention may also be essential after hatching, since the young often require frequent feeding, warmth and protection. For many bird species, one parent is insufficient to rear a brood of young successfully; when the parent leaves the nest to feed itself, or to gather food for the young, the eggs or chicks are likely to die of cold or through predation. In such circumstances, it is more adaptive for a male to remain with a female and share parental care with her, than it is to allocate his reproductive effort to obtaining additional matings. No matter how many females he inseminates, if those females cannot rear their young alone, he will leave no offspring. Thus, hostile environments impose a selection pressure that favours males that remain with, and participate in parental care with, one female and select against those that do not.

Even when the environment is such that a single parent can raise young on its own, it is to the advantage of each parent if the other shares in parental care, since the energy demands on itself will be reduced. In most birds, females are the primary care-givers to the young; what factors can enhance the probablity that males will also care for the young? As explained above, the nature of the environment can be a potent external force in this respect, but there are also aspects of social behaviour that can play a part. Consider just two factors among many. First, it is quite common among birds that breed in dense colonies, such as many seabirds, for females to show reproductive synchrony, such that they all mate and start to lay eggs around the same day. A consequence of this is that a male, having mated with one female, has a low probability of subsequently finding another female that has not mated. Thus, in the trade-off between staying with one female and seeking other matings, the benefits of seeking additional matings may not outweigh the costs, in terms of reduced hatching success, of leaving the first female. Secondly, if a male leaves a female, she may mate with another male, with the result that the male that leaves may father few if any of her eggs. When more than one male inseminates a female, there is competition among their sperm to fertilise the female's eggs. This sperm competition is an important factor in the evolution of sexual relationships (Parker, 1970a; Smith, 1984; Birkhead, 1988). In some species, the sperm from two or more males simply mix in the female's reproductive tract so that, if two males inseminate a female, each fathers, on average, only half her progeny. In many species, however, the structure of the female's tract is such that sperm from the last male to inseminate her fertilise all or most of her eggs. This phenomenon of 'last-male paternity' often makes it more adaptive for males to remain with a female that they have mated with and to guard her against the sexual attentions of other males than to seek matings with additional females (Birkhead & Møller, 1991).

These arguments incorporate only a few of the factors that have shaped the evolution of the enormous diversity of relationships that exist between male and female animals. They do however, make an important general point: the relationship between males and females involves both co-operation and a conflict of interests. Neither sex will leave any progeny if they do not co-operate to the extent of mating, at the very least. A very hostile environment may impose conditions under which young can only survive if both parents care for them. In other situations, however, one sex may behave in certain ways that alter the balance of interests of the other. Thus, breeding synchrony and the form of the female's reproductive tract may make it more adaptive for males to stay with a female than to seek further matings.

6.5 The sex ratio and sex change

In the great majority of sexually reproducing organisms that exist as separate sexes (a condition called gonochoristic in animals, dioecious in plants), sons and daughters are produced in roughly equal numbers, giving a 1:1 sex ratio. In genetic terms this is readily explained by the chromosomal mechanisms of sex determination in which one sex (males in mammals, females in birds) is heterogametic (XY), the other is homo-gametic (XX). Such a system apparently makes it inevitable that, at least at conception, sons and daughters will be produced in equal numbers. There are, however, examples of species in which the sex ratio is biased towards one sex, suggesting that selection can operate on the sex ratio in some circumstances. A more general reason why a 1:1 sex ratio is not inevitable is that there are factors determining that it is not necessarily adaptive for the sexes to be produced in equal numbers.

Consider a specific example, the elephant seal (*Mirounga angustirostris*). Females produce one pup each year and make a massive contribution to the rearing of their pups in the form of large quantities of highly nutritious milk. Males play no part in parental care, but fight in a particularly violent way for sexual access to females (Le Boeuf, 1974). Only the largest and most powerful males mate with females; they establish the status of 'harem masters' with exclusive access to a large number of females. The outcome of this competition among males is that, whereas all females that survive to breeding age produce progeny, only 10% of males breed during their lifetime. From the point of view of the species, or of a population, a 1:1 sex ratio thus seems maladaptive. Many more young would be recruited into the population if the male to female ratio was 1:9. No effort would then be wasted in producing sons that fail to

breed. Natural selection does not, however, act on species or on populations, but on individuals, and so we must seek to explain the evolution of the sex ratio in terms of whether it is adaptive for individuals to produce more sons or more daughters. Such an explanation was formulated by R. A. Fisher (1930).

Suppose that a situation arises in a population of elephant seals in which females are twice as numerous as males. A parent that produces x daughters will have xy grandchildren, where y is the average expected number of young produced in a female's lifetime (in elephant seals $y =$ one pup per year for up to 10 years). A parent that produces x sons will have $2xy$ grandchildren, because each son will, on average, inseminate two females. As a result, it is more adaptive to produce sons than daughters in a female-biased population, and the frequency of sons will tend to increase as a result. The fact that only 10% of sons actually reproduce in the elephant seal mating system is irrelevant; it is the average reproductive success of males that is important. Likewise, it will be more adaptive to produce daughters than sons in a male-biased population. Selection thus acts against any departure from a 1:1 sex ratio; put another way, a 1:1 sex ratio is said to be evolutionarily stable, a biased sex ratio is not (see Chapter 5).

This arguments holds true only if the effort required of a parent to produce a son is the same as that required to produce a daughter. In elephant seals this is the case: young of each sex are the same size at weaning and require the same amount of parental care. If, however, female progeny 'cost' a parent twice as much to produce as sons (for example), that parent has the option, given that the resources it can allocate to reproduction are finite, of producing *n* daughters or 2*n* sons, or some equivalent combination of sons and daughters. If the population sex ratio is 1:1, a parent producing 2*n* sons would have, on average, twice as many grandchildren as one producing *n* daughters and so would have higher fitness. In this situation a 1:1 sex ratio is thus not evolutionarily stable and would evolve towards one in which males are more numerous than females. The evolutionarily stable sex ratio of females to males would be 1:2. Thus, a more accurate statement of Fisher's argument is that parents should expend equal amounts of effort on producing sons and daughters (Clutton-Brock, 1991).

If the XX/XY mechanism of sex determination always produces sons and daughters in a 1:1 ratio, there will be no variance on which selection could act, even if it were adaptive for parents to produce more progeny of one sex then the other. It is not, at present, clear to what extent this is an important constraint on the evolution of the sex ratio. There is, however, increasing evidence that biased sex ratios do occur in nature and that these arise by other than genetic mechanisms. In an American bird, the common grackle (*Quiscalus quiscula*), male and female eggs are produced in roughly equal numbers but, by the time they fledge, females outnumber males by 1.6:1 (Howe, 1977). In this species males, at fledging, are approximately 20% heavier than females; consequently, they require a greater energetic investment on the part of their parents. It appears that the parents devote roughly equal amounts of parental care to male and female progeny with the result that mortality among male nestlings, which have a greater energetic requirement, is higher than among females.

In an experimental study of wood rats (*Neotoma floridana*), McClure (1981) kept a control group of mothers on an unrestricted diet and an experimental group on a diet that provided only 70% to 90% of the nourishment required to maintain the body weight of non-reproductive females of comparable size. In both groups the sex ratio at birth was 1:1

but, whereas this was maintained until the time of weaning in the fully fed control group, the sex ratio in the food-restricted group fell steadily until, by 20 days after birth, it was three males to seven females. This change resulted from the food-restricted mothers allowing their sons less opportunity to suckle than their daughters. The young of mothers kept on the restricted diet were small at all ages, including full maturity, than those of fully fed mothers. The small adult size of daughters did not reduce their eventual reproductive success, as measured by litter size, but small sons were less able than the sons of fully fed mothers to obtain and defend territories and thus did suffer reduced reproductive success in comparison with the sons of fully fed mothers. The adaptive explanation proposed for the discrimination shown by food-deprived mothers to their male progeny is that, because their sons have poor reproductive prospects, mothers on a restricted diet devote less parental care to them.

Baboons live in matriarchal societies, in which the females of a given age within a group are socially dominant to males of the same age. Within each sex, there is also a hierarchy, with larger individuals being dominant over smaller ones. As a result, larger females enjoy greater access to high quality food than smaller females or males and so have larger progeny that have a greater chance of surviving. Individual young, as a result of these size differences, tend to 'inherit' (because of this phenotypic effect, not because of a genetic effect) the status of their mother: high-ranking females tend to have progeny that subsequently themselves attain high rank. In addition, high-ranking females produce more daughters than sons; low-ranking females produce more sons (Clutton-Brock, 1991). The mechanisms underlying this sex ratio bias, in a species that has genetic sex determination, is not known.

Perhaps the most remarkable example of sex ratio bias is that reported by Burley (1986) in captive zebra finches (*Taeniopygia guttata*). Individuals of both sexes can be made more or less attractive to the opposite sex by giving them particular colours of leg ring (red is especially attractive). In long-term breeding experiments, birds with attractive leg rings produced more young of the same sex, whereas those with unattractive rings produced more opposite-sexed offspring.

These examples suggest that, although there may be no departure from a 1:1 sex ratio at birth (the primary sex ratio), the sex ratio in a population can be distorted by unequal allocation by parents of their parental care to sons and daughters. One group of animals in which the primary sex ratio is manipulated, though in quite a different way, is the hymenopteran insects, such as bees. Females determine the sex of their progeny by controlling whether or not their eggs are fertilized. Unfertilized eggs become haploid males, fertilized eggs become diploid females. Many hymenoptera show strongly biased sex ratios with a much higher proportion of females than males; in honeybees for example, the sex ratio within a hive is one male to three females (Hamilton, 1967; Bull, 1981).

It can be argued that, in a variable environment, a female should manipulate the sex ratio among her progeny according to whether her sons or her daughters are more likely to maximize their reproductive potential under the prevailing conditions. The basis of this prediction is that, if females do this, they will maximize the number of their grand-progeny, i.e.: maximize their fitness. Such a model has been developed by Charnov *et al.* (1981) and has been applied by them specifically to parasitic wasps, the female of which lays single eggs on the corpses of prey which vary in size and on which the growing wasp larva feeds. The size of the young wasp that eventually develops from each egg depends on the size of the prey on which it is laid. Because eggs are large and energetically expensive to produce, the fecundity of a female is likely to be a function of her size: the larger she is, the more eggs she will produce. This dependency of fecundity on body size will generally be less marked in males; even the smallest male will produce enough sperm to fertilize several females.

Sex allocation theory predicts that female wasps should apportion male and female eggs among prey of varying size. They should lay female eggs on larger prey, male eggs on smaller prey. In a laboratory experiment, this prediction was tested using the parasitic wasp *Lariophagus distinguendus*, which attacks granary weevils and in which progeny from eggs that were laid on large prey are larger than those laid on small prey (Charnov *et al.*, 1981). The results are shown in Fig. 6.3. As predicted, females laid a greater proportion of female eggs when the available prey were large than when they were small.

In the solitary bee *Osmia lignaria propinqua*, females produce equal numbers of each sex over the course of a breeding season, but produce a preponderance of daughters early in the season and switch to producing sons later. This appears to be related to a greater availability of flowers that provide food early in the season. When food is abundant, progeny can attain large size and will achieve high fecundity if they are female. Later, when food is scarce, they will be small and can be more successful as males (Torchio & Tepedino, 1980).

In a refinement of sex-allocation theory, Trivers and Willard (1973) suggest that, where there is greater variance in the reproductive success

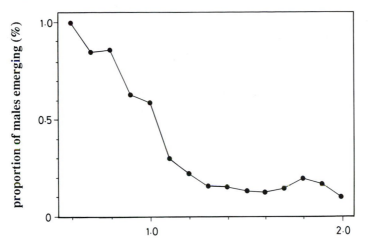

size of host (mm)

Fig. 6.3. The relationship between the size of insect hosts and the sex of parasitic wasps (*Lariophagus*) laid on them by their mother. (Modified from Charnov *et al.*, 1981.)

of males than of females, mothers should preferentially produce some sons if they themselves are in good condition, since the sons of poor-condition mothers are less likely to achieve high reproductive success. Conversely, poor-condition mothers should produce daughters. This prediction has been supported in studies of red deer (*Cervus elephas*), a species in which male mating success is strongly skewed towards larger males. Hinds with high dominance rank produce a larger proportion of sons than low-ranking mothers (Clutton-Brock *et al.*, 1986).

6.5.1 Environmental sex determination

Perhaps the most puzzling form of departure from a 1:1 sex ratio is that which results from environmental effects. This environmental sex determination is best known among reptiles, in several species of which the gender of an individual is determined by the temperature at which it undergoes development in the egg. In the leopard gecko (*Eublephoris macularius*), for example, eggs reared at a temperature of 32°C yield progeny of which 80% are male and 20% are female. In contrast, eggs reared at 26°C result in 100% female offspring. A 1:1 sex ratio results if the eggs are reared at 29°C (Fig. 6.4; Crews, 1988). These results are due to a real effect of

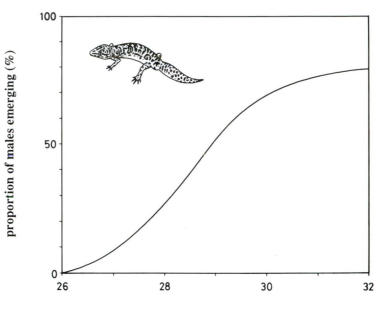

Fig. 6.4. The effect of temperature during development on the sex ratio in the leopard gecko (*Eublepharis macularius*). (Modified from Crews, 1988.)

temperature on development, the nature of which is not known, and cannot be attributed to differential mortality of males and females at different temperatures. Whatever the temperature at which it is reared, an individual gecko is unequivocally male or female; hermaphrodites or individuals of indeterminate sex are never found and there are no physiological differences between 'hot' females, reared at temperatures that produce a preponderance of males, and 'cold' females. It appears that temperature during development activates some kind of switch mechanism that determines whether an individual will become male or female. In crocodiles such as *Alligator mississippiensis*, temperature-dependent sex determination is even more dramatic: a temperature of 30°C during development produces 100% females, 34°C and above produces 100% males (Ferguson & Joanen, 1983). In another group of reptiles, the chelonians (turtles and tortoises), in contrast, high temperatures produce females, low temperatures males (Bull, 1983; Janzen & Paukstis, 1991). In those reptiles that show environmental sex determination, the effect is on the primary sex ratio, i.e. that among progeny at birth. In the other examples discussed above, females manipulated the sex ratio of their progeny in a variety of ways to produce a biased population, but in the context of a primary sex ratio that is 1:1.

The ecological, and therefore evolutionary significance of environmental sex determination is not yet clear. The phenomenon has been studied intensively in the laboratory, where temperature can be controlled, but there are rather few data on sex ratios in natural populations. It is known, however, that the laboratory results found in *Alligator mississippiensis* are consistent with what happens in nature. Crocodiles and alligators lay their eggs in large nests consisting of mounds of vegetation; the temperature within nests varies from one to another, hotter nests produce a higher proportion of males than cooler nests, and, in natural populations, there is a preponderance of females, suggesting that the majority of nests are cool (Ferguson & Joanen, 1983). It remains to be seen whether female crocodiles, and other reptiles, can adaptively manipulate the sex ratio among their progeny by controlling the temperature within their nests.

6.5.2 Hermaphrodites and animals that change sex

Hermaphrodites are animals that are capable of producing both male and female gametes. True hermaphrodites are those that can produce both at the same time, a condition found in a wide variety of animals, including platyhelminthes, molluscs, crustaceans and fishes. The most likely selection pressure favouring hermaphroditism is difficulty in finding a mate (Ghiselin, 1969). Many hermaphrodites are sessile animals that cannot move about in search of mates, animals that live at very low population densities that have a low probability of finding a mate, or internal parasites, which typically live in an environment that is very isolated from conspecifics. For such animals, the ability to fertilize one's own eggs in the absence of a mating partner is a great advantage. In addition, they do not bear the costs of mating, since they do not have to seek or compete for mates and are not exposed to the risks that attend mating. On the other hand, they must bear the cost of inbreeding, of which self-fertilization is the most extreme form. Self-fertilization will only be adaptive for hermaphrodites that are self-fertile and which do not carry a large number of lethal recessive genes.

Even hermaphrodites, that are not self-fertile, are at an advantage over gonochoristic animals that live at low densities, because any two individuals that meet one another will be able to mate, since each produces both male and female gametes. Mating in many hermaphrodites involves a mutual exchange of gametes, each individual passing sperm or eggs to the other. In the polychaete worm *Ophryotrochia diadema*, mating partners exchange eggs with one another, a process called 'egg-trading' (Sella, 1985, 1988), whereas in the sea slug *Navanax inermis*, they trade sperm (Leonard & Lukowiak, 1991).

Among plants, many adaptations exist that prevent self-fertilization. There are, however, a number of plants, such as violets (*Viola* spp.), that possess flowers that never open, insuring that self-fertilization occurs. Much less is known about the incidence of self-fertilization in animals. Some gastropod snails, such as *Helix* and *Cepaea*, are completely self-sterile, whereas another, *Rumina*, typically fertilizes itself in nature. The white-lipped land snail *Triodopsis albolabris* will never fertilize itself if kept in pairs, but will do so after being kept in individual isolation for several months. The reproductive success of cross-fertilizing pairs is 86 times greater than that of self-fertilizing individuals, suggesting that, for this species at least, self-fertilization incurs severe fitness costs that will make it adaptive only when no partners are available (McCracken & Brussard, 1980).

There are many hermaphrodites that are not sessile or parasitic and which do not live at low densities, most notably several species of coral reef fish. One species, the black hamlet (*Hypoplectrus nigricans*), employs its dual sexuality in a remarkable way (Fischer, 1981). Each individual has a large ovary and a relatively minute testis. When spawning, they form pairs which last for a day or more and the two partners take turns to adopt male and female roles. One fish produces a batch of eggs which its partner fertilizes externally. The other fish then produces a batch of eggs which the first fish fertilizes and this alternating pattern is repeated many times until both fish have exhausted their egg supply. The fertilized eggs are released into the plankton. The two partners in effect 'trade' eggs with one another, in return for having them fertilized by their mate. This system does not incur one of the major costs of sex, the production of sons, since all individuals are fully capable of functioning as females. Furthermore, because each fish needs only a small testis to produce sufficient sperm to fertilize the eggs of its mate, the allocation of resources to the production of male capability is much less than it is in gonochoristic species, in which, as discussed above, it is typically equal to the allocation to female capability. The black hamlet, by egg-trading and cross-fertilizing with another fish, gains all the benefits of sexual reproduction, but bears very little of the costs. It is curious that this reproductive pattern is known for only a very few species.

Several species of fish consist of individuals that are capable of producing both male and female gametes, but not at the same time. This condition is called serial hermaphroditism. Some are protandrous, beginning life as males and becoming female later in life, others are protogynous and change sex in the reverse direction. The existence of these two patterns raises the question: under what conditions is it better to be male or to be female? In both protandrous and protogynous species, sex change typically occurs when an individual attains a certain size, suggesting that body size is an important determining factor. The exact role of body size in determining the direction of sex change depends on the mating system of the species (Ghiselin, 1969; Warner *et al.*, 1975; Warner, 1984).

Fishes normally continue to grow after they have begun to reproduce. For both sexes, an individual's fecundity usually increases with its size because its gonads will grow along with the rest of its body. Thus larger males will produce more sperm than smaller individuals, larger females more eggs. For females, because they produce relatively very large gametes, fecundity is strictly limited and will typically show a linear increase with body size (Fig. 6.5). For males, however, sperm can be produced very cheaply and in huge numbers so that even a tiny male has a very large reproductive potential, in terms of the number of eggs he can fertilize. If, however, the mating system of a particular species of fish is such that larger males dominate and prevent smaller ones from mating, a common pattern, only larger males will be able to realize their reproductive potential (Fig. 6.5a). In this situation it is adaptive for an individual to start breeding as a female and to continue to do so until it has reached a critical size at which it can compete effectively with other males. This is protogyny and is characterized by males being larger and less numerous than females.

If mating is random, however, and there is no advantage to being a large male, small males can produce enough sperm to fertilize many eggs, many more in fact than they could produce eggs. In this situation, what limits a male's reproductive success is his ability to find females to mate with, and variance among males in terms of mating success is not related to their size (Fig. 6.5b). In this situation, an individual should start its reproductive life as a male as it can expect to fertilize more eggs than it could produce as a female of comparable size. It will change to being a female when it has grown large enough to be able to produce large numbers of eggs. This is protandry, in which males are typically smaller than females.

There is another factor to be considered here, and that is the sex ratio in the population. Here we can apply the same kind of argument that was

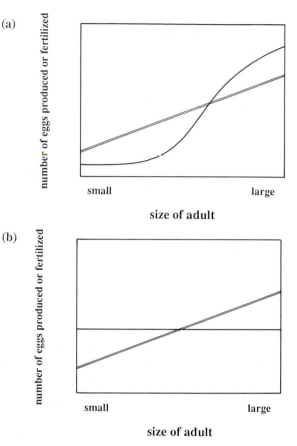

Fig. 6.5. Hypothetical relationships between reproductive success and body size in male (single line) and female (double line) fishes that can change sex. In both cases female reproductive success (fecundity) is strongly related to body size. In (a) males can only achieve high mating success when they are large and it is therefore advantageous for individuals to be protogynous. In (b) small males have the same mating success as large males and it is advantageous for individuals to be protandrous. (Modified from Warner, 1984.)

developed above, based on the idea that, in a given population, there will be a particular sex ratio that is evolutionarily stable. If, in a population of protandrous fishes, males are very abundant relative to females, average male reproductive success will be relatively low, because there are few eggs available for them to fertilize. It will then be adaptive for one or more individuals to change into females, since they will then produce more progeny than they would if they remained male. Likewise, in a protogynous species in which females become relatively very common, an individual may gain higher reproductive success by changing into a male. This hypothesis has been tested experimentally in a number of protogynous species. If one or more large males are removed from a population, one or more of the larger females typically turn into males and take their place.

In serial hermaphrodites each individual has the option of being male or female. Which sex it is more adaptive for it to be, that is, which yields the higher fitness benefit, depends not only on properties intrinsic to the individual, such as its size, but also on what other individuals are doing. Thus, the sex ratio in a population, which at first sight may seem to be an attribute best understood at the population level, is a consequence of selection operating on the behaviour of individuals.

6.6 Sexual selection

The many differences between males and females, particularly in morphology, reproductive potential and behaviour, that are based ultimately on anisogamy, create selection pressures that act only on one sex. Selection of characters that give certain individuals an advantage over others of the same sex in obtaining successful matings is called sexual selection (Partridge & Halliday, 1984). Because variance in mating success is typically greater among males than among females, sexual selection has generally been regarded as a process that affects males but, recently, there has been an increasing awareness that females often compete with one another for matings (Ahnesjo et al., 1993). For example, female moorhens (*Gallinula chloropus*) compete for males on the basis of their ability to incubate eggs (Petrie, 1983a), female mormon crickets (*Anabrus simplex*) compete for nutritious spermatophores produced by males (Gwynne, 1981), and female dendrobatid frogs compete for those males that best care for their eggs (Summers, 1992). The stereotype of aggressive males competing for essentially passive females is increasingly being shown to be inappropriate.

There has long been debate as to whether sexual selection is a distinct process from natural selection. This arises because extravagant male characters, used in fighting for or attracting females, are apparently costly in terms of male survival. Some authors (e.g. Mayr, 1972) suggest that this makes sexual selection a distinct process, others (e.g. Endler, 1986; Halliday, 1978, 1990a) that it does not. It is typical of phenotypic characters of many kinds, whether or not they are involved in sexual competition, that they confer both fitness costs and benefits on individuals. Studies of the costs of sexual ornaments and behaviour are very important

because there is an increasing body of theory and empirical evidence based on the hypothesis that female choice selects for male ability to meet such costs (Andersson, 1982, 1986; Kodric-Brown & Brown, 1984; Halliday, 1987). Atmar (1991) emphasizes physiological costs of male mating activity in suggesting that the existence of males may be adaptive for a species (see Section 6.2.1).

Conventionally, two forms of sexual selection have been differentiated, intrasexual and intersexual selection (Halliday, 1978; Bradbury & Davies, 1987). Intrasexual selection refers to selection arising from fighting between members of one sex, usually males, that leads to the evolution of large body size, fighting ability and weapons such as horns. Intersexual selection refers to selection arising when males compete to attract females and leads to the evolution of conspicuous colours, structures, odours and other displays that attract females. In some instances, this distinction is quite clear and there are numerous empirical studies demonstrating the evolutionary consequences of the two types of process. For example, the large size of male elephant seals is clearly attributable to male combat (Le Boeuf, 1974) and the enormous train of the peacock (*Pavo cristatus*) has evolved through female choice (Petrie *et al.*, 1991). There are many instances, however, where the distinction is not clear and there are theoretical grounds for regarding it as invalid (Halliday, 1990a). As Brown

(1983) has pointed out, both forms of selection represent competition between male genotypes and are thus evolutionarily equivalent; males displaying to a female are as much in competition with one another as they would be if they were fighting.

The precise nature of the selection pressures acting on males as a result of competition for mates becomes particularly unclear in species in which female choice favours characters that also confer an advantage in aggressive interactions among males. Examples include large body size in fishes, such as the mottled sculpin (*Cottus bairdi*) (Downhower *et al.*, 1983), and the black throat patch of the house sparrow (*Passer domesticus*) (Møller, 1987, 1988). Since these characters give individual males an advantage both in male–male and in male–female interactions, their evolution cannot be attributed exclusively to either intra- or intersexual selection.

Determining the adaptive value of sex-related characters is a complex process because of the diversity of selection pressures involved. A fruitful approach has been to identify, for a given character such as male body size, 'episodes' of selection and, for each episode, to determine the relationship between variation in that character and variation in a specific component of fitness (Wade & Arnold, 1980). The slope of the regression between the two variables provides an index of the intensity of selection. Application of this method to Howard's (1979) data on bullfrogs (*Rana catesbeiana*) showed that large male size is favoured by selection, primarily because larger males acquire more mates, and there is only very weak selection arising from two other effects, that larger males tend to pair with larger, more fecund females, and that larger males hold superior territories in which eggs are more likely to survive (see also Arnold & Wade, 1984a, 1984b; Koenig & Albano, 1986).

While direct competition between males is generally exemplified by overt fighting, it often also takes a more subtle form and there is a huge variety of 'dirty tricks' whereby males may gain an advantage over their rivals. Among salamanders, several species show sexual interference, in which a male mimics female behaviour so as to induce a rival male to engage in an unproductive and energetically costly 'mating' (Halliday, 1990b). Among insects, males of several species apply secretions to females during mating; these render them unattractive, or apparently unreceptive to other males (Thornhill & Alcock, 1983).

Females may benefit from competition between males if the outcome is that the fittest males are those that win. In some species, females actively incite males to compete (Montgomerie & Thornhill, 1989). Examples include the female jungle fowl (*Gallus gallus spadiceus*), which calls when most fertile (Thornhill, 1988), and the female spider *Linyphia litigiosa*, which attracts to her web a succession of males that must fight one another to stay there until she is ready to mate (Watson, 1990).

Probably the most widespread form of non-combative sexual competition among males is sperm competition. This arises when the sperm from two or more males compete to fertilize a female's eggs. Sperm competition is common in animals, such as many fishes, with external fertilization; in many species males dart towards a mating pair and shed sperm on the eggs as they emerge, e.g. the labrid fishes *Symphodus tinca* and *S. ocellagus* (van den Berghe *et al.*, 1989). In animals with internal fertilization the occurrence of sperm competition depends on a number of aspects of female reproductive biology: there must be a delay between mating and fertilization, sperm must be stored within the female's reproductive tract, and a female must mate with more than one male. Sperm competition and its evolutionary consequences are generally discussed in terms of selection acting on males, but the fact that the females play a crucial role in providing the conditions under which it can occur suggests that it may have selective advantages for them, a consideration that has been neglected (Eberhard, 1990). The existence of sperm competition or, more accurately, the possibility that it may occur, has led to the evolution of a diversity of male adaptations that prevent it or which counter its effects (Smith, 1984; Birkhead & Møller, 1991). These include increased investment in sperm production by males and various forms of mate-guarding behaviour by which males prevent other males from mating with a female before and/or after they do.

Competition among males for matings can be intense if the fitness differential between winners and losers is very large. Consequently, there is strong selection for less competitive males to evolve alternative mating strategies, forms of behaviour by which they may achieve some measure of reproductive success. These are discussed further in Chapter 5.

Finally, while selection in relation to sex is generally associated with mating systems in which one or both sexes is polygamous (has several mating partners, see Section 6.7), it is wrong to assume that it is an unimportant influence in monogamous species. While there may be little or no variation among individuals in terms of mating success, there may still be considerable variation among individuals, of either sex, in terms of their reproductive potential. Selection favours adaptations that enable individuals to mate with the fitter members of the opposite sex, promoting competition for mates and leading to sexual dimorphism in morphology and behaviour and other sex-related adaptations (see Halliday, 1978; Kirkpatrick *et al.*, 1990, for reviews).

6.7 Sexual relationships: mating systems

Across differential species of animals, there is considerable diversity in the distribution of mating among individuals; the various kinds of distribution are called mating systems (Emlen & Oring, 1977; Davies, 1991). The main categories of mating systems are:

Monogamy: individuals mate exclusively with one partner.
Polygamy: individuals mate with two or more partners. This is divided into:
  Polygyny: an individual male mates with two or more females.
  Polyandry: an individual female mates with two or more males.

Before using these terms, some caveats are in order. First, the term 'system' can be misleading, since it implies some external constraint on the way animals behave. In human societies, monogamy may be prescribed and polygamy allowed or prohibited by legal and religious constraints, but among other animals the pattern of mating is a consequence of the behaviour of individuals. Second, while it is common practice to describe a *species* as monogamous or polygamous, these terms should properly be applied to *individuals*. Because most sexual species have a 1:1 sex ratio (i.e. equal numbers of males and females), only monogamy can be the rule for all individuals in a species. In so-called polygamous species, some individuals will have several mates, some one, and some none at all. Third, it is necessary to bear in mind the time-scale under consideration. In a species in which individuals breed several times, a female may be monogamous each year, but with a different male; over the course of her lifetime she is, in effect, polyandrous. Bearing these caveats in mind, the rest of this section considers the factors that promote monogamy in some species, polygyny or polyandry in others.

6.7.1 Monogamy

Monogamy is more common among birds than in any other group of animals and it is estimated that about 90% of bird species are mono-gamous; it is very rare among other vertebrates. The prevalence of monogamy among birds is probably due, as discussed earlier, to the fact that most of them produce young which, both as eggs and chicks, require considerable parental care if they are to survive. Eggs require constant incubation and defence against predators; chicks require warmth, protection and, in many species, frequent feeding. Only in habitats in which it is warm, food is abundant, and predators are few is one parent able to rear the young alone. If females are assisted by a male, they can realize a greater proportion of their reproductive potential. If males engage in parental care of the young they have fathered, they gain greater reproductive success than if they do not.

Some birds maintain their pair-bonds over several years. In the kittiwake (*Rissa tridactyla*), the breeding success of individuals and pairs has been carefully monitored for many years and it has been possible to show that, the longer a pair stays together, the higher is their reproductive success (Coulson, 1966). This provides an additional advantage for individuals that maintain a monogamous relationship.

6.7.2 Resource-based polygyny

This type of mating system occurs where males control access by females to resources, such as food or nest sites, that are essential to the reproductive effort of females, and where competition among males results in resources being divided unequally among them. As a result, males that successfully defend a large share of resources attract and mate with many females; those that hold a small share obtain few or no matings.

If a resource occurs in patches, it is likely that some males will hold territories that are rich in it, while others will hold territories that are deficient. A particularly clear example of resource-based polygyny is provided by a bird, the orange-rumped honeyguide (*Indicator xanthonotus*) of Nepal (Cronin & Sherman, 1976). In this species, beeswax obtained by raiding the nests of giant honeybees is an essential component of the diet of both sexes. Bees nests are scarce; they also occur only on cliff faces and so are patchily distributed. Males do not show any parental care but focus all their reproductive effort around bee nests, where a small proportion of them successfully establish territories. Only females and non-breeding immature birds are allowed into these territories and males copulate with all mature females that enter their territories to eat wax. Females are unresponsive to the courtship displays of non-territorial males.

6.7.3 Female-defence polygyny

In some species males compete for females directly, not by competing for resources, and successful males establish a group or 'harem' of females.

The formation of such groups is facilitated if females show a predisposition to gather in groups for some reason other than mating. For example, female elephant seals leave the water once a year, hauling themselves out onto a beach to give birth to their pups (Le Boeuf, 1974). Because suitable beaches are few and far between, large numbers of females tend to gather in confined areas. Mating occurs immediately after a female has given birth, at the same site, which thus becomes the focus of intense competition among males to gather groups of females. Fights among male elephant seals are extremely violent and only the largest and most powerful males become 'harem masters'. This intense selection for fighting ability has favoured large size in males, which are three times heavier than females. In a single season, only a third of males mate at all, and less than 10% of males fertilize nearly 90% of the females. Fighting incurs high costs for males: some die immediately after a season in which they have been harem masters and few achieve this status for more than 3 years, whereas females may breed for up to 10 years in succession. Some males never mate during their lives.

A male's success in maintaining exclusive access to a group of females depends not only on his ability to compete with rival males, but also on his ability to control the movements of females. Red deer stags spend all their time during the rut (mating period) in competing with rivals by roaring and, occasionally, fighting, and in chivvying their hinds like a sheep-dog to keep them in a compact group. Nevertheless, females frequently wander away and join other harems (Clutton-Brock *et al.*, 1982). The rarity of this kind of polygyny among birds is probably partly due to the fact that they are highly mobile and that males cannot easily control the movements of females.

6.7.4 Leks

A quite different form of polygyny occurs when males do not defend or control either resources or females. Instead, males gather in dense clusters, called leks, where they defend very small territories. These contain no resources of importance to females but simply provide sites on which males display to and mate with females that visit the lek. Males display vigorously to females and it is typical of lek species that males have very elaborate plumage: the peacock is the classic example. Lek species have been studied very intensively, largely because of the opportunities they provide to study female choice (Vehrencamp & Bradbury, 1984).

6.7.5 Polyandry

Polyandry, in which individual females simultaneously have long-lasting mating relationships with several males, is a very rare mating system. For this reason, and because it contradicts the general proposition that males are expected to be the polygamous sex, it has been intensively studied. It is exemplified by a water-bird, the American jacana or lily-trotter (*Jacana spinosa*) (Jenni, 1974). Jacanas live on lily-covered lakes, where their greatly elongated toes enable them to walk on floating vegetation. Each female defends a large territory within which there are several smaller male territories, each containing a floating nest that is attended and defended by a particular male. She moves around her territory, mating from time to time with all the males, and laying eggs in each nest. Each male defends and carries out all the incubation of the eggs in his nest. The female is thus emancipated from parental care and allocates all her reproductive effort to mating, egg laying and defending her territory against other females. She is considerably larger (50–75% heavier) than the males, is dominant over them and will stop any fights that break out between them.

The critical factor in the evolution of polyandry is the nature of the breeding habitat. Suitable breeding sites are scarce and nests are subject to heavy predation. These conditions favour the ability of females to lay large numbers of eggs, both to exploit such suitable habitats as do exist and to replace eggs that are lost to predators. Selection has thus favoured two aspects of female reproductive biology that enhance their egg-laying capacity, large body size and a shift in resource allocation away from parental care and towards egg production. Selection for large size is accentuated by the advantage that it gives females in competition for breeding sites with other females. It has led to females being larger than males and thus able to dominate them.

This mating system is clearly highly advantageous to the female since she is able to achieve much higher reproductive success than if she had to care for her young. What are the selective pressures acting on males that have led to their behaving as they do? As discussed under monogamy, an environment that is hostile to egg survival will generally favour parental care by males, and polyandry most probably evolved from a monogamous mating pattern. A high level of predation requires the ability to replace lost eggs; only females can do this, shifting the allocation of parental care towards males and favouring large size in females. When a female is large enough to dominate and control a male, his options become restricted. To seek further matings he would have to leave his nest, exposing it to predation and himself to attack by the female.

There remains the question of why males within a female's territory tolerate one another. This is related to the pattern of mating and egg laying shown by the female and introduces a critical factor in the evolution of this and other mating systems, 'certainty of paternity'. The female mates with all the males in her group before and during egg laying. Each male thus guards and incubates a clutch of eggs, only some of which are likely to be fertilized by him; on the other hand, some eggs that he has fertilized are being cared for by other males in the group. Natural selection will favour males that allocate paternal care only to those progeny that they themelves have fathered and will select against those that care for progeny that do not carry their genes. In the jacana, however, males cannot distinguish between eggs that they have fathered and those of other males, because particular episodes of mating by the female are not linked to particular episodes of egg laying. Selection thus favours their caring for a clutch of eggs, some of which are probably theirs, and not disrupting the breeding effort of other males that are caring for eggs, some of which are also theirs.

6.7.6 Conflicts of interest

Within a population, it will usually be the case that individuals will vary in their mating strategy: some will be monogamous, others polygamous. As a result, individuals will vary considerably in terms of their fitness and there will be a conflict of interest between the sexes. For a female it may be better if her mate is monogamous, but for him better to be polygamous. The nature of this conflict between the sexes is well illustrated by Davies' study of dunnocks (Davies, 1985; 1992; Davies & Houston, 1986), discussed in detail in Chapter 5.

6.7.7 Experimental tests of mating system theory

The arguments presented above about the environmental conditions that favour one kind of mating system rather than another are based on observational evidence. Information about mating patterns is related to the nature of the environment in which different species live. These hypotheses can, however, be tested by carrying out appropriate experiments.

Polygyny

Mating system theory suggests that polygyny occurs where males defend resources that are essential to the reproductive effort of females and where there is variation among males in the quantity of such resources that they hold. This hypothesis can be tested by enhancing the resources held by selected individual males; if the hypothesis is correct, such males should attract more females. This kind of experiment has been carried out on the red-winged blackbird (*Agelaius phoeniceus*) by Ewald and Rohwer (1982). They placed either a single feeder containing protein-rich food, or several feeders, in the territories of particular males and compared the number of females that they attracted with the number attracted by control males that did not have feeders. They found that males with a single feeder did attract more females than control birds, but that males with several feeders did not. What happened in the latter case was that several additional males moved in so that what had originally been a single territory containing several feeders became a number of very small territories. These results suggest that males are attracted to and compete for areas where resources are rich and which will, therefore, be attractive to females.

In polygynous systems, individual females face a choice between joining one or more females already settled on a good territory or being the sole partner of a male holding a poor territory, a problem addressed by the polygyny-threshold model (see Chapter 5).

Monogamy

Mating system theory argues that monogamy is favoured when the environment is such that both parents can rear young more successfully together than if males are polygamous. This hypothesis has been tested in a number of birds by removing the male from an established breeding pair after mating has occurred (Bart & Tornes, 1989; Mock & Fujioka, 1990). If the hypothesis is correct, the reproductive success of a female whose mate has been removed should show a decrease, compared with that of a female whose mate has not been removed.

Greenlaw and Post (1985) carried out a male-removal experiment on the American seaside sparrow (*Ammodramus maritimus*) and found that the number of young successfully reared by experimental females was 33% less than the number reared by control females. Bart and Tornes (1989), working with house wrens (*Troglodytes aedon*), removed males on four

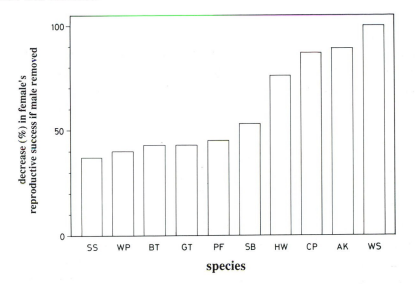

Fig. 6.6. Maximum values, obtained from mate-removal experiments, for the contribution of male birds to female reproductive success in ten bird species. SS: seaside sparrow (*Ammodramus maritimus*), WP: willow ptarmigan (*Lagopus lagopus*), BT: blue tit (*Parus caeruleus*), GT: great tit (*Parus major*), PF: pied flycatcher (*Ficedula hypoleuca*), SB: snow bunting (*Plectrophenax nivalis*), HW: house wren (*Troglodytes aedon*), CP: common pigeon (*Columba livia*), AK: American kestrel (*Falco sparverius*), WS, western sandpiper (*Calidris mauri*). (Modified from Mock & Fujioka, 1990.)

separate occasions and found that nestling survival decreased by 63% in one trial, but showed no reduction in the other three trials. The trial in which a reduction was found was conducted when climatic conditions were especially severe. In both these species, the presence of a male increases a female's reproductive success. In the house wren study, the variation in the results suggests that this effect only operates when the environment is particularly hostile.

Similar studies have been carried out on several species, and there is a great deal of variation between species in the strength of this effect (Fig. 6.6). There is a range from the seaside sparrow, in which male removal leads to a decrease of 33% in fledged offspring, to the western sandpiper, in which the figure is 100%.

In some male-removal studies, researchers have closely observed 'widowed' females to see how, in behavioural terms, they compensate for the loss of their mate. For example, Whillans and Falls (1990) removed the males from pairs of white-throated sparrows (*Zonotrichia albicollis*) when their chicks were 6 days old and found that females increased the frequency with which they left the nest to collect food and that, in consequence, they spent less time brooding their chicks. In this experiment, male removal did not reduce fledging success but the chicks in experimental nests weighed significantly less than those in control nests.

6.8 Summary and conclusion

The selection pressures that maintain sexual reproduction in natural populations are unclear. A number of costs and benefits, for both sexual and asexual reproduction, can be identified, acting both at the species and the individual level. It is not clear, however, what the relative values of these costs and benefits are in nature. Outbreeding sex and asexual reproduction by mitotic parthenogenesis represent the extremes of a continuum of reproductive modes that yield varying degrees of both genetic variation between parents and progeny and heterozygosity within a population. For sexual species there may be an optimal level of outbreeding. For a given species, evolution towards an optimal mode of reproduction may be countered by the costs of changing from one mode to another.

The existence of sexual reproduction makes inevitable the evolution of anisogamy through disruptive selection on gametes. In turn, anisogamy provides the basis for a diversity of differences between males and females, in morphology, behaviour and, especially in the allocation of parental care. Although genetic mechanisms produce males and females in equal numbers, there is accumulating evidence that, in a variety of contexts, parents can preferentially produce sons or daughters in an adaptive manner.

Sex differences lead to different selection pressures acting on males and females. Both males and females may compete with members of the same sex for access to the other sex, with the result that sex differences have been influenced by sexual selection. Animals show a variety of mating systems. States such as monogamy and polygamy are best regarded as properties of individuals rather than of species and, within a single population, individuals vary with respect to these states. At the population level, the pattern of mating is best understood, as in the case of dunnocks, as the product of both conflict and co-operation between males and females.

7

Kinship and altruism

P. J. B. SLATER

7.1 Introduction

Of all the misunderstandings that surrounded the theory of evolution in the century after *The Origin of Species* was published, that concerning the level at which natural selection acts was the most persistent and pervasive. It is easy to find examples, even in the writings of distinguished biologists, of animal characterisics being referred to as for 'the good of the species'. Sometimes this may have been what they intended to say, but perhaps it was more often sloppy wording or written in the belief that what was good for the individual was much the same thing as what was good for the species. Wynne-Edwards (1961) did a service by putting such arguments more explicitly. Even though it is now generally accepted that he was wrong, the alternative viewpoint that he presented acted as an important and useful stimulus to the field. It was his belief that much of the behaviour of animals could only really be understood in terms of natural selection favouring the group or species to which they belonged. Why else did some animals, such as the sterile castes of social insects, forego breeding altogether? Why else were animals within a group so often altruistic to one another and so seldom aggressive? He interpreted these and a wealth of other observations in terms of this theory, and argued that it was much better able to account for the social behaviour of animals than the alternative that each animal behaved purely in its own interests. For example, he argued that many of the communal displays of animals, such as leks and roosts, the dawn chorus and even the vertical migration of plankton, could be interpreted as mechanisms to allow groups of animals to assess population size so that their reproductive efforts could be raised or lowered appropriately. Unlike humans, animals did not over-exploit resources, he claimed, but were prudent farmers. If in danger of diminishing their food supply, they would reproduce less so that supply and demand were neatly matched. To back up his case, his book provided numerous examples of animals behaving for the good of others rather than in their own selfish interests: the phenomenon of altruism, which appeared so hard to account for in any other way.

These views provided a challenge that was quickly taken up by many others. The problem in accepting them was at a very basic level. They depended on selection acting at the group level, groups of animals that behaved in such ways out-competing those that did not, yet it was hard to see how this process could override the effects of individual selection except under very unusual and special circumstances. It may be for the good of the group that a particular animal does not reproduce but, if it has genes that make it breed despite this, it will pass these on to more offspring than will others that only breed when it is in the interests of the group, and they will thus spread through the population. As Richard Dawkins (1976) put it, in his characteristically pithy style: 'Even while the group is going slowly and inexorably downhill, selfish individuals prosper in the short term at the expense of altruists'. As individuals reproduce and die much more rapidly than groups divide and become extinct, selection at the level of the individual must normally override that at the level of the group.

So, how then does one account for phenomena such as sterility and altuism if animals are not behaving for the good of their group? Most of the answers to Wynne-Edwards' challenge seem to lie in considering the effects of kinship, which have been increasingly studied over the three decades since his book was published. It is with these effects that the first part of this chapter is concerned. We will then go on to consider how kinship is assessed – by both animals and biologists. Lastly, we will ask whether kinship is all there is to altruism, or do unrelated animals show it to one another?

7.2 The level at which selection acts

J. B. S. Haldane, in a famous remark, commented that he would lay down his life for at least two brothers or eight first cousins. Whether because he was years ahead of his time or because his mode of publication was unconventional (he is said to have expounded the view over a drink in the Orange Tree Pub in the Euston Road), the importance of this point went unremarked for several decades. The emergence of kin selection theory, as it has come to be called, did not really occur until two papers by Hamilton (1964a, b), which are now viewed as landmarks, although again it took some time before they became widely recognised. How does this theory differ from the idea that selection acts purely at the level of the individual?

It is correct that selection does act mainly on individuals, because it is individuals that vary in their ability to survive and produce. No one would even wonder why an animal should help its own offspring, because leaving more children or grandchildren is clearly advantageous in terms of natural selection. But what Hamilton did was to extend this argument to other relatives by taking it down to the genetic level. An animal shares half of its genes with each of its children, and also with each of its parents and siblings. It shares one-eighth of them with each first cousin (hence Haldane's remark) and a progressively lower proportion with other relatives as the relationship gets weaker. Just as helping its offspring assists the spread of its genes, so too does helping its more distant relatives, though to a lesser extent.

In understanding these arguments, it is useful to grasp the concept of the coefficient of relatedness (r). This varies from 0 to 1 and represents the extent to which individuals are related to each other. It can be defined in two different ways, which come to the same thing.

1. The proportion of one individual's genes (or alleles) that are identical by descent with those present in another.
2. The probability that a particular gene (or allele) possessed by one individual is identical by descent with one present in another.

Including 'for reasons of common descent' in these definitions may seem curious, but it is important. Many genes are widespread within a species, and animals may share the majority of their genes without necessarily being closely related. However, while siblings may share 99% of their genes, they only have 50% in common by reason of their shared parentage. These are the genes that are 'identical by descent' and it is these that are relevant to assessing kinship. Using an ESS argument, Dawkins (1979) shows that a population of universal altruists can be invaded by individuals that are altruistic only to kin. The 'identical by descent' provision introduced by Hamilton is thus an important one.

Table 7.1a gives values of r for various degrees of relatedness in a population of diploid animals such as humans, and Figure 7.1 illustrates relationships in the form of a pedigree diagram. It is assumed that the population is outbred: inbred animals share a higher proportion of their genes. For example, a brother–sister mating gives $r = 0.75$. The figures are also averages: the exact amount of sharing between individuals may

Table 7.1. *Coefficients of relatedness (r) between various*
kin pairs in diploid animals (e.g. mammals) and
haplodiploid animals (e.g. hymenopterans)

(a) Diploid	
Parent < > offspring	0.50
Full siblings	0.50
Half siblings	0.25
Identical twins	1.00
Grandparent < > grandchild	0.25
Uncle/aunt < > nephew/niece	0.25
First cousins	0.125
(b) Haplodiploid	
Mother < > daughter	0.50
Father > daughter	1.00
Daughter > father	0.50
Mother > son	0.50
Son > mother	1.00
Brothers	0.50
Sisters	0.75
Brother > sister	0.50
Sister > brother	0.25

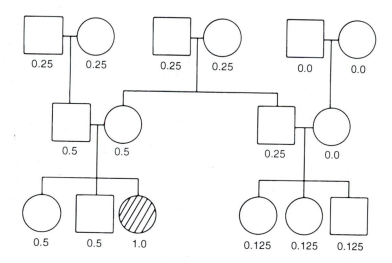

Fig. 7.1. A pedigree diagram illustrating relationships within a family. Squares and circles indicate males and females respectively. The figures are coefficients of relatedness between the female, shaded, and all the other individuals. Within the diagram she has two siblings, two parents, four grandparents, an uncle and three first cousins. (Adapted from Shields, 1984.)

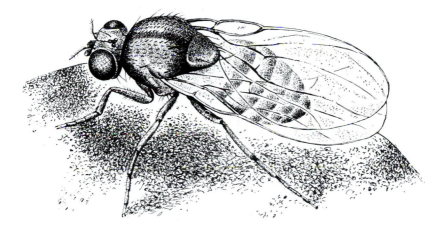

vary according to which particular chromosomes each inherits, the extent of cross-overs, etc. Its variability may also depend on the number of chromosomes a species has, a feature that differs quite a bit between them. In species with few chromosomes, such as *Drosophila* in which there are only four pairs, two gametes from the same parent may often have a very high proportion of their genes in common by chance, or a very low one. In humans, however, with 23 pairs of chromosomes, the proportion must usually be very close to 50%, as we would expect at random.

Coefficients of relatedness are not so simple in some species. Persistent inbreeding can lead them to rise progressively towards an asymptote of 1.0, and can also make them very hard to calculate. Imagine working out the r between the parent great tits and their offspring at the bottom of Figure 7.2! In species with a short generation time such as this, it is also quite possible for a male to be mated to his great grandmother. Such problems involve diploid species, but other complexities arise where the genetic system is haplodiploid, as it is in hymenopterans (Table 7.1b). Here, males have only one set of chromosomes whereas females have two. Males are formed from unfertilised eggs and so receive all their genes from their mother. This has the interesting effect that r is asymmetrical. Every gene a male has he got from his mother ($r = 1.0$), but she only passed half her genes (one allele of each pair) on to him ($r = 0.5$). It also introduces complexities in the case of siblings. A father gives each of his daughters all of his genes, so that half their genes are exactly the same. As to the rest, each gets half of her mother's genes and on average they will therefore share 50% of these with each other. Thus, for sisters $r = 0.75$. On similar reasoning $r = 0.25$ between a sister and

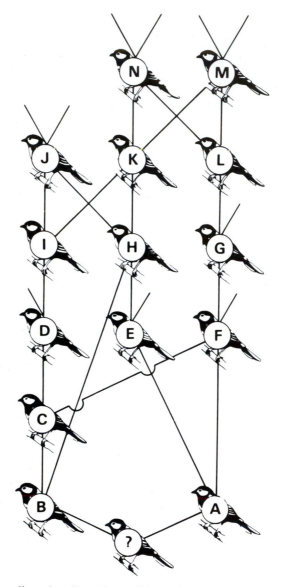

Fig. 7.2. A complicated pedigree in a wild population of great tits (*Parus major*) from the island of Vlieland in the Netherlands. All birds are at least as closely related as grandparents to A and B. (From van Noordwijk & Scharloo, 1981.)

her brother (he has none of the genes she received from her father). This curious genetic system thus leads sisters to be more closely related to each other than they are to their brothers; they are also more closely related than they are to their mothers or, if they are fertile, to their offspring.

Following on from such considerations, there are two ways in which an animal can enhance transmission of copies of its own genes into the next generation. The first ('direct') way is by reproducing itself, thus increasing its Darwinian fitness in a straightforward way. The second ('indirect') way is by helping relatives. Hamilton coined the term 'inclusive fitness' to take account of both these components.

How important is this indirect component? In many animal breeding systems, some of which we will consider below, it may be a crucial factor. An obvious case is in those social insects where workers do not reproduce at all and their inclusive fitness consists entirely of the indirect component. In many other cases the indirect component may be unimportant as the animals simply do not help relatives. How then should inclusive fitness be measured? Where no helping is involved this is simple as it amounts to the average number of offspring of the individuals of a particular genotype that survive to breed, or for a single individual the number of surviving offspring that that animal has. Where helping is involved, the calculation for a single individual is much more complicated, and quite incorrect methods have often been recommended (see Grafen, 1982, and the particularly lucid account of these issues by Dawkins, 1986). The problem is that, as in a balance sheet, the same credit must not appear twice. If the inclusive fitness of one animal is augmented by the help it gives to a relative in raising its offspring, that of the relative must be lowered by an equivalent amount. Some frightening calculations might result, and the information on which to base them will almost never be available. However, looked at in terms of genotypes rather than individuals, the problem becomes simpler. Inclusive fitness can be defined exactly as where no helping is involved: the average number of offspring of that genotype that survive to breed. This does not mean that helping is unimportant. What it does mean is that there is a simple way of comparing between the success of animals with helping genes and those without. Does a queen social insect with 99 sterile workers to help her produce more offspring on average than could 100 solitary fertile individuals? That question is much easier to answer than it is to calculate what the inclusive fitness of the queen is and what is that of each of her workers.

7.3 Implications of kin selection for behaviour

Hamilton argued that the helping of relatives was beneficial if the following equation was satisfied:

$$b/c > 1/r$$

where b is the fitness benefit to the relative and c the fitness cost to the helper. Thus, the more distant the relationship, the greater the benefit has to be in relation to the cost. From this we would predict that animals will favour closer relatives over more distant ones for any given act of helping, and also that help to more distant individuals would only occur where great benefits could be bought at rather little cost. A few case histories will help to indicate the extent to which these predictions are met.

Sterility in social insects

Why does it pay a female honey bee to become a sterile worker rather than a fertile queen? The answer to this question seems to lie in the close genetic relatedness amongst individuals in a honey bee hive. For example, their haplodiploid genetic system may provide one clue. As females are more closely related to their sisters than to their offspring, it is more advantageous for them to rear a fertile sister than a fertile daughter. However, this cannot be all there is to it, as rather equivalent systems exist in termites and in naked male rats (*Heterocephalus glaber*), which are diploid. The key seems nevertheless to lie in the closeness of the relationship between the reproductive female and her helpers. In termites it has been suggested that the pattern of inbreeding within colonies, followed by outbreeding when new colonies are set up, may also lead individuals to be more closely related to their siblings than to their offspring (Bartz, 1979). Mole rats live in dense colonies surrounded by inhospitable terrain, so that dispersal from one colony to another is very difficult and colonies are certainly highly inbred: the mean $r = 0.81$, and their level of inbreeding is the highest for any mammal yet studied (Reeve et al., 1990).

Helpers at the nest in birds

There are at least 200 bird species in which the young are fed by more individuals than just the breeding pair. In most such cases the helpers are

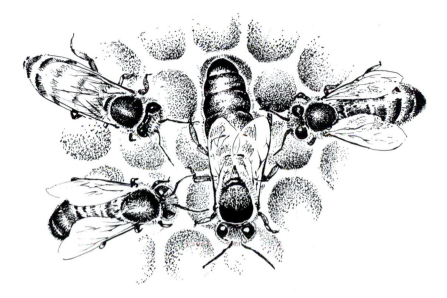

close relatives and they are often offspring of that pair from previous broods. In the white-fronted bee-eaters (*Merops bullockoides*) studied by Emlen (1990), for example, there is a highly significant tendency for birds to help relatives, and 45% of helping is directed to rearing full siblings. On average $r = 0.33$ between helpers and the nestlings they are feeding (Emlen & Wrege, 1988). Why should birds help their parents rather than moving off to find mates and set up breeding territories on their own? The first point is that, as with all diploid animals, a bee-eater is as closely related to its siblings as to its offspring ($r = 0.5$), so benefits will accrue equally from rearing one of either. Secondly, bee-eaters are colonial and it is not easy for two birds to find sufficient food for their brood near the colony, so that helpers may be crucial to breeding success. For each helper a pair rears, on average, half an extra chick (Fig. 7.3). On the other hand, a young bird that sets up on its own with a mate has great difficulty in rearing young at first because it has no older birds to help it.

The balance of advantage between helping and breeding oneself may involve many factors, such as the availability of mates and breeding sites, and whether one's parents are still alive. In white-fronted bee-eaters, kinship appears to be the predominant factor leading to helping and determining which pair an individual helps (Emlen & Wrege, 1988). However, there may also be direct benefits to the individual helper, and these may be of crucial importance in some species. For

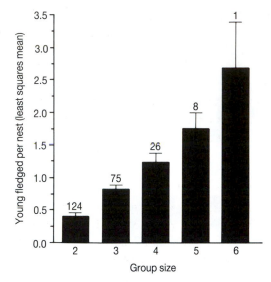

Fig. 7.3. The effect of group size on fledging success in a population of white-fronted bee-eaters (*Merops bullockoides*) at Baharini in Kenya. The values are least squares means (±s.e.) from an analysis of covariance in which the effects of season and clutch size were controlled. (From Emlen, 1990.)

example, helpers gain experience that may enhance their later breeding success. Birds that help may also be in the best position to replace the pair member of their own sex if it should die, and so achieve 'promotion' with the group to full breeding status. Even more directly, it is also not uncommon for female helpers to lay an egg in the nest at which they are helping. Thus, seen in terms of costs and benefits, there are a number of reasons why birds will sometimes help rather than finding a mate and having offspring of their own. Although these reasons undoubtedly vary between species (see the reviews in Stacey & Koenig, 1990), kinship is an important factor in many of them, including the white-fronted bee-eater.

Weaning conflict in mammals

Until the advent of kin selection theory, the conflict between mothers and their infants at the time of weaning, and indeed parent–offspring conflict generally, was hard to understand. At that time infants will often continue to demand attention while their mothers increasingly drive them away (Hinde, 1974). Trivers (1974) put this phenomenon in a theoretical context by pointing out that the optimal time for weaning is earlier for mothers than for their infants. The mother is related to both her present infant and her next one by $r = 0.5$ and, at a time when the present infant is reasonably self-sufficient, it is to her advantage to move on to investing in the next one (lactation often being incompatible with pregnancy). On the other hand, the infant is related to itself by $r = 1.0$ (it *is* itself), but to its sibling by only 0.5 if they share a father or 0.25 if they do not. Thus the costs and benefits for it are different and the optimal time for weaning will be later. An understanding of kin selection theory thus helps to account for this behaviour.

Alarm calling in ground squirrels

The production of alarm calls by animals has always been something of a brain-teaser. Why should an animal that sees a predator make a noise that is likely to draw the predator's attention to it? There is evidence in birds that the costs may often be minimised: the sounds produced may be very hard to locate and may, at least in some cases, even be inaudible to the predator (Klump, Kretzschmar & Curio, 1986). There must nevertheless be benefits to account for their evolution.

Of the many ideas put forward to account for alarm calling, kin selection is one of the most likely. In line with this, evidence for Belding's ground squirrels (*Spermophilus beldingi*) suggests that calling depends very much on the number of relatives present in an animal's group (Sherman, 1977, 1980). This species is subject to a great deal of predation, from animals such as coyotes and American badgers, and the calls are conspicuous and easy to locate, so the benefits of calling must be considerable. In ground squirrels, as in most mammals, young females tend to stay in the group of their birth. On the other hand, young males transfer to other groups and continue to move from year to year thereafter. Males call less than females, as one would expect because they have fewer relatives around them. Females that do move between groups also call less than those that are resident and, amongst residents, those with close relatives present in the group call more than those without.

Examples such as these leave no doubt that kin selection theory can account for many phenomena that would be hard to understand at the strictly individual level. These include some of the most tricky cases, such as worker sterility, and also many cases of altruism, even if the 'altruism' becomes more apparent than real when we take account of the effect of kinship. In its everyday sense, the word altruism means doing something of benefit to another individual at a cost to oneself. But it is questionable whether the word should be applied to many animal examples, where kin selection theory may tell us that benefits outweigh costs at the level of the genes.

7.4 Kin recognition in animals: how and why?

That kinship influences behaviour does not necessarily mean that animals recognise kin. Selection may favour altruism where the probability of an animal being a relative is high: for example, birds feed the young in their nests though sometimes, as when the chick is a cuckoo, they are not related to them at all. But there is now considerable evidence that animals discriminate between kin and non-kin and behave differently towards the two, indeed the literature on kin recognition has grown phenomenally in recent years (see reviews by Hepper, 1986, 1991; Barnard, 1990). Here we will consider the function of kin recognition and the mechanisms that underlie it.

7.4.1 Why recognise kin?

Despite the large amount of recent work on kin recognition, there is little evidence on its functional significance (Blaustein *et al.*, 1991). Furthermore, although it is easy to give theoretical reasons why it may be adaptive, as will become clear, many of the advantages that it may bring could be achieved in other ways.

There are two main reasons why it might be beneficial to animals to behave differently towards kin and non-kin. First, as discussed above, helping relatives (or nepotism) may enhance the indirect component of inclusive fitness. We might therefore expect mechanisms to arise that ensure that animals discriminate between their relatives and others so that they give help to the former but not the latter. Some evidence also suggests that animals of similar genotypes may compete more than dissimilar ones, and kin recognition may allow animals to avoid such competition (Barnard, 1990).

As far as helping relatives is concerned, an interesting point is made by Altmann (1979). He points out that, where animals are surrounded by individuals of varying degrees of relatedness, they should not simply apportion help according to r, but should help close relatives in preference to more distant ones. If help, regardless of the amount received, is simply translated into offspring by the recipient, then it pays the donor only to help the closest available relative. But if more help brings diminishing returns, some reallocation of help to more distant relatives may be advantageous, although this help will not pay so well as its proceeds must be devalued by the more distant *r*. Put another way, if the offspring of a sibling and those of a cousin would benefit equally from the help an animal can give, it should devote its help to those of the sibling. But the animal may have no siblings, or may have a cousin struggling to raise a brood to whom help might be more than four times as beneficial as it would be to a sibling that was rearing its brood without difficulty. In this case the help would be better allocated to the cousin, even though it has only one-quarter of the shared genes that the helper has with its sibling. This situation is unlikely to arise often. In any case, the most important point from this argument is that, from the aid-giving viewpoint, the recognition of close relatives is that most likely to be advantageous.

The second benefit of kin recognition is in mate choice. Inbreeding may lead animals to be homozygous at an increasing number of loci and thus to the expression of deleterious recessive mutations. Morton, Crow & Muller (1956) estimated that each human carries the equivalent of 3–5 lethal recessive alleles, so that it is not surprising that the adverse effects of inbreeding are well documented (Cavalli-Sforza & Bodmer, 1971). 'Inbreeding depression' has also been described in many other species. For example, in white-footed mice (*Peromyscus leucopus*) inbreeding leads to smaller litter size and lower offspring weight at weaning (Keane, 1990); the former is also true of domestic mice (Barnard & Fitzsimons, 1989). In great tits (*Parus major*) inbreeding leads to more hatching failure (van Noordwijk & Scharloo, 1981) and higher nestling mortality (Greenwood, Harvey & Perrins, 1978).

The results point to the benefits of inbreeding avoidance. One way in which this might be achieved is by animals recognising close relatives and not mating with them. However, much the same result can be achieved by differential dispersal. In mammals males usually disperse from their birth site further than females, whereas in birds the opposite is the case (Greenwood, 1980). Kin recognition is thus not a prerequisite for outbreeding. In laboratory tests incest avoidance is sometimes not found, and this may be because closely related animals would not normally meet in the wild (Duncan & Bird, 1989).

As discussed in Chapter 6, outbreeding may also be deleterious. At an extreme, animals selecting mates very different from themselves may be in danger of mating with the wrong species, with the many adverse consequences that hybridisation entails. But, even at a less drastic level, complexes of genes which are co-adapted or which adapt animals to their particular local environment, may be broken up by outbreeding (Shields,

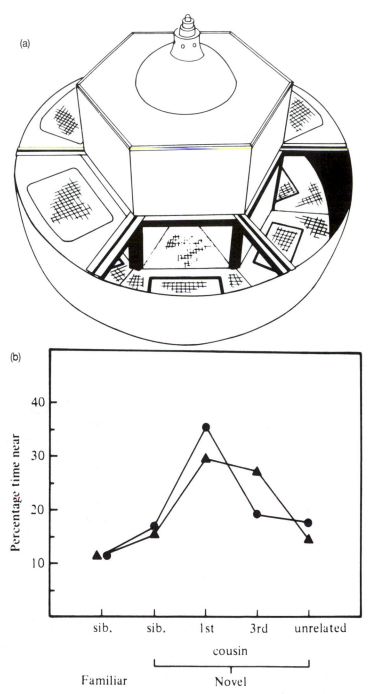

(a)

(b)

Percentage time near

40

30

20

10

sib.   sib.   1st   3rd   unrelated

cousin

Familiar   Novel

For caption, see opposite.

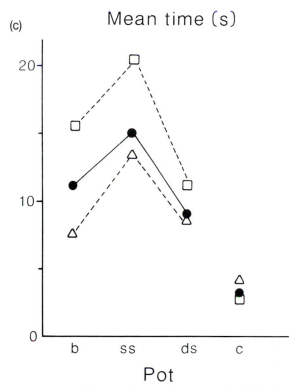

(c)

Mean time (s)

Pot

Fig. 7.4. (a) Apparatus used for testing preferences of adult Japanese quail (*Coturnix c. japonica*). Single birds are put singly behind one-way screens in the illuminated inner compartments while the test bird can move freely around the unlit outer part of the apparatus. A pedal outside each window records the time spent there. (b) The mean percentage time spent by adult Japanese quail near members of the opposite sex that were familiar siblings, novel siblings, novel first cousins, novel third cousins, or novel unrelated individuals. Male results are shown as triangles, female results as circles. (c) Mean time in a 5-minute test spent by female mice of two strains (squares and triangles; black circles are overall means) investigating each of four different pots containing sawdust which was unsoiled control (c), or from the floor of the cage of a male who was of a different strain (ds), of the same strain but not a brother (ss), or was a brother (b). (Parts (a) and (b) from Bateson, (1982) *Nature*, 295, 236–7; (c) after Gilder & Slater (1978) *Nature*, 274, 364–5. Reproduced with permission from Macmillan Magazines Ltd.)

1984). The combined effect of selection against both this and inbreeding depression may lead to what Bateson (1978) labelled 'optimal outbreeding', animals choosing mates that are neither too closely nor too distantly related to themselves. In his own work on quail (*Coturnix c. japonica*), Bateson (1982) has found birds to prefer the company of first cousins to that of either third cousins or brothers (Fig. 7.4a), and birds mated to first cousins also breed rather earlier (Bateson, 1988b). As well as finding inbreeding depression in white-footed mice, Keane (1990) obtained some rather weaker evidence for deleterious effects of outbreeding: at weaning the young of unrelated pairs are rather lighter than those of first cousin pairs. In line with both sets of results, he found some tendency for oestrous females to prefer males of an intermediate degree of relatedness. Although this effect was not a strong one, the finding is in line with earlier work on laboratory mice (e.g. Gilder & Slater, 1978; Fig. 7.4b).

Although most kin recognition systems are probably adaptive in the context of either helping or mate choice, not all of them are likely to be explained within these two categories. A perplexing example is that of social aggregation in tadpoles of various species (e.g. Blaustein & O'Hara, 1986; Waldman, 1981), where the animals have been found to form kin groups. Several advantages have been suggested for this, including the idea that it may lead to optimal outbreeding in adulthood (Blaustein & Waldman, 1992); it could also be a case of kin grouping as a byproduct of some other mechanism, a possibility discussed further in the next section.

Despite such difficult examples, there seem to be good reasons why it might be advantageous for animals to behave differently towards kin and non-kin. However, such differences in behaviour might arise without any recognition as such: differences in dispersal between the sexes may lead animals normally to mate with non-kin; animals may help others with which they are familiar, and which are thus likely to be kin, without necessarily recognising kin as such. To be sure that some recognition process is involved, we need experiments such as those of Bateson and of Keane described above, which clearly demonstrate that animals of different degrees of relatedness are treated differently. Many such experiments have now been performed, and they point to a variety of different kin recognition mechanisms.

7.4.2 How are kin recognised?

In a recent paper, Grafen (1990) threw down the gauntlet, by claiming that only one of the many papers on the subject (that by Grosberg & Quinn, 1986) had demonstrated kin recognition satisfactorily. However, his definition of kin recognition was a very restrictive one: 'recognition by genetic similarity detection'. In whatever way recognition works, the process must involve matching to some standard which, he argues, 'must be genetic if it is kin that are recognised, as kin are defined by genetic similarity'. Most other cases that have been described, he claims, are byproducts of species, group or individual recognition. For example, animals may learn the characteristics of their species by imprinting on their mother early in life. Although the function of the system may be species recognition, the standard is acquired from relatives, and the young animal may later respond more to kin than to non-kin as a result, but without this having been selected for as such.

This critique sounds a cautionary note amidst the host of claims that different species 'recognise kin'. In most of these the animals are far from 'reading *r*', were that possible; in few are the demonstrations in natural situations and in few has the selective advantage of kin recognition been demonstrated. Nevertheless, as Stuart (1991) argues cogently in reply to Grafen, many systems using acquired standards. and involving group or individual recognition, may have fitness benefits associated with that recognition that 'typically flow among kin'. To exclude them from 'kin recognition' just because they do not involve a special mechanism is unduly restrictive.

A broader definition of kin recognition proposed by Waldman, Frumhoff & Sherman (1988), is 'the processes by which individuals assess the genetic relatedness of conspecifics to themselves or others based on their perception of traits expressed by or associated with these individuals'. Being internal, such processes are unobservable, but they can be inferred if animals treat kin differently from non-kin, in other words if they show kin discrimination.

How might such discrimination be achieved? Discrimination systems have been classified in various different ways in reviews of the subject (e.g. Holmes & Sherman, 1983; Hepper, 1986; Barnard, 1990), but they all propose several categories which are not necessarily mutually exclusive. We will consider these under four headings.

Context-based discrimination

Parent birds and rodents may be quite undiscriminating about providing care to young in their nests, even if these belong to another species (e.g. Denenberg, Hudgens & Zarrow, 1964). Some male rodents which normally kill pups they encounter, are less likely to do so after a period of living with and mating with a female (e.g. Elwood & Ostermeyer, 1984). In these cases, either the location of the young or the state of the adult makes it probable that the young encountered are relatives, and behaviour is

adjusted accordingly, without any more sophisticated discrimination
system coming into play.

Discrimination based on familiarity

This and the subsequent two categories all involve comparison of cues
(sometimes referred to as 'signatures') from a conspecific with some
standard. There are several ways in which this may be done (Fig. 7.5).
Animals may respond differently to familiar and to unfamiliar individuals,
familiarity being normally associated with kinship. Such a system can be
demonstrated experimentally where animals treat individuals with which
they have been reared differently from those they have not encountered
before, regardless of whether they are relatives or non-relatives. For
example, spiny mice (*Acomys cahirinus*) prefer to associate with siblings
rather than with non-siblings, but this is based on experience as they do
not discriminate against non-siblings with which they have previously
been housed (Porter, Wurick & Pankey, 1978). Early experience is also
important in the sibling recognition of tadpoles of the toad (*Bufo
americanus*) as animals reared with siblings and non-siblings do not
discriminate between them, while those reared only in sibling groups do
show a preference (Waldman, 1981). Both of these preferences are
olfactory (Porter *et al.*, 1978; Waldman, 1985a), whereas that of tadpoles
appears not to be because cues converge amongst animals that live in
association (Waldman, 1985b); in spiny mice shared diet seems to
accentuate the effect (Porter & Blaustein, 1989).

Phenotype matching

This mechanism involves the development of a broader standard, which
may be based on experience with parents, siblings, or indeed self, and
against which others are compared (Fig. 7.5). Individuals showing low
discrepancy are treated as kin. Unlike the last category, the prediction
here is that animals will discriminate between unfamiliar kin and unfamiliar
non-kin. An example of such an effect is that sibling mice are less
aggressive to one another than are non-siblings, even if they have never
encountered each other before (Kareem & Barnard, 1986). Any characteristic that is more similar in kin than in non-kin might be used to make
such a discrimination: the quail studied by Bateson (1982) appear to use
visual cues; vocal signals learned from relatives have also been suggested
as having a possible role here, though evidence is currently weak

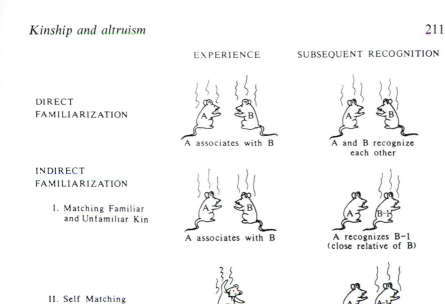

Fig. 7.5. Three ways in which experience might affect recognition in animals, illustrated with the example of smell in mice. (From Porter & Blaustein (1989). Reproduced with permission from Blackwell Scientific Publications Ltd.)

(McGregor, 1989). The most convincing evidence comes from discrimination using smell in rodents (e.g. Kareem & Barnard, 1986). In such cases, while animals may learn the standard that they use, the standard itself may be more or less closely based on relatedness and so correlated with r.

Genetic systems

Two rather different systems come under this heading, the first with and the second without a possible role in kin discrimination.

1. Genes providing kinship markers. Mice prefer to mate with individuals differing from themselves in the major histocompatibility complex (MHC) of the genome (Yamazaki *et al.*, 1976). This group of loci is highly variable and controls the production of cell surface molecules, which differ between individuals as a defence mechanism against viruses which might otherwise mutate to mimic such characteristics and so get past the immune system. These molecules also contribute to urine odour and it is on this that the discrimination is based (Singh,

Brown & Roser, 1987). The mating preference may give offspring greater resistance to disease but, as the MHC difference is greater in non-kin than in kin, this genetic difference may also allow kin recognition and so avoidance of mating between close kin (Egid & Brown, 1989; Potts, Manning & Wakeland, 1991). Indeed, its associated smells may be amongst those used in phenotype matching.

2. 'Green beard' genes. This is a hypothetical system which might lead animals to behave altruistically towards each other, and was originally described as the 'green beard' effect by Dawkins (1976). If a gene favouring altruism towards others that possess it happens to have a phenotypic readout that can be sensed (such as a green beard), it will spread rapidly in the population. However, this is not an example of kin discrimination because selection will favour helping any individual with that particular gene regardless of their relatedness. Nor is it likely that a gene would happen to have these two unrelated effects in the first place.

What can we conclude about kin recognition? It is now a much researched subject and some interesting and exciting results have emerged. However, there are more theoretical possibilities than there are firm conclusions. Most crucially, it remains far from clear the extent to which the many discrimination systems described have their benefits through kin and, if they do so, just what these benefits are.

7.5 Measuring relatedness in animals

Given the important place that kinship has in current evolutionary theory, and also all the recent research on how it affects behaviour, it is obviously important that we are able to assess as accurately as possible how closely related animals are to one another. This is not a simple matter. In many species females mate with more than one male for each brood they produce. Furthermore, the proportion of young fathered by each is not necessarily related to the frequency with which each mates. In some species first male advantage is well established (e.g. Schwagmeyer & Foltz, 1990; Watson, 1991), while in others the most recent male to mate fathers most if not all of the young (e.g. Birkhead *et al.*, 1990). Even in species which form monogamous pairs, such as many birds, females may show extra-pair copulation, and 'egg-dumping', birds laying in a nest other than their own, is also well documented (Andersson, 1984). The attentive guarding of mates by males, and of nests by established pairs, may minimise the risk of these eventualities, but they are relatively common. The long period of courtship before mating and egg laying in many birds may be accounted for by the need for paternity assurance on the part of males. In some species sperm can survive for weeks in the female reproductive tract (Birkhead, 1988), so that only a long period of mate guarding before egg laying can ensure that young are not fathered by any earlier mates the female may have had. All of these phenomena point to the difficulty of establishing relatedness by observation alone.

The correlation of phenotypic traits between individuals has allowed some estimates of extra-pair paternity and intraspecific brood parasitism. In the lesser snow goose (*Anser caerulescens hyperboreus*) there are two morphs, and the frequencies of these among the offspring of different pairs allowed it to be estimated that 5.6% of goslings resulted from nest parasitism and 2.4% from extra-pair fertilisation (Lank *et al.*, 1989). In indigo buntings (*Passerina cyanea*) offspring wing length is more similar to that of the mother than to her mate, and parent-offspring regressions suggest that as many as 40% of young may not be fathered by the pair male (Payne & Payne, 1989). This is especially interesting as observational data suggested that only 2% of copulations were outside the pair, indicating how poor a guide these can be to paternity.

Until recently, the main biochemical method for analysing genetic similarities between individuals was gel electrophoresis using polymorphic enzymes. Incompatibility between parents and offspring at one or more loci could indicate where extra-pair copulations or egg-dumping have occurred. However, this method (while cheap) is both laborious and very uncertain. The more loci that are examined the more reliable it becomes, but the possibility that one individual is the parent of another cannot often be excluded with confidence.

In place of such methods, the field has now been revolutionised by the advent of DNA fingerprinting. This technique was developed only recently, by Jeffreys, Wilson & Thein (1985), but its use in studies of behaviour and evolution was appreciated as quickly as were its more publicised legal uses in forensic science and in cases of contested paternity. Though each test is quite time consuming and costly to carry out, the method has the great merit that it can identify mismatches between parents and offspring with extremely high reliability. The chance of getting it wrong, and thinking that individuals are parent and offspring when they are not, is less than 1 in 10 000. So far, much of the work using this technique has been with birds, as they are unlike mammals in having red blood cells with nuclei, and thus containing DNA. A small blood sample will thus provide sufficient DNA for the test.

The procedure is based on the existence of 'minisatellites' in the DNA of the genome, consisting of the same short sequence repeated many times. While all these sequences are identical within an individual, they are highly variable ('hypervariable') between them. The first stage of the finger-printing process consists of splitting them into sections using an enzyme that cuts where it identifies a particular short series of bases. For example, the Hae III enzyme cuts between the guanine and the cytosine wherever the sequence GGCC occurs. For each individual this then cuts a strand of DNA into many fragments. Although the full procedure from this stage need not concern us, in essence the fragments are separated using gel electrophoresis and a short DNA probe, which has been radioactively labelled so that it can be seen on an autoradiogram, is bound to them. The resulting autoradiogram shows a series of bars unique to that individual, rather like the bar codes on goods in a supermarket, each bar being produced by the DNA fragment or fragments that have migrated that distance up the gel.

The reason why minisatellite sequences are so variable is that they are, for some reason, subject to a very high mutation rate. Close relatives share many sequences and thus bars on the fingerprint, but the variability in the sequencing means that the chances of all of them being the same (except in identical twins) are minute. Indeed, even close relatives in an inbred line can be distinguished easily in humans. However, almost all the bars in the radiogram of a young animal should be present either in that of its mother or its father or both. Where it has several bars that fail to match one or other there is good evidence that it is not that individual's offspring. Figure 7.6 shows how clearly one can identify cases where animals are not parent and offspring from analysis of band sharing (Graves *et al.*, 1992). Five chicks in Figure 7.6(b)(i) have a very low rate of bar-sharing with the male at their nest, and are therefore examples of extra-pair paternity. One of these five chicks is also that in Figure 7.6(b)(ii), which shares few bars with its apparent mother, and it is therefore related to neither of the birds tending it: its presence in that nest must be due to egg-dumping.

So far, the major use of DNA fingerprinting has been in identifying cases of extra-pair paternity and egg-dumping or, in more complex breeding systems, determining just which male is the father of a particular offspring (Burke *et al.*, 1989). Amongst birds, while there are species in which extra-pair paternity appears to be negligible (Gyllensten, Jakobsson

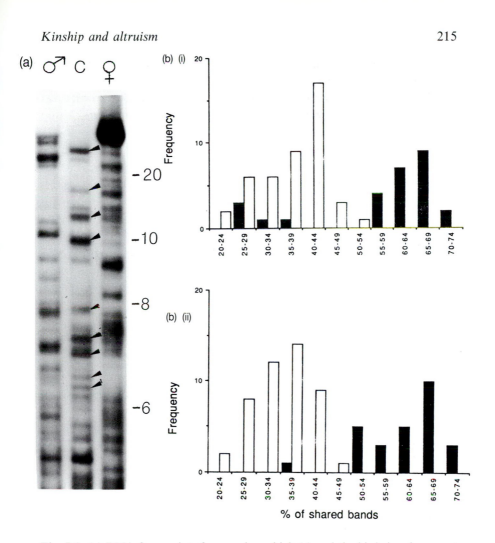

Fig. 7.6. (a) DNA fingerprints from a shag chick (c) and the birds in whose nest it was reared. Arrows indicate bands derived from the chick that are not identical to any bands in either parent. The chick shares over half its bands with the female, but only 25% with the male, thus indicating its extra-pair paternity. The scale on the right gives the length of the DNA fragments in kilobases. (b) The distribution of the percentage of bands shared between young shags and adults. (i) The percentage shared between each offspring and the male at its nest (closed bars) compared with two other, presumably unrelated, males (open bars). (ii) That between each offspring and the female at its nest (closed bars) and two other adult females from the colony (open bars). (From Graves *et al.*, 1992.).

& Temrin, 1990), it ranges up to around 40% in some others (Westneat, 1990). Intraspecific brood parasitism has been found to account for up to 11% of offspring (Birkhead *et al.*, 1990). These high levels indicate just how important it is to consider such processes in assessing kinship amongst animals.

The technique of DNA fingerprinting is thus proving very useful in detecting cases where animals that appear on observational evidence to be parents and offspring are not closely related at all. If the technique is applied to groups of animals, it may also be possible to determine who within the group is really the parent of one of these young. As well as helping towards the understanding of breeding systems, information of this sort will enable biologists to refine measures of relatedness within social groups. While the technique cannot be used to identify relationships other than very close ones, in the long term such relationships derive from those identified between parents and offspring. So once these are known, estimates of r between less closely related individuals in the population can be made, and this can be done with far greater assurance than would be permitted by any other technique.

7.6 Altruism without kinship: does it exist?

Maynard Smith (1982a) defines an altruistic trait as one 'which, in some sense, lowers the fitness of the individual displaying it, but increases the fitness of some other members of the same species'. If the fitness we are talking of is inclusive fitness, then helping between relatives is not altruism, because it has indirect benefits for the individual showing it. Is there then such a thing as altruism between non-kin?

Trivers (1971) was the first to argue that altruistic relationships could arise between unrelated animals, through a process he referred to as 'reciprocal altruism'. This is much the same as co-operation between animals, but with a time lag so that, instead of each gaining an immediate benefit, one benefits on one occasion, the other on another. There are various prerequisites before such a system will arise.

1. Animals must be reasonably long lived and live in stable groups so that the same individuals encounter each other repeatedly in situations where altruism can be displayed.
2. Animals must be able to recognise each other as individuals and detect when an individual 'cheats' (accepts altruism without displaying it in return). Cheaters receive the benefits without bearing any of the costs:

if they cannot be detected and discriminated against, genes favouring cheating will spread through the population and altruists will be excluded.

3. The cost of the altruistic act must be low in relation to the benefit the recipient receives from it. The higher the cost, the more certainty there must be that an opportunity for reciprocation will arise.

The species which most obviously meets these requirements is our own. Indeed, a high proportion of human social interactions are based on reciprocal altruism. But, as far as other species are concerned, convincing examples have been rather rarely described. The same two crop up repeatedly in discussion of the subject. First, the study of vampire bats (*Desmodus rotundus*) by Wilkinson (1984) in which he found that an animal that had fed one night would share its blood meal with one that had not, with the favour returned when their situations were later reversed. Sharing was sometimes, but not always, between kin. Second is the co-operation of pairs of males in baboons, one of which distracts the α individual while the other copulates with his consort (Packer, 1978). Again, the two males take different roles on different occasions. This example is rather more controversial and both Bercovitch (1988) and Noë (1990) have criticised Packer's interpretation, arguing that the individuals involved are acting purely selfishly.

There are a few other examples that seem to fit the definition of reciprocal altruism. Not all helpers at the nest in birds are kin, and Ligon & Ligon (1978) argue that those in the green woodhoopoe (*Phoeniculus purpureus*) gain benefits through reciprocal altruism. Helpers tend to become breeders in the group where they previously helped, and their success is critically dependent on having helpers of their own. By helping breeders to rear their young they are ensuring a supply of helpers when they become breeders themselves. Their altruism in rearing the young of others is reciprocated when those young help them to rear their own. Another example comes from the work of Seyfarth & Cheney (1984) on mutual grooming in vervet monkeys (*Cercopithecus aethiops*). While much of the grooming in a group is shown between kin, they found that there are other grooming pairs. Playback of a call used by the monkeys to solicit aid in disputes was more likely to yield a response from a non-relative if that animal had recently been groomed by the caller, suggesting that such animals have reciprocal relationships.

Despite these examples, there is no doubt that instances of reciprocal altruism are rare. One reason for this is that animals in social groups are

Player B

		Cooperate	Defect
	Cooperate	**R=3** Reward for mutual cooperation	**S=0** Sucker's pay-off
Player A			
	Defect	**T=5** Temptation to defect	**P=1** Punishment for mutual defection

Fig. 7.7. Pay-off matrix for the prisoner's dilemma game. The pay-offs shown are those to player A with illustrative numerical values. (After Axelrod & Hamilton, 1981.)

usually related to each other and, without long-term study or good knowledge of relatedness, the effects of kinship are hard to exclude. Indeed, one reason for the rarity of reciprocal altruism may be that kinship is virtually essential for relationships of this sort to get started.

When unrelated animals first meet, how should they behave towards each other? They are faced with a classic game theory problem known as the prisoner's dilemma, illustrated in Figure 7.7. The game is defined by the payoffs in the matrix being ordered $T > R > P > S$. The greatest payoff (T) comes to the cheater that receives help without giving it. Next greatest (R) comes when the two co-operate with each other. The third most profitable (P) is when neither assists the other, and the fourth (S) is the sucker's payoff: the animal that gives help without receiving it bears all of the costs and gets none of the benefits.

If $T > R$ how then can co-operation get off the ground? This situation has been examined by Axelrod & Hamilton (1981). They found two ESSs (evolutionarily stable strategies: see Chapter 5) where animals encountered each other repeatedly. One was where neither co-operated with the other (both gain P). The other was the so-called tit-for-tat strategy. In this an animal co-operates the first time it meets another individual and on subsequent encounters it does what that individual did on the previous occasion. Of course, if all animals play the game this way, they end up co-operating the whole time and each gains R. There is one proviso: the benefit of tit-for-tat depends on a reasonably high likelihood that the animals will meet again. If this is sufficiently low, it will pay an animal to defect when it meets a tit-for-tat opponent as it will get the higher reward, T.

In a neat test of these ideas, Axelrod (1984) played numerous different possible strategies, which had been suggested to him by colleagues, against each other in a computer simulation. Despite the sophistication and complexity of many of these strategies, tit-for-tat proved the consistent winner. This then suggests that, where animals meet each other repeatedly, there are good reasons to believe that co-operation and reciprocal altruism will evolve. However, despite these strong theoretical predictions, the relevance of tit-for-tat to the lives of animals remains somewhat contentious. The evolutionary stability of tit-for-tat has been questioned (e.g. Boyd & Lorberbaum, 1987), and Boyd & Richerson (1989) have also suggested that tit-for-tat is only likely to arise in groups larger than two under very restricted circumstances. One difficulty is in seeing how a tit-for-tat relationship might get started. When two animals meet only once, the ESS for each of them is to defect, yet tit-for-tat requires that they co-operate on the first encounter. A possibility here is that tit-for-tat first arose among kin in groups and then, once established, spread to interactions between unrelated animals.

Tit-for-tat has three key characteristics (see Dugatkin, 1991).

1. It is 'nice'. It involves co-operation on the first encounter.
2. It is 'retaliatory'. If one animal defects, the other does so in response the next time they meet.
3. It is 'forgiving'. Because it has a short memory, co-operation can easily become established even with a partner that defected in the past.

Is there any evidence that animals actually show behaviour of this sort? Perhaps the best tests have been those by Milinski (e.g. 1987) and Dugatkin (e.g. 1988) on the inspection visits that fish pay to a model predator. When a predator appears, small fish such as sticklebacks and guppies shoal more closely but will, either singly or in small groups, make occasional dashes towards it, before retreating back to their shoal. In one situation Milinski (1987) simulated co-operation by mounting a mirror alongside a stickleback's tank so that it appeared to be accompanied by a co-operating companion as it swam towards the predator. In another a shorter mirror was placed at an angle so that the animal's image appeared to defect, moving further away and finally disappearing, as it approached the predator. The stickleback spent more time close to the predator in the first of these conditions, which Milinski interpreted as evidence that it was co-operating in the way that a tit-for-tat rule would suggest. This may well be so, but alternative and more parsimonious explanations, such as that the fish tend to shoal more closely in the presence of a predator (Masters & Waite, 1990; Lazarus & Metcalfe, 1990), seem just as plausible. In another experiment, Milinski *et al.* (1990) found that sticklebacks kept in groups of four tended to make inspection visits towards a predator in the same pairs. This suggests that they can recognise each other as individuals, and is certainly consistent with co-operation between them. Guppies, too, remember particular partners, at least for a few hours, and are more likely to associate with one that has been co-operative in the past (Dugatkin & Alfieri, 1991). Here again we have results consistent with tit-for-tat. However, interpretation of these fish data in terms of this strategy is not yet compelling. It is, for example, questionable whether fish approach predators preferentially in pairs, as tit-for-tat would predict, rather than singly or in larger groups (see Turner & Robinson, 1992; with reply by Milinski, 1992).

Whitehead (1987) has also performed an experiment which he interpreted in terms of tit-for-tat. Using a loudspeaker, he played calls recorded from a male of their species to groups of mantled howler monkeys. In some tests he played a sequence of distant calls followed by close ones, so simulating an approaching troop, whereas he reversed this order in others, as if the sound came from a group that was retreating. The calls that appeared to get closer attracted the troop, whereas the retreating ones led the troop to withdraw. He argues that these responses imply co-operation between groups: retreat leads to retreat and approach to approach, just as the tit-for-tat strategy would suggest.

These results, and these of Milinski and his co-workers, are certainly compatible with tit-for-tat, but they do not provide a very strong test of the idea. Recent theoretical work also suggests that tit-for-tat may not be the ultimate in successful strategies. It was simply the best that Axelrod's contacts suggested. A more subtle (and biological) approach is to run simulations in which strategies evolve, with more successful ones spreading and less successful ones becoming extinct. Such studies point to approaches that can be even more successful. One of these is 'generous tit-for-tat', which is like tit-for-tat, but forgives defection on one-third of occasions (Nowak & Sigmund, 1992). Another is labelled 'Pavlov' or win–stay, lose–shift; in this case, an animal winning one of the higher payoffs that occurs when a partner co-operates (R or T) stays with its previous strategy for the next round, whereas one gaining only S or P shifts to the alternative one (Nowak & Sigmund, 1993).

More experimental work is certainly needed on this topic as the theoretical arguments have, for the moment, run ahead of the data, The theory suggests that altruism may arise in unrelated animals in situations where there is a good chance that it will be reciprocated. But it also indicates that the precise conditions required may be rather restrictive. This may account for the fact that very few examples have so far been described. But it also highlights the need for more studies of real animals in the real world.

7.7 Conclusion

This chapter has examined some of the most exciting topics on the interface between behaviour and evolution. There is no doubt that kin selection theory has, over the past 30 years, proved a most powerful tool in our interpretation of behaviour. It has highlighted the importance of kinship in groups of animals and led to a great growth of research into whether and how animals recognise their kin. If we are to understand their interactions with each other, we too must recognise their kin relationships, and here tremendous scope is offered by the developing tools of molecular biology. But, beyond kinship, does altruism really exist? Animals certainly co-operate with one another, and a few good examples of reciprocal altruism have been described, but in all cases each individual's inclusive fitness benefits, at least in the long term, so use of the word 'altruism' is something of a misnomer. Just how co-operation evolves in animals that are not genetically related to one another, the final topic of the chapter, is of great theoretical interest and demands more experimental studies. After two decades in which 'selfishness' was the dominant notion in interpreting the interactions of animals with one another, perhaps ideas such as tit-for-tat start to provide us with a sound theoretical basis for a more co-operative social order.

8
The evolution of intelligence

R. W. BYRNE

8.1 Intelligence

How can one possibly hope to understand the evolution of such an abstract quality as intelligence? The task is a hard one, but not impossible; and many of the problems are common to the study of other types of behaviour. In order to trace the evolution of any behaviour, one needs to use the *comparative method*. In this, this phylogenetic history of a behaviour is inferred from its pattern of occurrence in surviving species; sometimes in addition the original selective value of a behaviour can be deduced from its adaptive consequences for the fitness of those species today. Often, this general approach can be supplemented with fossil records that show concrete results of the behaviour. However, in the case of intelligence, these signs are minimal and problematic to interpret: they consist just of stone artifacts and cave paintings, and even these date from only the last two million years.

So, what one has to do is: (1) find reliable differences in the intelligence of extant animal species and how these differences affect the species' ability to survive under different circumstances; (2) deduce from this pattern the likely intelligence of hypothetical extinct ancestors – whose existence is inferred from a reliable, modern evolutionary taxonomy; and then (3) look for plausible selection pressures that could have favoured these evolutionary changes. For instance, if all carnivores are found to be more intelligent than other mammals, one could infer that something in the lives of the common ancestor of modern carnivores promoted intelligence, perhaps the need to hunt; we could test this idea by asking whether more intelligent carnivores *do* hunt more successfully. The emphasis, unfortunately, is on the word 'if'. In practice, demonstrating real differences in intelligence between species is difficult and a matter of controversy. This chapter reviews the attempt and what it has found, before considering theories of the evolutionary origin of intelligence.

223

8.1.1 Meanings of 'intelligence'

Perhaps the main reason for difficulty is a lack of consensus about what the term 'intelligence' means, when applied to animals. When researchers come to formalize people's everyday understanding of the term's meaning, they often arrive at a definition appropriate to variation *within* but not *between* species: for instance, 'the aggregate or global capacity of the individual to act purposefully, think rationally, and to deal effectively with his environment' (Wechsler, 1944). If 'purposefully' is equated with 'goal directedly', and 'rationally' with 'optimally', this definition is suspiciously close to that of biological adaptedness. This resemblance is explicit in some definitions: 'the faculty of adapting onself to circumstances' (Binet & Simon, 1915). Small wonder then, that some biologists have doubted whether degree of intelligence is an appropriate measure by which to compare animals. All extant animals are evidently, by their very survival, adapted to their environment – but in very different ways. These ways are fascinating, but calling them 'intelligence' adds nothing.

This understandable view is in unholy alliance with that legacy of pre-Darwinian arrogance, the assumption that animals 'merely' learn, but only humans are intelligent. On this view, human intelligence is unique (not simply large), and if anything it is a consequence of verbal communication, since that is also unique to humans; any use of the term for animal skills is simply anthropomorphism. Now, in the end it may turn out that the gulf between human intelligence and that of any other animals is a very large one; but to assume that there is no comparison is an approach that will appeal only to a creationist. Since this argument is a 'dead end', is there a better definition of intelligence to help us out?

A central problem that must first be addressed is that many performances of animals *look* intelligent, but there is every reason to suppose that their development is under tight genetic guidance; an obvious example is the nest-construction skill of social wasps and bees. Furthermore, there is no real chance that the search for diagnostic actions with which to deduce variation in degree of intelligence can be simplified by first separating all actions into discrete categories, 'innate' and 'learned', and then only using the latter; most animal behaviour is influenced by both genes and learning (see Bateson, 1983a). For instance, passerine birds such as chaffinches (*Fringilla coelebs*) are predisposed to copy songs that match a particular, but loosely specified, pattern (see Slater, 1983). Youngsters remember very exactly the song that they hear; months later, when they are adult and it is spring, they successively improve their attempts at singing by comparing their own song with this memory. Is song, then, innate or learned?

Learning and genetic predisposition are intricately meshed in development, and the same is even likely to be true for skilled human behaviour. For instance, it would seem that all adults, male and female, are predisposed to talk to babies (and pet animals, and even occasionally house-plants) in a particular way: 'What's that you've got there? A nice, red ball, is it? Isn't that pretty? Wouldn't it be fun to drop it?' and so on. This speech is singularly low in novel information, which is largely obvious from context anyway. Yet it is flamboyantly rich and complex in syntax – the largely unconscious, everyday grammar that allows people to structure a limited set of words into phrases and sentences with an unlimited range of meanings. This kind of talk also has an exaggerated intonation pattern, making both the syntactic structure and the individual sounds of speech (phonemes) more obvious than in normal adult speech. It is, in short, an ideal vehicle for helping the child (if not the house-plant) to learn phonology and grammar, although mothers and other caretakers who talk this way are unaware that their speech has this special function in language development. If similar analyses apply to such disparate organisms as human babies and young chaffinches, is there any need to go beyond this sort of explanation – channelling of simple, individual learning processes by genetic predispositions – in order to explain differences between animals? Why bring in 'intelligence'?

There are two reasons. The first is that psychologists have less confidence than in the past that learning is well understood. Traditional learning theory held two convenient beliefs: learning processes are identical in all situations and organisms (and therefore best studied in a laboratory animal, where experimental control is greatest), and that by just these processes animals can learn anything (see review by Roper, 1983). As I will sketch below, learning now seems to differ from species to species, and not all information is equally learnable; this is particularly true in social learning contexts. It is not just that different species are channelled into *different* knowledge-bases according to their needs; species differ in overall flexibility of learning. To most people, intelligence includes the ability to learn: so variation in flexibility might reasonably be ascribed to differences in 'intelligence'. This view of intelligence is matched by another type of definition of human intelligence, one that stresses *understanding*, for instance, 'grasping the essentials in a situation and responding appropriately to them' (Heim, 1970).

The second reason is that intelligence cannot be entirely equated with learning ability, even if social learning is included. In humans intelligence certainly means more than this: terms like 'thinking clearly', 'solving difficult problems' and 'reasoning well' recur in attempts to define the ability. These abilities, in combination with a large and unrestricted memory, add up to efficient *computational* functioning of a 'Turing machine'; that is, a universal computer that can in principle solve any well-posed problem. While Turing machine capabilities would not be straightforward to demonstrate in non-verbal animals, they cannot be ruled out by definition and the frequent use of the computer metaphor for the function of an animal's brain (e.g. Dawkins, 1976a) shows that the idea that some animal brains compute is at least plausible. With 'hardware' capabilities that allow computing, a machine's (and perhaps a brain's) 'software' of plans and programs becomes crucial.

8.1.2 *Animal intelligence as a specialization*

Intelligence may include, then: learning an unrestricted range of information; applying this knowledge in other and perhaps novel situations; profiting from the skills of others; and thinking, reasoning, or planning novel tactics. What is clear about all these (possible) components of animal intelligence is that they contribute to general purpose skills, not highly specialized ones. This may help understanding of what intelligence is *for*. All these abilities are what one should expect in generalist animals, and intelligence should most benefit extreme generalists, species adapted to exploit continually changing environments, since they must cope with novelty to survive. Creatures adapted to unvarying environments, however intricate, would be much better off with reliable, if inflexible, methods: genetically coded strategies, or genetically channelled learning of a narrow range of information. In a given species, then, one might expect a trade-off between special purpose, hard-wired abilities serving particular biological needs, and general purpose 'intelligence' which can be applied widely. Those inborn tendencies that allow reliable learning of songs by passerine birds and grammar by children, however remarkable and well adapted, imply *restrictions* on generalized learning capability.

Viewing the flexibility of intelligence as an *adaptation* to a life of change and unpredictability, just as being able to run fast or breathe infrequently are adaptations to quite different selection pressures, shows that there is nothing unscientific about calling one animal intelligent and another less so. The difficulty is only an empirical one: do data exist that enable us to reject the null hypothesis that all non-verbal animals have equal intelligence (Macphail, 1985)?

In trying to answer this question, I will often speak as if animals mentally represented information (about the world, or about their behaviour). Behaviourists have long pointed out that this way of talking is unnecessary: one can always replace such phrasings with ones that describe merely behaviour and its history and context – the observables, which can be studied. This is true, but not necessarily virtuous; presumably one can just as easily run the maxim in reverse, and replace all behavioural histories with ones that hypothesize mental representations. This cognitive version has the advantage that it is much easier for people to understand; and, provided that our hypotheses are the most parsimonious for our observable data, should give no cause for alarm. It also enables comparisons in equivalent terms to be made between animal ethology and cognitive psychology – where mental representations are routinely inferred on the basis of observed behaviour.

In representational terms, intelligence must involve an animal's ability to:

1. gain knowledge (and whether there are constraints on the type of knowledge it can represent, and the circumstances from which it can extract knowledge) from interactions with the environment and other individuals;
2. use its knowledge to organize effective behaviour, in familiar and novel contexts;
3. deal with problems, using (if it is able) 'thinking', 'reasoning' or 'planning' – in fact, any ability to put together separate pieces of knowledge to create novel action.

8.1.3 Traditional explanations of animal abilities

The traditional story of how animals gain knowledge and solve problems was a simple one; this is a brief résumé. There are three strands: learning, imitation and insight. Learning happens by an associative process. This takes two different forms, but in each of them an initially neutral event becomes associated with a valuable reward or unpleasant punishment, a 'reinforcer'. In *classical conditioning*, the neutral event is presented by an experimenter, or simply occurs naturally in the world, shortly before a reinforcer on a number of occasions; afterwards, the neutral event alone brings on the reaction appropriate to the reinforcer – such as salivating to food. In *instrumental conditioning*, the neutral event is a behavioural act by the animal itself; since the animal has to discover for itself that a certain behaviour will be followed by a reinforcer, this is often called trial-and-error learning. The same rules (or 'laws') apply in all situations, so the tight control of laboratory experiments can be used to unmask them. With the mechanisms of conditioning, an animal associates any pair of events that it can detect, and by building up a mass of associations it will actually come to behave appropriately.

Social influences have little place in this picture, but *imitation* by animals has long been recognized, and usually regarded as a problem getting in the way of testing whether animals have any *insight*, a more powerful intellectual ability than 'mere' learning. Since birds like parrots and hill mynas (*Gracula religiosa*) can imitate so perfectly, even when they clearly have no understanding of the human words they copy, imitation has traditionally been regarded as a 'cheap' way of appearing clever, a sham of intelligence. Many animals can imitate, to judge from the way

behaviour spreads culturally; a well-known example concerns the spread among species of small birds (mostly tits *Parus* spp.) of the habit of tearing off the tops of milk bottles to obtain cream (Fisher & Hinde, 1949; more recently, tits in Britain learned to peck holes in metal foil tops, before being finally defeated by cartons!). What most animals lack is *insight*, and insight learning has been treated as diagnostic of intelligence. The classic demonstration that at least one species of animal can show insight was by Köhler (1925). He gave the puzzle of raking in a reward with sticks to a chimpanzee (*Pan troglodytes*); in one version, no stick was long enough. After many attempts, the chimpanzee gave up, and Köhler stopping observing. A few minutes later the keeper excitedly reported that the chimpanzee had succeeded. The animal had been playing with the sticks, and happened to join two sticks together; then he suddenly jumped up and used the joined bar to solve the puzzle – apparently after a sudden realization that he now had a long enough tool.

These views of associative learning, imitation and insight have all had to be modified in recent years. The current picture is a more exciting one for those interested in the evolutionary precursors of human intelligence; I will therefore sketch these recent changes before considering theories of the selection pressures that caused the evolutionary increase of intelligence in some species.

8.2 Individual learning channelled by genetical constraints

It is now known that not all events are equally easy to associate. Garcia and Koelling (1966) gave rats water that was novel either in its taste (sweet or salt) or in that a flashing light and clicking noise always accompanied drinking. They paired these with aversive reinforcement, either by illness (in fact brought on by X-rays) or by painful electric shock. After novel-tasting water, illness causes avoidance of the taste in future, but shock causes none. After 'bright–noisy' water, shock causes aversion but illness none. Further, Garcia, Ervin and Koelling (1966) showed that this aversion to novel drink or food can be caused by illness up to 12 hours after consumption. It seems that rats are specially prone to associate feeling ill with the last novel food they ate or liquid they drank: a wonderful mechanism for an explorative, generalist animal to learn which items are poisonous. Humans too seem prone to this specific kind of food-aversion learning; most people have experienced 'going off' some unusual food that preceded nausea, even when they know very well that the nausea was caused by some infective illness. Garcia's results have led to a plethora of research on the specificity of reinforcers in classical conditioning, and equivalent favoured pairings of some events over others have been found in instrumental conditioning (see Roper, 1983). To give just one example, hamsters (*Mesocricetus auratus*) readily learn to rear up or scrabble more if an experimenter pairs this with food rewards, yet fail completely to increase the rate of washing their faces when this behaviour is reinforced instead (Shettleworth, 1975).

It is clear from this work that animals are equipped to learn some things better than others. Genetic predispositions have even been blamed for our own absurd tendency to learn fear of (harmless) spiders and snakes, yet failure to learn appropriate fear of motor cars and electricity. Of course, early ancestors of humans in Africa encountered dangerous snakes and spiders, but no cars or power lines. In fact, most apparent constraints on learning generally make sense in an evolutionary timescale: for instance, it is plausible that at no point in their respective evolutionary histories would a rat have been at risk of poisoning by bright–noisy water, or a hamster have been able to affect its fitness by scratching more. As noted by Gould and Marler (1984), the usual term 'constraints' can be a misleading one since it suggests evolution of incompetences, whereas in reality what have evolved are tendencies to learn efficiently just those specific kinds of information that have favoured survival.

These cases may all be viewed as *channelling of selective attention by genetic predisposition*, towards certain aspects of the environment and not others. Putting it this way highlights common ground with the way some passerine birds learn songs: there, the chick has a predisposition to listen to, remember and copy sounds that approximately resemble the song of its own species. A similar parallel can be made with imprinting. The young of precocial birds approach conspicuous objects as soon as they are steady enough on their feet, learn their characteristics, and so narrow down their preferences to (eventually) one particular conspicuous object. Usually this is the mother bird in the natural condition, but sometimes it is a rotating flashing object in the ethologist's laboratory (see Bateson, 1973). In all these cases, the organism is equipped with predispositions and preferences that increase the chance that useful, adaptive information will be acquired by learning.

The mechanisms of learning themselves, classical and instrumental conditioning, are best regarded as devices for ensuring that animals learn adaptive information. This interpretation is particularly suggested by the finding that reliable temporal association is not enough for conditioning to occur. If one element is a *redundant* predictor of another, the two remain unassociated. For instance, suppose a red light has been paired with food repeatedly, so that it alone evokes salivation in a dog; later, both red and green lights are repeatedly paired with food, but the (redundant) green light will not come to evoke salivation on its own (this is called 'blocking'; see Roper, 1983). Classical conditioning makes evolutionary sense as a correlation-learning device (Dickinson, 1980); with it, the animal will automatically come to learn the pattern of event probabilities in the environment. The animal would indeed seldom be misled if it treated these correlations as *causal*, although this would not be strictly correct. Certainly, humans suffer from a tendency to see cause in mere correlation, perhaps an evolutionary consequence of the effectiveness of classical conditioning. Instrumental conditioning has been likened to evolution by natural selection (Skinner, 1981), since it is an automatic process by which behaviour patterns with beneficial consequences are selectively increased in frequency, by reinforcement, among the many types of behaviour produced by the animal. Instrumental conditioning functions automatically to record the significant results of exploration and experiment, ensuring that an animal profits from its experience and need not repeat its errors.

While precisely *what* an animal focuses on to learn can be heavily channelled by its predispositions, the rules of associative learning seem to apply rather generally (Roper, 1983). The many examples of apparently remarkable learning feats in otherwise unremarkable animals, such as the ability of some fledglings to learn the axis of rotation and star configuration of the night sky for later use in navigation (Emlen, 1970), or the ability of elephants (*Elephas maximus*) to remember 600 other elephants as individuals (Moss & Poole, 1983), may be best explained by attention

channelling rather than a special kind of learning. This has the advantage of making it quite clear that describing these feats as 'intelligent', or comparing such differences in what animals learn on a unitary scale of 'intelligence', would be inappropriate.

8.3 Social learning: mechanisms other than imitation

From being regarded as a cheap trick to mimic intelligence, imitation has been promoted to a sign of remarkable intellectual ability – and one perhaps absent in non-humans. How has this come about?

8.3.1 *The demise of imitation as a general explanation of social learning*

When humans want to acquire a new and complex skill, they watch and imitate a skilled performer. It used to be assumed that animals too used this strategy, but a few doubts about the strength of evidence for animal imitation have recently snowballed. Once the claims were tested with tight experiments, time after time animals showed no evidence of imitation! In the case of birds learning to remove milk-bottle tops to obtain cream, for instance, naive birds shown a bottle with pecked top learned just as quickly as those which saw a bird actually pecking a top and gaining cream (Sherry & Galef, 1984).

Questioning the role of imitation has far-reaching implications. It used to be assumed that the famous cultural behaviour of Japanese macaques (*Macaca fuscata*) was transmitted by imitation. To give a classic example, a monkey in the Koshima troop, Imo, in September 1953 began to wash sand off a dirty sweet potato (a food used by researchers since 1952 to provision the monkeys), and over the next 6 years this habit spread to 57% of the troop. Most of this spread was to young animals, especially to Imo's own relatives (Kawai, 1965); other behaviour showed similar patterns. However, Green (1975) pointed out that all the special cultural differences involved food processing or – in the case of the vocalizations that he was studying – food calling; and provisioning gives the opportunity for unintentional conditioning by the human provisioners. Perhaps monkeys that called dramatically, or fed in ways considered cute by people, were consequently fed a little more than average. If so, the cultural spread observed might have been a human artifact, based on reinforcement of chance variations in behaviour.

But surely such an unusual behaviour as food washing is highly unlikely to occur by chance exploration? Maybe not. Visalberghi and Fragaszy

(1990a) gave a social group of capuchin monkeys (*Cebus apella*) dirty food under close observation in the laboratory; each monkey learned food washing within 2 hours. But these monkeys showed no sign of watching what others did, and instead assembled the whole pattern in a piecemeal way. Apparently, separate components were reinforced by a history of rewarding chance events – such as finding clean food under water. A similar demonstration of the surprising ease with which a 'clever' behaviour can be learned came from the study of an ability known in wild common chimpanzees: making a sponge out of chewed leaves in order to get drinks from a cavity. Kitahara-Frisch and Norikoshi (1982) found that two zoo-born chimpanzees quite rapidly discovered the technique, with no need of a model to copy.

Mitchell (1987), Galef (1988) and Visalberghi and Fragaszy (1990b) have reviewed claims of imitation in animals, and found most of the evidence wanting – even in monkeys, so famed as imitators. Most of the data could easily result from special kinds of associative learning, which can *mimic* imitation; some reviewers even suggest that imitation is restricted among animals to the case of birds' learning of sounds, and only humans can imitate actions.

8.3.2 Social learning without imitation

What other mechanisms are used to 'explain away' the frequent social learning that used to be called imitation? Explanations are of several types, sharing the fact that the act must *already* be a part of the animal's repertoire before it can be affected (Fig. 8.1). None of them can account for the acquisition of really novel behaviour without trial-and-error learning – whereas imitation can – but sometimes they may underlie learning that looks surprisingly like imitation.

Stimulus enhancement

The commonest explanation of apparent imitation is *stimulus enhancement* (Spence, 1937), in which the probability of approaching or contacting something in the environment is increased by the presence of a conspecific interacting with it. Once an animal's responses are limited to a small part of the environment or a particular object in an experiment, Spence realized, the chances of it discovering by trial and error the correct technique to gain reward are much increased. He did not explicitly specify that the consequences of the performer's behaviour (such as whether it

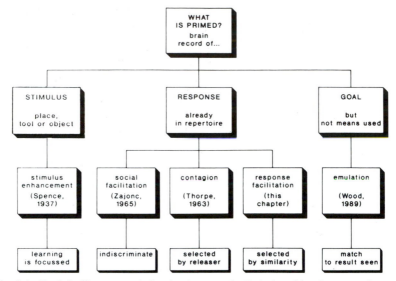

Fig. 8.1. Social effects on existing brain records. Relationships between the more important social effects on learning that have been proposed to underlie behavioural copying that at first sight appears to be intelligent imitation, but which in fact may result from simpler mechanisms.

could be seen to be rewarded) were important, nor whether 'negative enhancement' would occur if the observed animal were instead punished, but these would greatly increase the power of enhancement to produce social learning. At its most general, then, stimulus enhancement could cause an increase or decrease in the probability of approaching or contacting parts of the environment, contingent on seeing a conspecific interacting with these parts and the consequence for it. The rapidity of learning that this allows might easily be mistaken for imitation of the conspecific's behaviour.

Such a mechanism is clearly of functional benefit to social species, and is probably the major factor in many cases of culturally transmitted behaviour of ecological importance. Cambefort (1981) found that juvenile baboons (*Papio ursinus*) learn the palatability of a new food rapidly once one animal had tried it, whereas propagation of this knowledge is slow in vervet monkeys (*Cercopithecus aethiops*) in the same habitat. The difference correlated with the spatial organization of the troops, young baboons tending to be found close to each other, and this may be one reason for baboons' greater flexibility of ecological niche. Red-winged blackbirds (*Agelaius phoeniceus*), which are a major pest of crops in the

USA, have been shown experimentally to acquire food preferences by observation of conspecifics feeding, and to learn to avoid foods which they have seen a conspecific eat before being sick (Mason & Reidinger, 1981, 1982). The benefits to such an opportunist species are obvious and unfortunate for farmers! In cats, observation of successful performance by a conspecific (especially by the mother, when the subject was a kitten) enabled rapid learning of a task which was learned slowly or not at all by individual trial-and-error learning (John, Chesler & Bartlett, 1968; Chesler, 1969). Squirrels (*Tamiasciurus hudsonicus*), which could watch an experienced squirrel feeding on hickory nuts (for them a novel food), after 6 weeks experience were using only half the time and energy to gain the same food as a group which had no model to watch (Weigl & Hanson, 1980). All these behavioural changes are likely to be due to stimulus enhancement rather than imitation, but their biological significance is considerable.

Avoidance of anything seen to be aversive to others is also useful. Mineka *et al.* (1984) showed rapid, strong and long-lasting avoidance of a snake in young rhesus monkeys (*Macaca mulatta*) that observed their wild-born parents display intense fear of snakes. The common observation of children acquiring their parents' fears (e.g. an irrational fear of spiders) no doubt taps the same phenomenon. Similar constraints apply to both in the type of object for which fear is easily learned: monkeys are much less likely to learn fear of flowers than of snakes (see Mineka & Cook, 1988), and children are notoriously slow to learn fear of mechanical perils. This so-called *observational conditioning* involves more than stimulus enhancement, since the young monkeys show signs of fear and not merely avoidance of the snake. The possibility that responses as well as stimuli may be acquired socially brings us to the next social phenomenon.

Response facilitation

Social effects that act directly on responses have usually been called social facilitation, but unfortunately several quite different phenomena have been given this same label. At one extreme, Zajonc (1965) has used *social facilitation* to describe his idea that the presence of conspecifics increases the behavioural activity of all of them, indiscriminately. Galef (1988) notes a lack of evidence for this kind of social facilitation; if it does happen, the selective pressure for its evolution is still unclear, since it could only aid social learning in the most unselective way. At the opposite extreme of specificity, Thorpe (1956) has proposed that social facilitation is restricted to a few rather stereotyped behaviour patterns, a phenomenon he also called *contagious behaviour*. Examples are the synchronized movement of a wader flock, thought to decrease vulnerability of any individual to predators (when part of a flock takes to the wing or turns in flight, the rest follow), and the way adult humans respond to the sight of a person yawning, by yawning themselves (Provine, 1989). A behaviour is 'released' by a particular stimulus, to use the classical ethological terms; when the releasing stimulus happens to be another animal behaving in the same way the result is contagion. This may sometimes be mistaken for imitation.

Babies soon after birth respond to an adult's smile or tongue protrusion with smiles or tongue protrusions of their own (Meltzoff & Moore, 1977; Meltzoff, 1988). If the babies' behaviour is interpreted as imitation (and it usually is), this raises the interesting question of how the child computes what act in its repertoire is 'the same' as the adult's, with no obvious way of comparison or matching. The puzzle is solved if facial expression mimicry in babies is a special case of contagion in neonate humans. Babies at birth are no doubt equipped with a number of functional responses to special stimuli; only when the responses and the stimuli which trigger them just happen to be 'the same' would this mimic imitation. On this hypothesis, babies are genetically equipped with (*releaser*) ⇒ (*action pattern*) linkages for several facial expressions, in each of which the releaser is the same expression performed by another. Such a genetic endowment is plausible: it would have clear function in controlling the behaviour of adults to the baby's advantage. If this view is correct, there are only a fixed few actions that a neonate human can mimic; however, this is hard to test when the baby's range of behaviours is anyway small.

The intellectual problem, of how an individual can recognize that a behaviour is 'the same' when performed by itself and by another, is much less acute when the behaviour does not involve the individual's face. Perrett *et al.* (1989) have found single neurons in monkey brains which respond to a specific action, for instance picking up or rotating an object, whether the action is done by another or by themselves. Presumably then, monkeys can be taught to execute simple actions which look to them the same as what they are shown. Oddly, this task seems difficult for monkeys (Visalberghi & Fragaszy, 1990b) though not perhaps for budgerigars (*Melopsittacus undulatus*) (Dawson & Foss, 1965). These birds saw a demonstrator use one of three techniques to open a food dish, and subsequently tended to use the same method. Even so, Galef, Manzig and Field (1986) found the phenomenon fragile, short-lasting and of marginal effect on their behaviour. More powerful evidence comes from Palameta

(1989), who showed naive pigeons a trained bird which used a specific technique to obtain food from essentially the same apparatus: the demonstrator either rotated or lifted up a small handle. When the subjects were tested with the apparatus arranged so that only one method would work, those birds that had seen the 'helpful' demonstration learned much more quickly and efficiently. This does not imply imitation, since it is quite possible that the simple actions were already part of the birds' repertoires, but is powerful evidence of the importance of social learning for a flexible generalist forager like a pigeon.

These experiments suggest the existence of a kind of social effect more selective than Zajonc's indiscriminate increase in all activities in social cirumstances, yet more general than Thorpe's contagious copying of a small set of stereotyped responses by means of linkages that are probably 'hard-wired'. I use the term *response facilitation* for this: the presence of a conspecific performing an act (often one resulting in reward) increases the probability of an animal which sees it doing the same. Response facilitation crucially differs from imitation in that only actions already in the repertoire can be facilitated. The sparse current evidence for response facilitation suggests that it may not be common in animals. Indeed, Spence (1937) briefly discusses the evidence for this kind of copying (he called it 'social facilitation', in yet a third sense of the term), which at that time came only from Köhler's chimpanzees' mimicry of humans.

If response facilitation and stimulus enhancement were to influence an animal on the same occasion, their combination would be a powerful one. Imagine that the animal observes a conspecific performing a particular act (one also in its own repertoire) to a certain object, in a certain place, and gaining food in the process. The observing animal would then be more likely to go to the location, contact the object and perform the same act – a sequence so intelligent-looking that it is understandably likely to be mistaken for imitation.

Only the acquisition of novel behaviour, therefore, can be used in diagnosis of imitation. This was recognized by Spence (1937), who urged adherence to Warden Jackson's (1935) six criteria in identifying imitation: the task should be 'novel and sufficiently complex', the response must appear immediately after the observation, practice must be excluded, the act must be 'substantially identical' to that observed, and enough instances must be recorded to rule out chance performances. Recent experimental work has carefully followed the last five criteria to the detriment of the first, using tasks that are not complex and which are not really novel for the animal. For instance, Whiten and Ham (1992) use a monkey's twist or pull of a bar to obtain reward as the action to be copied. This leaves open the possibility of solution by response facilitation, yet they refer to this sort of copying as imitation. In other words, a rather straightforward piece of matching, which resembles contagious behaviour in form, is equated to the kind of imitation familiar in humans, where quite novel behaviour can be assembled by observation of a skilled model. (In fact, they failed to find any evidence for the phenomenon, whatever its name, but this negative result may have a number of causes). Lumping these categories together seems unfortunate when the theoretical implications of imitation and response facilitation are so different, and is avoided here by narrowing the term 'imitation' slightly from its everyday meaning and restricting it to the copying of novel actions.[1]

Emulation

Finally, it has been suggested that an animal's goals may be influenced by seeing and desiring the results of another animal's actions. This has been called *emulation*, a term introduced by Wood (1989) to describe human behaviour, for instance where a child sees a ball hit into a net and desires to copy this result, but uses its own idiosyncratic way of hitting to do so. Tomasello *et al.* (1987) used this idea of emulation – copying the results of other animal's behaviour but not its methods – to account for their puzzling experimental results. Young chimpanzees watched a model perform a two-stage raking task with a tool to obtain a food reward, and, while they did not copy the two-stage process (so gave no unequivocal evidence of imitation), they did tend to use the rake more than controls seeming to realize that the object had a use as a tool (Tomasello *et al.*, 1987). To term this 'emulation' seems an extension to Wood's original usage: the *use* of a tool is a strange sort of 'result' to become a goal for the observer, but Whiten and Ham (1992) similarly describe the results as due to emulation (in their version, 'goal emulation', in which use of the tool is perceived as a *goal* and thus copied). Stimulus enhancement of an object (tool) in the specific context of an action (raking) would seem a possible alternative explanation, though of course the actions could also result from inaccurate imitation. Despite the lack of a clear experimental demonstration of emulation, its possibility must be borne in mind when evaluating evidence for imitation. Tomasello (1990) cogently argued that many of the elaborate tool-using behaviours of wild chimpanzees (*Macrotermes* termite-fishing, *Campanotus* ant-fishing, *Dorylus* ant-dipping, use of hammer and anvil stones to crack *Panda* nuts, etc.) could be learned by emulation of adults' results, rather than imitation of their methods as often assumed.

8.3.3 Modelling social influences that affect existing brain records

The great attraction of associative learning as a mechanism is its extreme computational simplicity. Yet I have now had to introduce both 'stimulus enhancement' and 'response facilitation', and perhaps even 'emulation', in order to understand data from social as well as individual learning: has all hope of simplicity been lost? Not necessarily. Computationally, learning which shows these effects can be modelled mechanistically with the single notion of *priming*, since all of them affect existing records in the brain.

Consider a system which chooses a location or object to go to (or stay at), a current goal to seek, and an act to perform on the basis of the highest salience or *activity level* of units that represent the possible choices. These activity levels would vary independently of each other according to the motivations and past experiences of the animal, but the corresponding goal, act and location units are incremented each time the animal sees a conspecific perform an act in some location with a desirable result. This system automatically generates the categories of behaviour I have described as stimulus enhancement, response facilitation and emulation. This would give an impression of highly 'intelligent' social learning in some circumstances, as we have seen.

What must be emphasized is that none of these priming mechanisms can give rise to novel and complex sequences of behaviour without trial-and-error learning: they can only affect existing brain records, whether of stimuli perceived, of goals desired, or of responses executed. To build up new behaviours by observation alone, imitation is required. However, just as trial-and-error learning may become more powerful when it is channelled by social effects, imitation may occur to greater effect when influenced by these simpler mechanisms, that themselves rely on no more than priming.

8.4 Imitative behaviour of animals

With behaviour that appears intelligent underwritten by such a simple notion as priming, it is not surprising that attempts to define *imitation* have – implicitly or explicitly – concentrated on excluding the 'lower-order' explanations for given data, as described in Section 8.3.2. Defining by exclusion has the risk of producing a heterogeneous category of hard-to-explain behaviour, as noted by Hinde (1970). This is perhaps why 'imitation' in animals now includes two very different things: vocal copying by various birds, which is easily demonstrated, strikingly accurate, yet not obviously related to their intelligence in other spheres; and copying motor skills from others, shown by very few animals.

8.4.1 Imitating sounds

Hill mynas readily copy human speech to such a close tolerance that the formants (moving bands of major energy that are produced by cavity resonance in the complex vocal tract) are reproduced faithfully – despite the bird's absence of any similar resonant cavities (Thorpe, 1967). Dowsett-Lemaire (1979) has shown that an average European marsh warbler (*Acrocephalus palustris*) mimics 78 species of other bird, many learned in sub-Saharan Africa during their very first winter migration; she has noted 212 species' songs and calls copied in all, and even this may be an underestimate – there are considerable problems identifying the marsh warblers' flamboyant and eclectic selection! In vocal mimicry by birds, imitation seems only to function to augment the repertoire; no more intelligence is implicated than in the chaffinch's tendency to copy songs heard as a juvenile. Slater, Ince and Colgan (1980) showed that chaffinches' acquisition of song types by imitation is well-simulated by random copying with a fixed error rate, and the propagation of song types is a nice example of cultural transmission where precisely *what* is copied appears to have no adaptive consequence.

8.4.2 Imitating actions

Imitation, in the sense of copying novel motor acts from the repertoire of another individual, is surely a very different phenomenon from these 'tape recording' skills. Does it indeed exist in animals? Ruling out the cases where apparent copying can be explained by some form of priming, including response facilitation and stimulus enhancement, then clear cases of this sort of imitation are few and far between, even in simian primates. In recent reviews of this issue, conclusions include the following statements: 'imitation is as yet unproved in monkeys' (Whiten & Ham, 1992): 'monkeys do not seem to be capable under common circumstances of learning tool use by imitation. The data for apes are scantier but suggest similar although less severe limitations' (Visalberghi & Fragaszy, 1990b);

'we are aware of no evidence that any species other than human beings is capable of copying the precise topography of a conspecific's behaviour in a sensory-motor task'; and 'there is no unequivocal experimental evidence in this or any other study that chimpanzees or other non-human primates are capable of imitatively copying' (Tomasello *et al.*, 1987).

8.4.3 Mental perspective taking

Why should it be difficult to copy a motor task? There are two main conceptual difficulties. Firstly, there is the problem, already noted, of recognizing whether one's behaviour is 'the same as' the actions of another. In vocal imitation, a simple test would suffice: matching one's own sounds against a memorized template or trace of those heard from another. This has long been suggested as an explanation of accurate song learning and vocal copying (see Slater, 1986), although strictly the birds would also have to compensate for differences in sound transmission through bone (in the case of own song) rather than air. By contrast, a motor act can look radically different from different viewpoints (most extreme when the act is a facial expression), so its imitation may involve more than simple comparison. This distinction is evidently a matter of degree: raking in a peanut does not look all that different when performed by one's right hand or by the hand of a conspecific sitting to one's right. But certainly some kinds of imitation require 'putting oneself into the shoes of another', in order to understand the physical viewpoint and goal orientation of another individual. This implies a need to understand mental states, in order to take the other's perspective.

The ability to comprehend what another individual knows, because it can see something invisible to oneself, has been specifically tested in rhesus monkeys and chimpanzees (Premack, 1988; Povinelli *et al.*, 1990, 1991). In these tests, the primate could choose the advice of one of two human helpers in deciding which container was baited with food. The act of baiting was hidden from the primate, but visible to one of the humans (the other was prevented from seeing by a screen in the way, by being out of the room or pulling a paper bag over the head at the crucial time, in different versions). Some, though not all, chimpanzees can reliably pick the helper who should know the right answer. Rhesus monkeys failed to pick the correct helper even after a long series of trials; however, negative results are always hard to interpret. Monkeys do not naturally point or spontaneously learn to do so in captivity, they do not as infants play extensively with objects like paper bags even if given the chance, and they interpret a stare at another individual as a threat. None of these provisos would aid learning in this task, and none applies to apes like the chimpanzee. Just as in human intelligence testing it is very difficult to devise a 'culture fair' test, in animal work it is hard to find a test that is 'species fair'.

8.4.4 Program-level imitation

Another case where imitation may imply computational sophistication is when it is used to acquire an already-structured skill *as a whole*. Lashley (1951) first drew attention to 'the problem of serial order in behaviour' (the fact that what superficially appear as strings of behavioural acts are better understood as hierarchically organized; see also Dawkins, 1976b), and the problem this raised for 'linear' theories, such as the chains of associations in behaviourist explanation.

Since behaviour is hierarchically organized, imitation can in principle occur at any level of the hierarchy: from slavish copying of each act in sequence to copying only the outline logical structure – from the motor-act level to the program-level (Fig. 8.2). These distinctions are not mentioned in most current definitions of imitation, such as 'B learns some aspect(s) of the intrinsic form of an act from A' (Whiten & Ham, 1992), or 'one animal reproducing in its behaviour of a conspecific' (Tomasello, 1990). However, it is usually clear that motor-act-level imitation is meant, from what is accepted as evidence for imitation, and Tomasello (1990, p. 284) explicitly stated 'true imitation' to be *imperson-ation*, a term introduced by Wood (1989) to describe the effort of children to 'behave as like [others] as possible' (p. 72).[2]

Most laboratory studies have used single-step tasks; however, the question of level only becomes interesting when the behaviour involves some logical complexity, such as requiring several subgoals to be achieved in the process of achieving the main goal (Fig. 8.3). With several subgoals, it is a real issue whether (a) the logical structure is copied (program-level imitation), but the motor-act sequences used in achieving each stage are not, (b) some motor-act sequences are copied in detail (impersonation), but an idiosyncratic logical organization is used, (c) the entire sequence is copied, which could involve either or both mechanisms.

To copy the logical, hierarchical structure of behaviour, the minimum required is to copy the sequence of *subgoals* reached along the way to successful performance; how to achieve each may then be learned by trial and error. The subgoal structure of behaviour is often 'visible' in its

Impersonation 1. 'Pick out a strand of green galium from the mass with any precision grip of the left hand and transfer the hold to a power grip by the other fingers of this hand; then repeat this cycle while still holding the picked strands in a power grip of the other fingers of the left hand, until the bundle is sufficiently large; then fold in any loose strands by using the right hand to bend in any loose strands while loosening and re-grasping the mass of stems in the left hand; or, if this is easier at the time, by letting go with finger and thumb of the left hand so that bundle is held only by other fingers, then rocking the hand to allow grasping of both loose and gripped strands by finger and thumb again; then repeating this process so that the other fingers grasp the bundle firmly, and repeating the whole cycle until all strands are held; then grip the bundle of galium loosely with half-open left hand, pick out any debris with pad-to-pad precision grip of the first finger and thumb of right hand, then grip the bundle tightly with left hand and eat by feeding into the mouth until full, then shearing off the rest by a molar bite, repeating when mouth empty again.'

2. 'Pick out a strand of green galium from the mass, then repeat this while still holding picked strands until the bundle is sufficiently large, then fold in any loose strands with the other hand (or rocking motion of the hand holding the bundle with repeated letting-go and re-grasping of strands if this is easier at the time); then grip bundle of galium loosely with half-open hand, pick out debris with first finger and thumb of other hand, then grip tightly and eat with shearing bites.'

Program-level 3. 'Repeatedly pick green strands of galium with one hand, then use other hand to fold in loose strands, then hold the bundle loosely with one hand and remove debris with other hand, then eat.'

4. 'Pick a bundle of galium, tidy it up, remove debris from it, then eat.'

5. 'Eat galium.'

Fig. 8.2. Levels of imitation. Hypothetical illustration of how a complex behavioural sequence could be 'imitated' at various levels, in hierarchical relation to each other. The sequence that it refers to is shown in flow-chart form in Fig. 8.5b. The points marked 'impersonation' and 'program-level' roughly indicate the two levels of imitation distinguished in this chapter; however, these levels are arbitrary, and a finely graded series of levels can be imagined extending above, below and between these points.

consequences, so the ability to understand *mental* states is not strictly necessary to imitate at the program-level. All that is required is (1) an ability to structure a sequence of goals, and (2) the ability to copy goals (perhaps by emulation).[3] Recently, researchers have endeavoured to categorize any copying actions as *either* impersonation *or* merely emulation; this may risk missing an important point. From the perspective of an

Fig. 8.3. Mother gorilla eats the pith of a common umbellifer, *Peucedanum linderi*. From the day of birth, a baby gorilla is showered with the feeding remains of its mothers diet – learning what is edible presents no problem. But when the foods need to be processed in several complex and interrelated stages, program-level imitation of the logical sequence would pay off handsomely.

animal or human learning new ways of structuring complex behaviour, program-level imitation is of major significance whether or not acts are slavishly impersonated at any stage in the process of building up the behaviour. The flexible 'ideal learner' would use program-level imitation to copy complex action structures, and acquire the fine details by using trial-and-error learning, emulation, or impersonation – whichever was easier. Wood, who introduced the distinction, commented that 'No doubt children do attempt to impersonate others ... the best may even go on to become professional impersonators. However, I think it is a mistake to assume that impersonation is the only or even the main intention driving so-called imitation' (p. 72). Children as young as 16 months evidently display program-level imitation: their imitation of adults' sequences of action require less observation and are more accurate if the sequences are of actions that are causally related, such as the steps in bathing a teddy-bear, and in this case any unrelated actions tend to be missed out in the imitations (Bauer & Mandler, 1989). Useful imitation in problem-solving requires both this ability to understand and copy the logical organization of programs of action, and – where these cannot be achieved by methods already known or easily derived by trial and error – to copy detailed action patterns from others. Program-level imitation and impersonation are thus complementary skills.

8.4.5 Evidence for computationally complex copying

With these conceptual distinctions in mind, what types of imitation have been recorded? Little evidence of *impersonation* has been found in experiments. Heyes, Dawson and Nokes (1992) allowed rats to see a demonstrator rat gaining food by either a left or right press of a joystick; the observers were in front of the demonstrator, so their 'left' was the demonstrator's 'right'. Before they were themselves tested, the joystick was rotated through 90° so that the place where it had finished after a left push was now the place where it would finish after a right push, ruling out stimulus enhancement. Nevertheless, the rats copied the direction of pushing. Response facilitation could be argued to explain this result, since their response was not novel. However, even this would need the rat to view responses as 'left' or 'right' from the point of view of the *demonstrator*; and the ability to take the viewpoint of another individual is often suggested to be the stumbling block preventing animals from impersonating. Palameta (1989) interprets the more rapid learning of specific feeding techniques he found in pigeons that watched demonstrations, mentioned above, as impersonation; however, as mentioned above, it is possible that the techniques were in the birds' repertoires already, and their specific usage simply facilitated by the demonstrations.

In the common chimpanzee, evidence of impersonation comes from the acquisition of a wide range of acts, some of them novel. Hayes and Hayes (1952) report the ability of Viki, reared in their home, to learn and respond appropriately to the spoken command 'do this'. The experimenter spoke the command and then, for instance, whirled on one foot, or stretched the mouth with two forefingers; Viki soon learned to impersonate each act immediately. Impersonation presumably also underlies the common observation of chimpanzees that have spent long periods with humans performing acts such as putting on lipstick, tying shoelaces, toothbrushing, and so on. If so, the fact that orangutans (*Pongo pygmaeus*), pygmy chimpanzees (*Pan paniscus*) and gorillas (*Gorilla gorilla*) have often shown similar behaviour would suggest that they too can impersonate.

There is similarly excellent observational evidence for imitation in bottlenose dolphins (*Tursiops aduncus*), which mimicked the behaviour of humans and a sealion using their aquarium pool (Taylor & Saayman, 1973). An adult dolphin impersonated the swimming mode, grooming actions and sleeping posture of the sealion, all movements entirely foreign to a dolphin. When a human blew a cloud of cigarette smoke at the pool's glass just as an infant dolphin looked out, 'she immediately swam off to her mother, returned and released a mouthful of milk which engulfed her head, giving much the same effect as had the cigarette smoke' (p. 291; Fig. 8.4). Another dolphin mimicked the scraping of the pool's observation window by a diver, even copying the sound of the air-demand valve on the SCUBA apparatus while releasing a stream of bubbles from his blowhole.

These and other carefully described impersonations are convincing, and of a different order of complexity to the rat's copying of a left rather than right joystick movement. (This is despite the rat's inherently greater similarity of sensory and motor apparatus to the human, variables often felt to underlie apparent phylogenetic differences in intelligent behaviour, e.g. by Macphail, 1985.) It is no coincidence that the strongest observational evidence of impersonation comes from apes and dolphins copying behaviour of *other* species. If instead an animal were apparently to mimic the behaviour of a conspecific, much greater evidence would be needed that the behaviour change was not really a result of normal maturation: a species-typical pattern emerging without any specific influence of the animal's experience of the form of the behaviour.

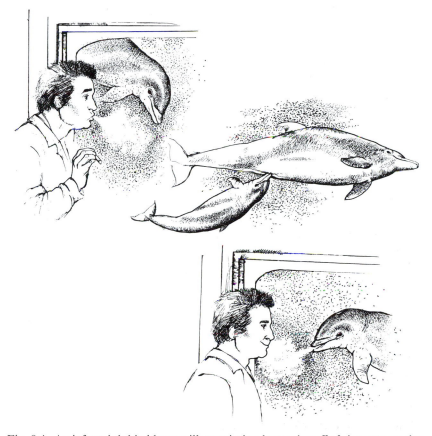

Fig. 8.4. An infant dolphin blows milk to mimic a human's puff of cigarette smoke. Here, the dolphin copies the final observed result of another's behaviour in its own way (emulation), but unlike most animals, bottle-nosed dolphins have also provided several compelling instances of impersonation, copying of the detailed action patterns observed.

The best evidence for *program-level imitation* is probably in the plant gathering of wild mountain gorillas (*G. g. beringei*), which shows complex, multi-stage processing skills, fully developed by the age of weaning (Byrne & Byrne, 1991, 1993). To eat stinging-nettle leaves, for instance, requires the following sequence: accumulation of a bundle of leaves, detaching the particularly virulent petioles from the more innocuous leaf blades, and then folding the blades so that the least stinging surface is presented to the lips (Fig. 8.5). Only the mother and the leading adult male tolerate infants near while feeding, and adults use their left and right hands in

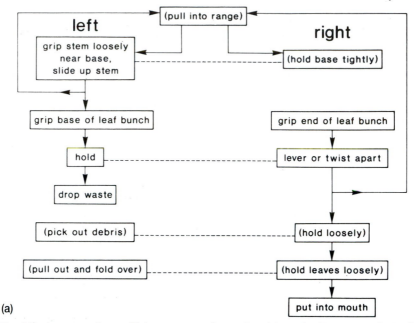

Fig. 8.5. A mountain gorilla's programs for eating (a) nettle (*Laportea alatipes*) leaves (*continued*).

unvarying roles during complex processing. Given this, the fact that there is no tendency for infants' hand preferences to match those of their only potential models (ibid) rules out impersonation. Trial-and-error acquisition seems highly implausible, given the task's complexity and structural uniformity across animals; genetically encoded action sequences similarly make no sense, as skills are specific to plants of very limited range. Thus, program-level imitation is the likely explanation, perhaps by emulation of each intermediate result in the sequence of processing stages.

To obtain clear evidence of program-level imitation a complex, multi-stage task is necessary, but unfortunately the natural lives of most animals and the experiments usually devised to test imitation are lacking in such complexity. The richer environment of humans gives more scope, and it is from human-reared apes that good evidence comes. Orangutans, during rehabilitation into the wild after illegal capture, have been noted reproducing rather elaborate human actions: chopping off weeds and then sweeping them up into a row of neat piles just as done by a camp assistant;

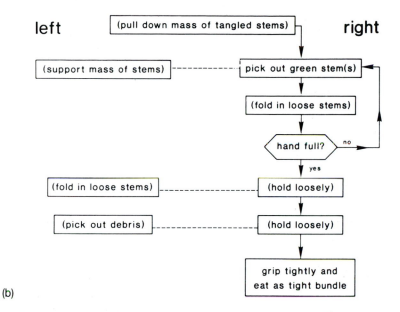

(b)

Fig. 8.5. (*continued*) (b) bedstraw (*Galium ruwenzoriense*) stems. The sequence of actions (which begins when an animal finds a food to eat) starts at the top and moves down: rectangular boxes show actions, described by the words in them, arranged to the left and right of the midline to indicate significant lateralities in the hand used (brackets show actions which are optional, depending on environmental conditions). Dotted lines indicate bilateral co-ordination between the separate actions of the two hands. Diamonds represent branch points, with the approximate criteria for the decision indicated in words in the diamond: thus a process may repeat or iterate until an appropriate size of handful is reached. The sequence ends with putting processed food in the mouth. (From Byrne and Byrne, 1993.)

attempting to siphon liquid with a hose, or to start a fire, following the techniques used by the camp cooks; tying a hammock between two trees (Russon & Galdikas, 1993). This program-level imitation may utilize emulation of the results of task stages (e.g. 1. weeds that are cut, 2. neat piles) but on the available evidence could also reflect impersonation of the complete sequences.

From the patchy current evidence, it therefore seems unlikely that any species apart from great apes and dolphins possess anything approaching the flexibility described above as 'ideal' for a social learner: program-level imitation that could use impersonation to copy details that were resistant to trial-and-error acquisition.

8.5 Understanding and insight

It will be clear that, for an animal to imitate by putting itself mentally in the place of another or by understanding the logical structure of another's behaviour, some sort of insight is implied; we have thus already begun consideration of what 'insight' might really mean. There seem to be at least four senses or components to insight.

8.5.1 Comprehension of mechanical function

Some animal abilities implying mechanical comprehension have already been mentioned: a chimpanzee's discovery that two sticks fitted together make a tool long enough for the problem in hand, and probably also a gorilla's learning that to fold nettle leaves enables them to be eaten less painfully.

Köhler's original work has since been qualified. Schiller (1952) showed that chimpanzees with no experience of the raking-stick problem will sometimes join sticks together in play even where there is no reward to be gained. The crucial behaviour may therefore not be novel in the experiment, but already part of the animal's repertoire: chimpanzees' solutions do not need to be computed entirely *de novo* from the mechanics of sticks, but appropriately deployed out of the repertoire of past experience. Object play (found in chimpanzees and humans, and in captive-reared orangutans and gorillas) might well have the biological function of building an augmented repertoire of possible solutions. Birch (1945) showed that chimpanzees which fail to solve the problem benefit greatly from a period of unstructured play with sticks, in that they are then likely to succeed in the task; a specific effect of the play is hard to pin down since the study lacked a control group.

That chimpanzees can generalize uses of an object between play and a problem-task suggests an appreciation of object properties at a rather abstract level. Visalberghi and Trinca (1987) have found that capuchin monkeys lack the chimpanzee's understanding of functional properties. They taught capuchins to use a thin tool to push foods out of a transparent tube, and then gave them a choice of a range of different objects. Monkeys made attempts with sticks far too thick to enter the tube, with sticks much too short to reach the clearly visible food, and even with a flexible chain! Apparently, capuchin monkeys have no 'insight' into *why* a certain stick is a useful tool, and fall back on 'blind' trial and error. Comprehension of mechanical function may thus be restricted among animals to the great apes.

8.5.2 Theory of mind

As well as insight into physical properties of objects, humans show an understanding that other individuals, like themselves, have mental states – beliefs, feelings, intentions, and so on. Although as humans we can never hope to know much about the private *minds* of other species, we can ask whether they – like us – treat other individuals as if they have mental states; those that do are said to have a *theory of mind* (Premack & Woodruff, 1978). Where imitation requires the animal to understand goals and intentions of another individual, this is already evidence for understanding the mental viewpoint of another individual. But having a theory of mind should allow much more. A constellation of superficially rather different skills is thought to depend on understanding of mental states: comprehending that an image reflected in a mirror is oneself, knowing that a conspecific meant to deceive, realizing that one's offspring would benefit from being taught a particular fact, empathizing with the predicament of an individual who has a problem, sympathizing with the grief of another, and worrying about malevolent intentions towards oneself. An underlying common variable linking these types of behaviour has been suggested several times (Gallup, 1982; Byrne, 1990; Whiten & Byrne, 1991; Jolly, 1991).

Most animals evidently lack a theory of mind – as has always been thought by scientists, since they saw no evidence of the diagnostic behaviour. However, increasingly some great apes *are* providing positive evidence. In addition to the evidence of cognitively complex imitation, self-recognition has been found in chimpanzees, orangutans and gorillas (Gallup, 1970; Suarez & Gallup, 1981; Patterson & Cohn, in press), whereas monkeys and elephants do understand how to use mirrors to look around corners but fail to recognize their own reflections (Anderson, 1984; Povinelli, 1989). Gorillas, orangutans, and both pygmy and common chimpanzees have given observational evidence of understanding that others may have deceitful intentions, and of what others can be made to think (Byrne & Whiten, 1990, 1991). This evidence includes cases of deception where a history of past reinforcement can be ruled out, counter-deception that shows an understanding of others' intent, and indignation at being deceived. Understanding of intent has also been shown experimentally: a chimpanzee who was deprived of a valued drink through an 'obvious accident' by one human caretaker, and by another who deliberately poured the drink on the ground, consistently chose the former to give it drinks in future (Povinelli, 1991). A young gorilla that

Fig. 8.6. A way for a young gorilla to solve a problem: requesting a human to
help. (Based on description by Gomez, 1991.)

initially solved the problem of getting to an out-of-reach goal by pushing
a chair or convenient human into position and climbing it, later switched
to requesting the human's help by the use of eye-contact and leading by
the hand, showing a practical understanding of the human as causal agent
(Gomez, 1991; Fig. 8.6). Common chimpanzees have occasionally been
seen to teach actions to their offspring by demonstration and hand-
moulding (Boesch, 1991; Fouts, Fouts & Van Cantfort, 1989), and to
choose appropriate solutions to problems presented to another individual
(Premack & Woodruff, 1978). Common chimpanzees also show empathy
with others' perspectives. As well as realizing that unobstructed vision

is necessary for accurate knowledge (the Premack and Povinelli experiments mentioned in Section 8.4.3), they apparently understand the role of another individual in a co-operative task. Povinelli and his co-workers devised a task in which one individual could see which of two handles would give food rewards when pulled, but could not reach the handle. To succeed at the task, this individual had to indicate the correct choice to a second individual, who could reach the handle but not see if it was correct – both thereby gained food rewards. Chimpanzees succeeded in either role, but more importantly were able – without more training – to assume the other's role when they were reversed (Povinelli, Nelson & Boysen, 1992). Monkeys, by contrast, showed no such immediate comprehension of their new role (Povinelli, Parks & Novak, 1992).

If great apes (and dolphins, to judge from their imitations), but not most other species, have a theory of mind, this will require a different kind of explanation to those discussed earlier in this chapter; it would not be sufficient to propose differences in attention bias as the source of such a contrast. Rather, it shows an ability (perhaps a restricted one) to *represent* the minds of others as data objects. Such representations would potentially allow computation of future behaviour from a starting point of different goals and beliefs than one's own, bringing us to the next sense of 'insight'.

8.5.3 Planning

The third sense in which animals might show insight is in terms of 'thinking', 'planning' or 'calculating': the derivation of novel solutions, or novel uses of old solutions, by mental operations. Minimally, this would involve concatenating knowledge learned in two previous circumstances, in order to deal with a third, novel circumstance.

Evidence for thinking in animals is understandably slight, in view of the practical difficulties of obtaining it. Köhler's chimpanzee is often said to have suddenly found his effective solution to the raking problem after a pause, and his insight was thus implied to result from thinking. However, inactivity does not necessarily imply thought; and, since play with the sticks filled the pause, the sudden solution was more 'noticed in his hands' than 'thought out'. Another potential avenue is the case where no learning history for a behaviour is plausible, yet the behaviour is built of two components, each learnable. For instance, chimpanzees characteristically behave with silence and cautious bushcraft when on their dangerous 'border patrols' into the range of other communities (Goodall, 1986).

When infants are distressed, they are immediately comforted; when adults (or a human observer) are carelessly noisy, they are threatened. This behaviour implies that the connection between noisiness of a party member and potential danger is well understood. Yet border patrols are rare and so risky that the 'punishment' for error may well be death; reinforcement learning thus seems unlikely to underlie the chimpanzees' knowledge. It is more likely that the connection is derived mentally. In a simpler sense, even laboratory rats can be said to exhibit concatenation of separate pieces of knowledge (Dickinson, 1980). Having learned to find a particular food in a hopper after a tone sounded, rats then experience illness (actually caused by other means) after eating this particular food; when the tone next sounds, they do not approach the hopper (Holland & Straub, 1979): they have put the two pieces of knowledge together.

8.5.4 *Abstraction and generalization*

The chimpanzee which realizes that a tactic in his play will serve to gain a food reward, and the person who realizes that a dancing ring in his dream will solve the structure of benzene (as Kekulé is said to have done), are generalizing from one situation to another. However, their generalizations are *abstract*, their solutions are not bound to their original context. The *level* of this abstraction may be an interesting component of intelligence. One artificial task that relies on seeing the logical connection between different problems is called 'learning set': an animal is given a series of problems to solve by trial and error, but the solutions are related by a rule, such as 'odd one out' or 'match the sample' (Harlow, 1949). Passingham (1981) has shown that primates learn these rules quicker than non-primates such as rats, cats or squirrels; and that rhesus monkeys far outperform small New World species like marmosets (*Callithrix* spp.). Rhesus themselves are outclassed by chimpanzees (Hayes, Thompson & Hayes, 1953). Passingham (1982) reviews other rule-learning tasks, showing that chimpanzees, rhesus monkeys, cebus monkeys, squirrel monkeys (*Saimiri* spp.) and marmosets, and finally strepsirhine primates such as the ring-tailed lemur (*Lemur catta*), can all learn rules; but their facility to do so decreases in that order. To notice such rules, animals must characterize the rewarded stimuli in abstract terms, as we do when we write the rules in English: 'odd one out' rather than just 'red ball'.

8.6 Evolution of intelligence

I have suggested that 'intelligence' has a number of facets. In many species, certain learning performances are channelled by constraints on what the individual attends to; learning is efficient but narrow in scope, and therefore lacks adaptability. Much learning is less constrained, such as the conditioning mechanisms for learning correlations in the environment and for recording the results of trial-and-error exploration, and these mechanisms are very general in animals. Even then, the way in which objects and events are represented – what is noticed, and whether in specific or abstract ways – affects the level at which knowledge can be generalized from one circumstance to another, and gives rise to species differences that are not just a consequence of different environmental needs. Mainly, these species differences have singled out simian primates from other animals; in other species, with the likely exception of dolphins, it makes little sense to describe differences as due to 'intelligence'. Much of the rest of this chapter, therefore, uses data from primates.

Certain simian primates, especially chimpanzees, show intelligence to a greater degree than others, for instance in understanding the mechanical properties of objects used as tools. The ability to represent the intentions of other individuals (and these of course can only be inferred, not observed) seems restricted to a very few species, and this has implications for those animals' ability to manipulate socially and to comprehend each other. Assembly of novel programs of behaviour from observation of the skills of others requires both empathy with other individuals' goals and intentions, and the ability to understand the logical structure of processes, in different cases; and these too are apparently restricted to a few species. These 'few species' usually equate to the great apes.

8.6.1 Why did some animals become intelligent?

In other words, what were the selection pressures that promoted intelligence? And what are the animals using it for now, that still confers biological advantage? Often these questions amount to the same thing, though the general-purpose nature of intelligence means that this is not necessarily so. There are two plausible theories.

Dealing with environmental complexity

Environmental challenges that have been suggested to be sufficiently demanding to select for intelligence have centred on efficient feeding. 'Feeding skill' amounts to two packages of ability.

Finding food This includes learning which items are edible, and getting to sources of them at each time of year, which involves a *cognitive map* of the known range. Although many species of animal probably have extensive spatial knowledge, this idea was first introduced to explain primate intelligence. Mackinnon (1978) noticed that an orangutan, when confronted with a poor crop yielded by its favourite fruit tree, chose a route through the featureless Bornean forest that took in a series of these trees – a route that was so economical in effort that the researcher could only match it with the aid of a carefully recorded map. Independently, Milton (1981) proposed mental mapping skill as the explanation for the large brain of the frugivorous spider monkey (*Ateles geoffroyi*) relative to that of the more folivorous howler monkey (*Alouatta palliata*). She suggested that to navigate individually around the spider monkeys' far larger range needs a good cognitive map, and a large brain would thus be necessary. Subsequent evidence of cognitive maps in primates has been sparse, but interesting. Chimpanzees in West Africa use round stones as hammers, and larger flat stones as anvils, to break open tough *Panda* nuts. Boesch and Boesch (1984) showed that these stones were in short supply, and the economical routes of chimpanzees when they found a source of nuts implied that they remembered where each stone last lay in the forest. Hamadryas baboons (*Papio hamadryas*) had long been noticed to mill around before they set off in the mornings: phalanxes of animals like 'arms' of an amoeba push out, only to collapse back into the mass, before finally the whole group pours out. When foraging, the group splits into much smaller parties, which rejoin later. Sigg and Stolba (1981) discovered that the 'arms', apparently thrown out haphazardly, function by 'pointing' at a possible place to meet up again later. When they set off to wander as small bands, their movements have no correlation to each other's route or to the direction of the eventual meeting place. Nevertheless, the orientation of the 'arm' along which they set off closely predicts the place where the troop later rests in the heat of the day. Baboons not only have cognitive maps, but they can label other animals' knowledge of large-scale space without using verbal descriptions.

Obtaining food This includes the capture of prey by *hunting*, and the skilled *tool use* shown by chimpanzees in insect-fishing techniques (Goodall, 1986). Surprisingly little is known of any intellectual skills that carnivores use in hunting, although big-game hunters have always claimed them to be considerable. Stander (in press) found 40% of lion hunts in Etosha, Namibia, to show co-operation of a kind that could not be explained as an accidental consequence of individually chosen strategies. When fanning out and circling, prior to driving the prey into an ambush by other, waiting lions, he found that 'all lions appear to watch both the prey and other pride members during the stalk as if to orient their own movements'. Clearly their prey could equally benefit from skilled evasion techniques, and Jerison (1973) has suggested that just such an evolutionary 'arms race' between predator and prey underlies the apparent increase in relative brain size of both carnivores and ungulates in the fossil record. Tool use is found in a number of birds and mammals (Beck, 1980), but only common chimpanzees fashion a range of different tools for separate purposes (McGrew, 1992). A suggestion of wider relevance is that any food processing where edible matter is *extracted* from where it cannot be seen is cognitively demanding (Parker & Gibson, 1979). Comparison of primates in signs of intelligence, according to whether or not they forage 'extractively' – especially if they use tools in the process – has been used to support this theory (ibid); in particular, the learning ability and relatively large brain of the capuchin monkey, a species notable for its destructive and extractive techniques of foraging, give it credence. However, many other mammals forage extractively (for instance, opening and even caching nuts to dig up later, probing for nectar, digging for water) and it is not yet clear that there is any systematic relationship between this and general intelligence.

Dealing with social companions

The 'environment' for a social animal includes its companions, in addition to other species and the physical world. Social companions are potential competitors for mates and food. Unlike the physical world, they change over time, often rapidly and interactively, and – having the same intelligence – they are likely to present peculiarly challenging problems to an animal. Nevertheless, for many reasons, animals of some species benefit from long-lasting associations. Humphrey (1976), making these points, argued that the resulting pressure, to maximize individual gains yet retain the benefits of group living, could have acted to select for a

Fig. 8.7. Caught in the act! The instant at which a young silverback male gorilla, Titus, was discovered, in the secret act of mating with the adult female Papoose, by the leading silverback of the group, Beetsme. Prior to this, Papoose had solicited Titus, and they suppressed normal copulatory calls during their mating. Beetsme is an old silverback, and it is likely that Titus will take over the group in time; in the meantime, the 'tactical deception' used to hide matings allows females a choice of mates.

social type of intelligence, poorly measured by the laboratory tasks of psychologists. Jolly (1966) had a similar insight, noting that the group-living ring-tailed lemur lacked the social complexity of any monkey (and was also, incidentally, poor at laboratory tasks, suggesting that what these measure is not wholly unrelated). Therefore, while social living could not result from high intelligence, she suggested that perhaps increased intelligence might be a consequence of social living.

Consistent with these ideas, social species of primate display both complexity of social manipulation and considerable knowledge of social information. Monkeys and apes of many species use behavioural tactics which rely on tactical deception (Byrne & Whiten, 1990). For instance, a female can obtain matings with a preferred but low-rank male by hiding, inducing the male to hide also, and by suppressing her mating vocalizations (Fig. 8.7); these 'tactics' may be learned from experience by trial and error, without insight into why they are effective (see Byrne, in press). Experimental playback of calls to vervet monkeys has shown that they are sensitive to the dominance ranks, patterns of close association, and group membership of their conspecifics (Cheney & Seyfarth, 1990). For instance, a different reaction is obtained to a 'submissive-to-dominant' call if it is uttered by a caller below or above the hearer in rank – since, of course, only in the latter case does the call signal the nearby presence of a higher-rank animal than the hearer. Given vervets' sensitivity to subtle *social* differences, then, their reaction to *environmental* events is striking: in fact, they give no reaction to finding the obvious (to a human) tracks of a python or the food cache of a leopard, their two major predators, and consequently have been noted to blunder upon the predators (ibid). Even in humans there is an echo of this pattern, with childrens' acquisition of the concept of rank and the principle of transitive inference (for instance, if A is stronger than B, and B stronger than C, then A must be stronger than C). Both are shown first in social judgements, only later in artificial contexts (Smith, 1988). However, interpreting 'failures' is always difficult: for instance, finding the track of a python does not tell one which way it went or when, and a reaction of alarm may be inappropriate.

8.6.2 Testing between theories: comparing animals' brain sizes

Finding a wealth of data consistent with the theory of a social origin of intelligence (see chapters in Byrne & Whiten, 1988) does not prove it correct. Social and technical skill are not independent in practice. The greatest sophistication in social manoeuvre and understanding of any animal is undoubtedly by the chimpanzee (de Waal, 1982; Menzel, 1974; Byrne & Whiten, 1991); yet this is just the primate species that uses tools the most, forages extractively, hunts regularly, inhabits the largest home ranges (Tutin, McGrew & Baldwin, 1983), and eats the largest variety of plant and animal foods known (Nishida & Uehara, 1980).

It is unfortunate that there is an almost complete lack of comparable data on social sophistication in non-primates such as the highly social mongooses, hyaenas, wild dogs (*Lycaon picta*), elephants and cetaceans. Also, apart from the case of hunting, the 'environmental challenge' theories are largely restricted in relevance to primates, since other species tend to have simpler diets. Most animals either use bacterial digestion to gain complete nutrition from plants, or eat other animals; there is little need of cognitive maps, or complex balancing of many dietary components, let alone tool use. The cognitive mapping abilities of various bird species are associated with hippocampal enlargement in the brain (Krebs, 1990), but not so far with signs of any more general intelligence.

In testing between theories, then, data from primates are of most use. There is no generally agreed metric of intelligence even between species within this group (see review by Warren, 1973). Instead, an indirect measure of intelligence, brain enlargement, must be used. The brain is energetically costly (Armstrong, 1985; Hofman, 1983a, 1983b), so brain enlargement must confer evolutionary advantage, and this is always assumed to be because of the benefits of intelligence.

To compare brain sizes properly, it is necessary to take account of two characteristics of growth and form. Larger animals, in general, have larger brains; this is not particularly surprising, since for most mammals, much of the brain is taken up with sensory and motor processes. Secondly, as the absolute size of living things changes, so the relative proportions of their parts is liable to change. In this case, absolutely larger animals have relatively smaller brains than one would expect from linearly scaling-up smaller animals. A technique that takes account of these difficulties is called allometric scaling. In allometric scaling, a double logarithmic plot of something, in this case brain size, is made against body size for the given group of animals. This forces the species points onto a straight line. Then one can see whether any particular animal in the group lies above the line (has a relatively larger brain than one might expect), on the line (has average size), or below the line (has a relatively small brain). This technique has many limitations (including doubt as to whether a straight line is the best fit to the log–log transformed data; see Deacon, 1990), but it is the one now used most often to compare animals' brain sizes. With this approach, one finds that humans have brains three times as large as we would expect from even a monkey of human size (Passingham, 1982), which of course tends to give us confidence in the relationship between relative brain volume and intelligence!

An underlying assumption of this method of working out relative brain enlargement is that the *extra* volume over that minimally required to service sensory and motor systems – in other words the 'computational part' – is some multiplicative fraction of total volume. So, 5% extra brain is assumed to be equally useful for intelligence in a 5-kg animal or a 50-kg animal, even though the extra brain tissue is far larger in the second case, and contains far more neurons. This is a very odd assumption for anyone used to computational machines, since these are chiefly limited in power by the number of their elements. One can instead make a different assumption, that of additive volumes, in which what matters is not the percentage of the total that is in excess, but the absolute volume (and the number of neurons) free for computation. This assumption is less tractable mathematically, as the standard of comparison is much less clear, and several versions have been tried: the amount of brain available beyond that for 'visceral function' (Jerison, 1973), the ratio of cortex to medulla volumes (Passingham, 1982), and the ratio of neocortex volume to the rest of the brain (Dunbar, 1992).

When either type of allometric scaling is applied to brain sizes of mammals, the primate order as a whole is found to be larger brained than most other groups (Jerison, 1973). But when the strepsirhine primates (the more primitive lemurs, lorises and galagos; Fig. 8.8) are partitioned from the rest, the strepsirhines emerge as a typical group of mammals (Passingham & Ettlinger, 1974). In other words, in general they have brains about the size one would expect from their body sizes. The monkeys and apes, however, have brains twice as large (or bodies half as big) as average mammals of their size.

In terms of 'brain volume above that expected from size', apes do not notably differ from monkeys by conventional allometry (Clutton-Brock & Harvey, 1980). On the basis of 'brain volume free for computing', all apes have brains relatively larger than those of most monkeys, however. The living apes also vary among themselves in apparent brain enlargement. A difficulty in interpreting this is that the apes also differ in diet type: a larger gut makes the abdomen larger but requires little brain expansion, so it is misleading to compare species adapted to different diets. Scaling against body length instead of weight helps a bit, but the length of primate bodies is still heavily influenced by gut size. Though all great apes have a broadly similar type of gut, they do vary in their adaptedness to coarse food, i.e. in the size of their large intestines. In fact, the gut sizes rank gorilla > orangutan > chimpanzee ≫ human, and this order is the inverse of their relative brain sizes. This suggests that selection on body size rather than brain size accounts for the variation (as argued also by Deacon, 1990).

The general relationship between home range size and relative brain enlargement of the primate order (Clutton-Brock & Harvey, 1980) has been used to support an origin of intellect in environmental complexity. However, this effect too may be an artifact of selection for big bodies in folivorous primates. Even in animals with simple stomachs, eating mature leaves requires a large gut for efficient fermentation, and hence a large body to support it; but, since mature leaves are relatively abundant in most primate habitats, only a small range is needed. By contrast, frugivory requires a larger range area for year-round access to a variety of fruit species, but the high sugar content allows digestion by a shorter gut,

Fig. 8.8. *Hapalemur aureus*, a strepsirhine primate. Lemurs, like this golden giant bamboo lemur, lorises and galagos make up the strepsirhines, which do not seem to be specialized for intelligence.

and thus smaller body size. This effect, rather than selection for big brains in species needing good cognitive maps, could equally cause the observed correlation. Recent analyses suggest that body size is labile in evolution relative to brain size (Shea, 1983; Deacon, 1990).

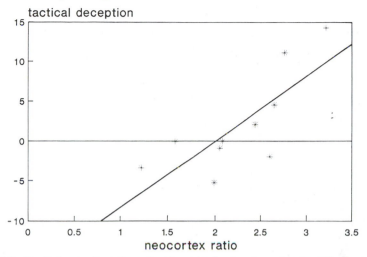

Fig. 8.9. Predicting tactical deception from brain size. An estimate of the frequency of use of tactical deception is plotted against the relative size of neocortex for each species. The estimate used here is the chi-square deviation, between the frequency of tactical deception *observed* and that *expected* on the null hypothesis that there are no species differences except in the number of studies undertaken. The correlation of 0.77 is significant at $p = 0.009$, one-way ANOVA $F(1, 8) = 11.89$. (From Byrne, 1993.)

If so, then one of the estimates of 'free computing space' will be a much better way of testing theories; of these, neocortical enlargement is the most straightforward. Sawaguchi and Kudo (1990), controlling for diet type, found a number of correlations between relative neocortical size and both group size and polygynous mating in primates. Now, Dunbar (1992) has tested directly between the two theories. Range area, day journey length and the amount of fruit in the diet were found to be unrelated to relative neocortex size, when body size effects are removed. However, group size, a rough-and-ready measure of social complexity, does correlate with neocortical enlargement. Dunbar goes on to propose that neocortical size limits the social complexity that an individual can cope with: when a group exceeds this limit, it becomes unstable and begins to fragment.

A specific test of the relevance of neocortex size to social manipulation can be made with the extensive but unsystematic data on primate tactical deception. In order to correct in a rough-and-ready way for variations in how much different species have been studied, Byrne and Whiten (1992) tested whether the number of cases reliably reported matched the number of observational studies undertaken. The pattern differed significantly from that expected by chance, largely as a result of high rates of deception in chimpanzees and baboons. Going beyond these findings, if the deviations are now plotted against Dunbar's estimate of neocortical enlargement (Fig. 8.9), they match – in fact the correlation of 0.77 is highly significant (Byrne, 1993). Despite the very approximate nature of this measure of social intelligence, its close relationship to neocortical enlargement suggests that we are at last beginning to get some hold on the elusive concept of primate intelligence.

While any relation of brain sizes to behaviour must still be regarded as tentative, it seems more than a coincidence that the neocortical enlargement of the primates, which is at its greatest in some of the largest species, the great apes, correlates with differences in intelligence. The enhanced ability of monkeys and apes to solve rule-based problems and to learn quickly in social contexts, together with the ability of apes (or some of them) to show insight into mechanical function, the logical structure of behaviour, and the intentions of other individuals, may all prove to be reliant on selection for localized brain enlargement. The current evidence suggests that the selection pressure for this was largely a social one.

8.7 Summary

There are conspicuous species differences in what animals learn and how efficiently they do so. To humans, it seems at first sight obvious that the species must differ in intelligence: research by psychology and ethology has shown instead that these differences are often a result of the genetical channelling of individual learning towards aspects of the environment crucial to survival. Individual learning mechanisms are themselves adapted to discovering useful information about the world, and seem very general in animals. In social species, learning can be further guided by the successes of companions, drawing attention to relevant environmental features or potentially useful patterns of action; these social influences lead to cultural differences in animal populations and a continuing need for a social context for development. For most species, however, there is little evidence to suggest variation in any general, computational intelligence.

A number of lines of evidence imply that simian primates, especially great apes, are specialized for intelligence; one implication of this is that the phylogenetic history of human intelligence, and perhaps the selective pressures that led to it, can be studied. (This view is in contrast to the belief popular in psychology, that human intelligence is very recent and inextricably linked to possession of language.) Intelligence in this sense includes several facets: the ability to represent complex entities, whether the complexity is physical, as in the comprehension of the mechanics of objects or the logical structure of behaviour, or mental, as in understanding the intentions, perspective and beliefs of other individuals; the ability to generalize at an abstract level and so to apply old skills to new situations; and the ability to think or plan – that is, to compute novel behavioural tactics on the basis of existing mental representations. Not all simian primates possess all these abilities, even in rudimentary ways.

In the primates, specialization for intelligence may be a result of selective pressures promoting skilled food-finding and food-processing, or selection promoting social skills; current evidence on the variation of brain size among primates favours the latter. Whether the apparent lack of any species specializing in general intelligence among other groups of animal is real, or some reflecton of the different way they have been studied, is still a moot point. The rapid learning and imitative abilities of dolphins are a pointer to a need for caution in claiming that only primates have evolved advanced intelligence.

Notes

1. For similar reasons, Tomasello, Kruger and Rarner (1991) use the term 'imitation learning' to restrict attention to acquisition of actions not previously in the repertoire and explicitly exclude copying of actions which are not new.
2. In everyday usage, impersonation could include mimicry of behaviours that the mimic can already perform; this could be achieved by response facilitation as well as by imitation. For reasons already discussed, this chapter restricts 'impersonation' to copying of actions not previously in the repertoire. Tomasello (1990) follows similar usage.
3. Equating 'true' imitation with impersonation, as some researchers have done, would suggest that program-level imitation (unless it can be shown to involve impersonation), has nothing to do with imitation at all. However, this would miss Lashley's point about the hierarchical nature of behaviour, and dismiss much that is of real significance in the ontogeny of human and animal behaviour.

9

Social structure and evolution

P. C. LEE

9.1 Introduction

The question of social life among animals has intrigued biologists for centuries. How do we define sociality? When is it likely to occur? What are the patterns of behaviour involved in being social? And what is its frequency in the animal kingdom? Social life requires associations between individuals, which potentially lead to interactions. Associations and interactions have costs as well as benefits to each participant. It is by addressing these costs and benefits that the questions posed above can be examined.

A definition of sociality is fraught with problems. A simple definition, that when two or more animals live together they constitute a social unit, is relatively straightforward but ignores the vast complexity of modes of living. Some animals are consistently social in that they form distinct and permanent units that are easily distinguished in time and space from other such units. A troop of baboons is such a social group. Here, individuals forage together for the majority of time, sleep in close proximity, mate and rear infants within the unit, exhibit and maintain friendly interactions within their group and potentially hostile interactions with non-group members. But what of colonially nesting birds? Aggregations of several thousand animals within a tiny area form a cohesive mass in time and space. While each pair will defend its nesting area, all the colony members can co-operate in spotting and possibly driving away predators (Kruuk, 1964; Birkhead, 1977). Some elements of sociality are clearly in play. Among the orangutans of Borneo and Sumatra, individuals forage separately, sleep alone, seldom mix with the opposite sex for longer than the time it takes to mate, and associations between individuals of the same sex have been observed for a maximum of several hours (see Gladikas, 1985; Mitani *et al.*, 1991). Orangutans could justifiably be called solitary.

Yet, each male's home range covers the separate foraging areas of two to three females, and roving ('unknown') males are forcibly evicted by the resident male. Females resist the advances of the stranger, while co-operating with their known males (Gladikas, 1979). Thus, a pattern of 'knowledge' of strangers and familiar animals exists above the day-to-day lack of association and interaction, and some form of sociality can be inferred. Other species such as sloths may also come together only for brief periods of mating while leading independent lives. Contacts between strange individuals tend to be mutually antagonistic, and the young disperse early. However, even in solitary species a period of parental care and association may be necessary. Among sloths, a close association between mother and offspring until well after weaning is necessary in order to establish dietary preferences (Montgomery & Sunquist, 1978). For other solitary species of birds and mammals, such parental investment may be prior to hatching or birth, and the offspring subsequently disperse to establish separate foraging or living areas.

These examples point to the problems inherent in an attempt to develop an all-encompassing definition of sociality. What unites social species is that some degree of interaction between associating individuals can be specified, and the interactions can be defined in terms of their nature and their frequency through time. The interactions then can be used to form the basis of distinctions between observed social systems, and can also be analysed in relation to the advantages and costs to individuals of their participation in specific aspects of social life.

9.2 Costs and benefits of grouping

In order to understand the evolution of sociality, it is necessary to determine the range of options in the nature and types of associations between individuals that is available and to discuss the resulting modes of sociality in relation to the advantages and costs of behaviour associated with grouping. These costs and benefits have been extensively discussed elsewhere (Bertram, 1978; Krebs & Davies, 1987; Wrangham, 1982; Wrangham & Rubenstein, 1986; Dunbar, 1988) and are only briefly summarised below (Table 9.1).

A discussion of the costs and benefits of sociality also needs to consider the differences between group formation (when and where individuals form groups), group size (when and where groups of different sizes are established), group composition (which individuals aggregate), and the stability and persistence of the group through time; all of these elements

Table 9.1. *Summary of costs and benefits of associations between individuals*

Functional category	Benefits	Costs
Predation	Dilution, defence, swamping; exchange of information; reduced time spent vigilant	Conspicuousness; larger area, higher area, higher contact probability
Foraging	Food defence; efficient food location; shared information as to quality, abundance and renewal; co-operative hunting	Reduction of individual intake on shared resources (competition); increased energy costs of foraging to cover a larger area or maintain a group
Information exchange	Opportunities for easy, quick and energetically inexpensive exchange of signals; enhanced assessment of status of signaller	Parasitism of signal producer by non-producer; increase in competition between signallers
Access to mates	Opportunities to gain access to mates, to assess mate quality, to complete mating; reduced search costs for mates	Competition for access to mates (direct or indirect through sperm competition); status differentiation in mating success; choice of mates leading to high variance in reproductive success; potential for infanticide to increase access to reproductive females
Access to helpers for rearing infants	Enhanced protection of vulnerable infants; reduced maternal energy costs and thus shorter intervals between reproduction; provisioning of infants with food	Potential for infanticide by non-parent helpers; delayed dispersal and reproduction by helpers
Disease transmission		Increased probability of morbidity through disease; higher parasite loads
Thermoregulation	Reduced costs of heat production when exposed to energetically costly cold stresses	

specify the nature of the sociality observed. The issues of group formation and group size are inextricably linked and a simple description of cost/benefit ratios does not necessarily separate the interacting pressures acting on size, stability and composition. Optimality modelling has attempted to resolve the problem (see Chapter 5; Sibly, 1983), and remains an area of extensive research. However, as Dunbar (1988) has noted, it is those costs and benefits which *maintain* groups that are of primary interest in discussions of the permanently social species. The descriptions presented below are brief summaries of the variety of benefits and costs of forming groups, and the pressures to maintain groups are discussed subsequently in relation to the specific types of social groups which result.

9.2.1 Benefits of grouping

1. *Protection from predation*

When individuals aggregate, each of them is less likely to be taken by a predator since the probability of the predator choosing one animal depends on the number of animals present if all individuals are equally attractive to or at risk from the predator. This is Hamilton's (1971) selfish herd principle. In practice, all animals are seldom equally at risk and thus some will be chosen disproportionately. Predators can choose vulnerable, inexperienced (e.g. young) animals or weakened animals (e.g. old, in poor condition, pregnant, injured) (FitzGibbon & Fanshawe, 1989; FitzGibbon, 1990). However, the chance of a predator taking its chosen prey is decreased by dilution, for example where many young are simultaneously present (wildebeest: Estes & Estes, 1979). Furthermore, if the animals in the group actively co-operate, they are more likely to spot the approach of a predator, to transmit this information by postures, chemical signals or vocalizations to others, and all group members are thus able to take evasive action at an earlier stage. The behaviour of individuals acting in concert can confuse the predator, or signal to the predator that it has been spotted and thus is wasting time and energy in further pursuit. They may even actively chase off or harass the predator, as in mobbing by birds.

These effects are not necessarily limited to members of the same species, and indeed, among African ungulates, mixed species groups of territorial males form when predation pressure is high (Gosling, 1986). A territorial male is unlikely to share his space with another potential competitor and, standing alone, he is a potential target for predators. If, however, he forms a group with males of different species, then each will be able to monopolise his females without interference while being better able to spot and avoid predators. Polyspecific associations among primates have also been proposed as potentially increasing group size in the face of predation pressure, whilst minimising intraspecific foraging competition that would occur in large groups of the same species (Cords, 1987; Peres, 1991).

2. *Increased foraging efficiency*

There are a number of foraging benefits that can be derived from associations between individuals. Firstly, animals in groups can spend more time feeding by sharing the time costs of vigilance for predators (ostrich: Bertram, 1980). Secondly, animals can co-operate to defend foods from other individuals or groups, thus increasing their intake (acorn woodpeckers: Hannon *et al.*, 1985; ants: Adams, 1990; capuchins: Robinson, 1988). Thirdly, they may be able to transmit information as to the location of foods (birds: Ward & Zahavi, 1973; Brown *et al.*, 1991; honey bees: von Frisch, 1967), the abundance or quality of foods (spider monkeys: Chapman & Lefebvre, 1990; rats: Galef, 1991), and means of exploiting foods (chimpanzees: Boesch, 1991). Finally, they can actively hunt prey with the assistance of others, as in many of the social carnivores (Kruuk, 1975; Bertram, 1978; Scheel & Packer, 1991) and in some fish (Major, 1978).

3. *Increased information exchange*

Both in relation to predators and to finding food, associations of animals facilitate the transmission of information. Close proximity increases the number of opportunities for observation and the rapidity of information procurement, while it decreases the time or energy costs of producing and dispersing the signal. Information about the availability and status of receptive females may also be more easily disseminated, and the quality of potential mates can be assessed (elephant seals: Cox & Le Boeuf, 1977; red deer: McComb, 1991).

4. *Access to mates*

When the sizes live separately, some mechanism is needed to bring males and females together to mate at the appropriate time of year (Halliday, 1981). While there is no need for permanent co-residence of the sexes, a period of interaction is obviously required to achieve mating. For some species, this period can be relatively long, or indeed continuous, while for others a mere matter of minutes is necessary. Aggregations of individuals during the mating period can ensure that both sexes are in close proximity when necessary, and furthermore provide an arena for the selection of mates of high quality. Leks or display areas among some birds (peacocks: Petrie *et al.*, 1991; sage grouse; Gibson & Bradbury, 1985) and ungulates (topi: Gosling & Petrie, 1990; kob: Buechner & Roth, 1974; Balmford & Turyaho, 1992) serve as focal points for congregations of both sexes. Beaches or grounds for hauling out among marine mammals provide both a stable platform for dependent young and access to mates for species that disperse widely and are unlikely to contact each other during the rest of the year (elephant seals: Le Boeuf, 1974; fur seals: Majluf, 1987). Aggregations of pods of whales (Whitehead *et al.*, 1991) and dolphins (Connor *et al.*, 1992) and large herds of elephants (Moss & Poole, 1983) may enhance the chances of finding mates from outside the small stable family unit, and thus reduce the search costs for species whose individuals are normally widely dispersed (Barnes, 1982).

If competition for access to mates occurs, then coalitions of individuals can be more effective in monopolising or controlling access (male lions: Packer *et al.*, 1988; cheetahs: Caro & Collins, 1987). Where numbers of coalition partners influence individual reproductive success (e.g. lions: Packer *et al.*, 1991), then larger coalitions will be advantageous and group sizes will increase.

5. *Thermoregulation*

Some species of mammals have difficulty maintaining body temperature in the face of environmental fluctuations. Specifically, small mammals such as bats (Trune & Slobodchikoff, 1976) and mole rats (Jarvis, 1981) may derive a thermal advantage from the body heat generated by others in close proximity. These animals tend to be relatively small in size and thus have a high ratio of surface area to body volume which leads to heat loss, and a correspondingly increased metabolic requirement. They also tend to roost or live in confined, enclosed areas such as burrows or caves which trap body heat. Other vulnerable individuals are small infants, which have a reduced capacity for thermoregulation due to the immaturity of their metabolism, compounded by the problems of small size. Mothers among some species of small mammals creche infants communally in a nest, as in the mouse lemur (Glatson, 1986), whilst the adults pursue a generally solitary lifestyle. A further advantage to communal roosting, as in bats or birds, or sharing a safe den site, burrows or nest, is the reduction of predator risk. In practice, separating these two effects of minimising environmental exposure and safety from predators is difficult.

6. *Access to helpers for infant-rearing*

Stable units among some species of birds, ranging from monogamous pairs (swans: Scott, 1988) to several males with a single female (dunnocks: Davies, 1990), through to the colonies of scrub jays (Woolfenden & Fitzpatrick, 1990), anis (Koford *et al.*, 1990) and bee-eaters (Emlen, 1990), provide helpers for sharing the extra energetic burdens of rearing infants. These extra costs have been estimated to be about four times the average metabolic requirements in some birds (Drent & Daan, 1980) and about 2–2.5 times the average for mammals (Oftedal, 1984). Thus any help a parent can obtain, either from its partner or from other individuals, will allow it either to rear more offspring within a single brood or litter, or to reproduce again earlier and thus produce a higher number of surviving offspring over its lifespan. Among the small-bodied callitrichid primates (marmosets and tamarins), females produce twins in each litter in contrast to the normal primate pattern of single births. In some species of tamarins, previous offspring remain in their mother's group for extended periods as non-reproductive helpers (Kleiman, 1985), or alternatively they may have two adult males in a group, both of whom will help carry and protect the twins (Terborgh & Goldizen, 1985). The form of sociality in these small primates appears to be strongly related to the need for helpers if infants are to be successfully reared. Among African elephants, nulliparous females within a matrilineal kin or family group also help protect and assist infants, and again, the number of allomothers affects calf survival (Lee, 1988). Similar effects have been observed in other mammals (social carnivores: MacDonald & Moehlman, 1982; jackals: Moehlman, 1979; mongooses: Rood, 1986; Rasa, 1989; coatis: Russell, 1983).

7. *Benefits of social transmission*

If the group consists of individuals of mixed ages, then information and skills as well as control of resources can be transmitted across generations. The acquired skills and experiences of the older individuals can be directly taught to younger animals, as in the case of chimpanzee tool technologies (Goodall, 1970; Boesch, 1991), or provided by example, as in information about the location of water and food during droughts among elephants (Moss, 1988). More common among mammals and birds is the transmission of critical foraging or breeding areas, such as territories, to the younger group members upon the death of the elders (acorn woodpeckers: Koenig & Stacey, 1990; foxes: Macdonald & Carr, 1989; gibbons: Leighton, 1987). The presence of many group members with a variety of skills, experiences and modes of behaviour is thought to enhance the learning of social skills during the period of development, contributing to the process of socialization and social cohesion (Altmann, 1980).

9.2.2 *Costs of grouping*

Each of the benefits mentioned above may have some associated costs. The extent of the costs, their distribution through the lifespan of the animal, and the possible behavioural strategies to minimise them can determine when groups will be stable, or when aggregations will be fluid and opportunistic. Some costs to being in groups are independent of the benefits to be derived. The costs of grouping can also be manipulated by individuals in those groups, providing the potential for some individuals to escape costs at the expense of others, who then must bear a disproportionate share. Under such conditions, some animals gain greater benefits through the imposition of extra costs on others. This internal cost–benefit analysis relates to the stability and structure of the group, and to the expression of sociality (see below).

1. *Predation*

While lone individuals may be less likely to be taken by a predator, a solitary animal can rely on escaping detection, through crypsis or through behaviour which minimises conspicuousness. A large group of animals is sometimes noisy, is more active and is easier to spot at a distance since it occupies a larger area; it is thus more obvious to predators. Predator detection of potential prey is enhanced when the prey exist in clumps.

Furthermore, while animals in groups may be able to share the time and energy costs of vigilance, they must still allocate some time to this activity, and this detracts from feeding time. Rather than feeding in a group, animals sometimes feed on foods in concealed places or at times when predators are less active, and minimise costs in this way, as is seen in the small, solitary nocturnal primates (potto: Charles-Dominique, 1977). Within a group, some individuals may spend less time being vigilant (ostrich: Bertram, 1980), or be less active in group defence (ground squirrels: Sherman, 1977), so that the burden of the costs of predator avoidance are not equally distributed among the group members.

2. *Competition for access to food*

Once animals are clumped in a group, individual food intake may be reduced. While individuals in the group may have gained information about the location or quality or food, that food must then be shared with all those exploiting it at the same time. Both direct 'contest' competition and indirect 'scramble' competition can potentially lead to differential intake among the group members (Isbell, 1991). When there is direct and immediate competition between animals for access to foods, some animals are likely to obtain more at the expense of others (hyaenas: Frank, 1986;

macaques: Dittus, 1979). Indirect competition, such as that related to the time costs of foraging, may increase as a larger area will need to be covered to satisfy the energy requirements of all the group members (geese: Prins *et al.*, 1980). Since individuals have different requirements due to body size, condition or reproductive status, and different energetic costs of locomotion, they may feed at different rates. Foraging in a co-ordinated body may reduce the intake of those who could move more rapidly, and cover a greater area, increasing intake relative to their nutritional requirements. The location of the individual within the foraging unit may determine its access to foods of different energetic or nutritional values. Peripheral animals can have reduced intake over central animals (fish: Major, 1978; monkeys: Wrangham, 1981), and peripheral animals may also be more exposed to predation.

3. *Access to mates*

While groupings enhance the likelihood of intersexual contact, they also increase the probability of intrasexual competition for access to mates (Halliday, 1981). Among many mammalian and avian species for which the costs of reproduction are higher for females than for males, females are the limiting sex in terms of offspring production (Trivers, 1972). Thus several males in association will be in direct competition for access to this limited resource, and this competition can have both immediate costs, in the form of time loss or energy depletion, and long-term mortality costs for the competitors. Bull elephants who have engaged in fights, and lost, are less likely to be reproductively active in the following year (Poole, 1989). Among mountain sheep, males who have mated have a higher mortality rate in the following winter (Geist, 1971). Some individuals are likely to be successful at the expense of others, producing asymmetries in reproductive success between individuals of the same sex. For females, direct intrasexual competition among group members similarly can alter the relative reproductive success of individuals, through the harassment of subordinates (baboons: Dunbar, 1989), or, at least potentially, through some females monopolising access to high quality males to the exclusion of other females (moorhens: Petrie, 1983b; elephants: Dublin, 1983).

4. *Disease transmission*

Many pathogens and parasites depend on dispersal from their host to another victim within a short period of time. Animals in close proximity facilitate the spread of disease and parasites (prairie dogs: Hoogland, 1979; humans: May & Anderson, 1987). These costs relate to increased mortality due to illness, and to condition loss which affects reproduction as well as long-term survival.

5. *Infanticide*

A further risk associated with group living is the killing of infants by other group members. A number of species show relatively high levels of infanticide (ground squirrels: Sherman, 1981; mice: Svare & Mann, 1981; langurs: Hrdy, 1974; gorillas: Stewart & Harcourt, 1987; lions: Packer *et al.*, 1988). Groups may form as defensive units to combat infanticide (see van Schaik & Dunbar, 1990), or they may increase the risks of infanticide by providing opportunities for the behaviour to occur. It is, however, a relatively rare occurrence, and may be a minor cost to group living rather than a major determinant of group structure, except in those cases where the costs of intrasexual competition have led to infanticide as a male reproductive strategy.

9.2.3 *Costs and benefits of maintaining groups*

The net result of a cost–benefit analysis of grouping appears to be an infinite range of solutions to the problems and advantages of social life, where no single factor can be paramount for all individuals of different ages and sexes. However, sociality can be viewed as the compromise between the demands of finding and defending food, minimising mortality through predation, and ensuring reproduction for each individual (Alexander, 1974). Once a group is formed, its maintenance depends on the balance attained between the conflicts over time, energy, survival and reproductive output for each participating individual. This truism provides little in the way of further understanding of the proximate mechanisms promoting sociality, and much of the current debate on the evolution of sociality centres on the conflicts between the costs to individuals of intra- versus intergroup competition for access to food or mates. For each cost or benefit presented above, there are behavioural mechanisms that produce a 'bias' in its distribution. Individuals can manipulate other group members so that they selectively accrue a greater benefit at a lesser cost, thus affecting the fitness of individuals differentially. Such differences between individuals will influence how long the group can persist, how cohesive it will be, and its composition.

As yet, there are few empirical means of examining the relative importance of the factors influencing the maintenance of social systems. Comparative studies of felids (Caro, 1989), canids (MacDonald, 1983; Moelhman, 1986), ungulates (Jarman, 1974), co-operatively breeding birds (Stacey & Koenig, 1990), and non-human primates (van Schaik, 1983; Dunbar, 1988) have pointed to the constraints of phylogeny, habitat and body size in determining modes of sociality, and have provided at least some data for suggesting appropriate tests (see van Schaik & Dunbar, 1990).

The maintenance of a group depends on factors such as its nature, whether it is stable or unstable; on the degree of structuring within it, on its age–sex class composition, the vagility (ability to disperse and distance covered during dispersal), and on the life history of the species. It is thus useful to attempt to categorise the kinds of groups observed before attempting to apply a cost–benefit analysis in order to explain their evolution.

9.3 Definitions of sociality

Functional definitions of sociality are difficult to determine for more than certain specific cases with present knowledge, but some levels of sociality can be defined primarily on the basis of the nature of the interactions between individuals. A continuum of interaction levels can thus be proposed, ranging from temporary associations and aggregations through to stable units, and to interactions between several such units within communities (Fig. 9.1). Each point along this continuum can result from different cost–benefit ratios in relation to the nature of the group, with a single underlying principle: individuals in association act to minimise costs while maximising benefits.

In this scheme, the different modes of sociality are characterised by the stability or persistence of the relationships between associating individuals. If the interactions at the basis of the relationships are non-existent or opportunistic, brief and transitory, then sociality is seldom present even though animals associate. If the interactions are differentiated and structured on the basis of the attributes of participants such as sex, age or power (physical size, strength, skill and motivation), are repeated over time between the same participants, and themselves have some content (are aggressive, affiliative, nurturing or sexual), then relationships result and sociality is present. This rather arbitrary exclusion of unstructured and temporary aggregations from inclusion as 'social' is not to diminish

Degree of association

Solitary
>   **Level 1:** Solitary for all activities
>   (sloth)

Gregarious species
Application of cost/benefit rules of thumb to formations of groups: groups
unstable in composition
>   **Level 2:** Aggregation for at least one activity
>   (seals)

>   **Level 3:** Aggregations for most activities
>   (Thompson's gazelle)

Social species
Complex rules with kinship, individual recognition, social maintenance through
interactions
>   **Level 4:** Aggregations for some activities
>   Stable associations for others
>   (African elephants, geese, Gelada baboons)
>   **Level 5:** Aggregations for some activities
>   Solitary for some
>   Stable associations for some
>   (bats)
>   **Level 6:** Stable associations for all activities
>   (macaques, guenons, wolves)
>   **Level 7:** Solitary activities embedded in stable associations
>   (lions, chimpanzees, ground squirrel)

Fig. 9.1. Levels of sociality, characterised by the degree of association between
individuals in different activities. The activities used are foraging, travel, sleep/rest,
rear infants and mate. Associations between mothers and offspring are not
included in the categories. An aggregation is defined as an association between
two or more individuals which is unlikely to be repeated through time. Examples
of species typically assigned to each level are given.

either the interest in these types of groups or their importance to the
participating individuals. Rather, it is to highlight the underlying principles
that transform temporary associations and their interactions into repeated,
ordered systems of interaction, and to distinguish between 'groups' and
'social groups'.

While it is relatively easy to provide examples of species exhibiting each
'level' of sociality, a more challenging task is to examine the relative
frequency of these options. An exhaustive survey of sociality in the animal
kingdom is beyond the scope of this chapter, but a number of speculations
can be made. Species that require a stable substrate for breeding or resting,
while foraging over relatively large areas, are predicted to fall into Level 2,

with unstable aggregations forming for sleeping, mating or infant-rearing purposes. Such species should be relatively frequent among those animals where there are physical limits to the habitats, as for example in the limited numbers or distribution of roosting trees or breeding cliffs among colonially breeding birds, and beaches or ice floes for pupping among the pinnipeds. Also in this group could be found those animals whose food supply alternates between extremely clumped and highly dispersed, with the result that aggregations form and break-up seasonally in relation to the food distribution (sparrows: Elgar, 1987; titmice: Grubb, 1987; minnows: Freeman & Grossman, 1992; hares: Monaghan & Metcalfe, 1985).

Species forced to move over large areas in response to changing food availability, and faced with high levels of predation, are predicted to form aggregations with variable membership for most activities (Level 3). Among mammals, such species should be relatively uncommon, since they may be large in body size in order to minimise the energy costs of migration and to face specific large-bodied predators. Large species tend to be less common than smaller species, and the combination of wide and fluctuating resource distribution with high predation pressure may be rare. It is interesting that many of the African ungulates tend to fall into this group, as a function of their large body size, foods which are seasonally productive and ephemeral, and digestive adaptations limiting food choice (see Estes, 1991; and below). Such species can be considered to be gregarious, interacting with constantly changing partners in relatively co-operative, but transitory ways. A useful distinction has been made between cases where animals aggregate temporarily for reasons associated with external pressures, such as predation or seasonal fluctuations in food availability, and where they associate through a mechanism of social attraction or social exchange (Freeman & Grossman, 1992). Where social attraction or exchange is likely, then a level of society above that of simple aggregation can be proposed.

When there is stability in associations and partners for interactions, marked sociality can exist. Of the four levels distinguished here, the most common should be Level 4. In this case, stable associations are maintained for some activities, while larger aggregations form for others. Thus, group size is variable in response to changing levels of resource availability, predation pressure and safe resting or breeding places. What underlies these groups is a stable core of associates or partners for some activities such as infant rearing. Avian species foraging in flocks and roosting communally, but with fixed partners for parenting, ungulate species which

fragment into small stable units and coalesce into migratory or mating units, depending on seasonality in resource distribution and quality, and primate species such as hamadryas baboons which forage as small units but coalesce in large bands to sleep in rare safe sites, are all examples. Level 5, with a mixture of aggregation, stable units within the aggregations and some solitary activities, may be uncommon, and limited to smaller species. In particular, feeding may be undertaken in a solitary mode, when resources are of high quality and dispersed, as in bats. The fixed membership, stable groups (Level 6) are also predicted to be relatively rare, and indeed may only be seen in the co-operatively hunting smaller carnivores and some primates. The final category (Level 7) of fission and fusion on the part of a stable group again can be predicted to be common, and dependent on the need to forage on dispersed foods in combination with opportunities to coalesce at larger food patches, and where mating relations tend to be confined to known partners within the community. In reality, such units are relatively rare and are confined to the extremes of the body size range among mammals, with few middle-sized species, and to the social insects such as honey bees.

The requirements for sociality in the continuum established above focus on persistence in association, discrimination between individuals involved in those associations, and in the nature of the resulting interactions (see also Wrangham & Rubenstein, 1986). The ability to recognise individuals, through vocal, visual or chemical cues, is fundamental to such stable social units. Some level of social maintenance is also critical, in that interactions are exchanged between these recognised individuals. As will be discussed below, the development of these attributes produces a number of species that can be termed social specialists, marked out by relatively large brains, high levels of communication, and abundant, diverse modes of social interaction. In addition, many such societies are based on some degree of genetic relatedness existing between individuals.

If interactions between individuals are a prerequisite for sociality, these can be split into those which are competitive and those which are co-operative. The reconciliation of individual needs in relation to competition and co-operation is the principle factor underlying the maintenance of sociality (Alexander, 1974). Three key factors affecting these basic types of interactions can be defined (Wrangham, 1982). These are *mutualism, kinship,* and *altruism.* The principles underlying these were presented in Chapter 7, and will be only generally stated here. Mutualism, where two participants both gain a fitness advantage from an interaction, is differentiated by Wrangham into two categories: non-interference mutualism, where the advantage is gained without either participant bearing a cost, and interference mutualism, where both gain but a third party suffers some costs as a result of the interaction.

The importance of kinship in the evolution of social behaviour has been stressed by many authors (see Chapter 7). Kinship underlies many of the stable systems of association, since the distribution of benefits is differentially allocated to relatives. However, if one individual bears a disproportionate share of the costs without any potential for avoiding or reducing those costs, then despotic systems can be produced even where co-operation is kin-biased (Vehrencamp, 1983). Such despotic systems may take the form of extreme hierarchical organisation, or of parental compulsion acting to produce specific behaviour among offspring (e.g. social insects). When the individual is willing to bear those costs, then its behaviour can be termed altruistic. The willingness to suffer costs can depend on the relatedness of the other (kin selection), or on whether interactions will be repeated over time (tit-for-tat models). As Trivers (1985) has pointed out, the relationship between costs and benefits to different participants can depend on the returns to both (Fig. 9.2), leading to systems of reciprocity.

Interference mutualism, tit-for-tat games, kin selection and reciprocity all have in common those elements that can be distinguished as social: they consist of interactions with a co-operative basis and associated costs, they occur repeatedly between the same individuals over time, and they tend to be stable in the ratio of costs and benefits between the participants over the short term. Furthermore, they include interactions within each sex as well as between the sexes, and can take into account characteristics of the individuals such as age or stage of development.

A final distinction can be made between sociality where each participant has some options in terms of interactions and responses to interactions, and those where options do not exist. Among the social insects, the fitness biases of different castes are fixed for all members of a caste. Individuals have no ability to alter their status and the very existence of sociality is obligate. Among most mammals or birds, if fitness biases become too extreme, the individual can disperse from the social group to avoid them (non-obligate sociality). If, however, as in the social insects, the costs of dispersal are higher (e.g. death) than those of remaining (e.g. no reproduction), then dispersal is unlikely. In contrast to the case of the social insects, for birds and mammals the costs of dispersal may not be fixed permanently, and thus there is always a chance for an individual to modify its status during its lifetime. These differences between obligate and

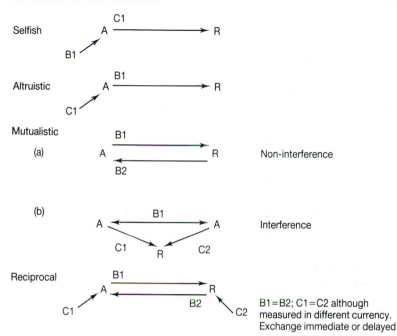

Fig. 9.2. The direction and exchange of intentions with the potential for affecting individual reproductive success. A = Actor; R = recipient; arrows note the direction of costs (C) and benefits (B) to the actors or recipients. (Modified from Trivers, 1985; Wrangham, 1982.)

non-obligate sociality are probably related to the body size, lifespan and brain size of the organism, demonstrating the importance of considering life history variables in relation to sociality.

9.4 Social systems

The ability to distinguish the participants in an interaction in terms of their individual characteristics and through the nature of those interactions provides a basis for descriptions of social systems (Hinde, 1976). At the heart of social systems are the interactions between individuals. Each individual has specific attributes, as does each interaction. A participant in an interaction may be male or female, old or young, of high rank or subordinate, and so on. Interactions can be defined in terms of their content, expressed as the form of the interaction and what elicits it, for example a copulation or a threat. The next stage is to specify the quality or intensity of that interaction – intense, moderate or mild – and its patterning through time – for example from infrequent to observed whenever the participants meet. When the same interaction, as character- ised by the categories above, is repeated between individuals, a relationship develops out of these interactions, which can also be defined by the same elements of content, quality and patterning. At levels of increasing social complexity, relationships themselves exist within a network of other interactions and other relationships, and are in turn modified as a result of the dynamic processes produced from such hierarchical interactions. A description of the relationships between individuals within a group provides knowledge as to the form of the social structure. The principles behind this approach appear complex and subjective, in that the ultimate description is only as good as the observations upon which it is built. However, in practice, the definitions of interactions and their subsequent assignment to relationships can be empirically tested and statistically validated (see Hinde, 1983, for a number of examples). The value of this approach is in its ability to reduce complexity to meaningful, recognisable, and quantifiable units.

The emphasis on relationships rather than on simple interactions has developed from long-term studies of a number of mammalian species, where the internal structure of the society has been identified. The importance of long-term studies lies in the ability to recognise the existence of relatedness between individuals emerging from dispersal and reproductive strategies and to assess the effects of kindship on the nature of the society that is observed (see Rubenstein & Wrangham, 1986; Clutton-Brock, 1988). The kinds of relationships known to exist include parental, sibling, sexual, affiliative, reconciliatory, and competitive (Hinde, 1983). Rather than simply categorising behaviour into finding mates, rearing offspring, gaining food and avoiding death, a relation- ship approach allows for the incorporation of individual character- istics, changes in reproductive status and age, and changes in group composition. These provide a deeper level of both description and explanation of variation in apparently similar societies. It has thus been possible to move away from simple descriptions of societies on the basis of male mating system (e.g. Crook & Gartlan, 1966), to a perspective that incorporates concepts such as role or alternative strategies (see Chapter 5) as well as determining the contribution of females to that structure. Although an approach based on defining relationships was initiated in studies of primates, it has been successfully extended to many non-primate species.

9.4.1 Social specialists

Among species with stable social units, the relationships between individuals can be used to define the internal structure of that society. Social structure is thus built up through individual actions in the context of co-operative and competitive relationships with others, and further by the relationships between relationships (Hinde, 1983). Each individual enters into an interaction with a history of such interactions at least partially dependent on its sex and age, and the sex and age of its partner. There are thus intrinsic differences between individuals (e.g. age, sex, strength, experience, capacity for learning, motivation, physiological capacities) and socially determined differences (such as reliability, predictability in exchange, the presence of relatives or allies). These intrinsic and socially influenced differences between individuals bias the outcome of interactions, strongly influencing those aspects of the social structure which are commonly examined.

One example of a social trait of abiding interest is that of systems of dominance. Fitness biases between individuals as a result of their status, taken to be dominant or subordinate, are firmly embedded in the literature, while the extent and indeed existence of such biases are also questioned. Hierarchical status per se is a property of interactions rather than individuals, and thus again can be related to proximate issues of relationships, although the experience of status, especially of extremes of rank, can influence the ways in which animals behave in a variety of interactions unrelated to dominance contests. Fish that lose fights are less active and have reduced levels of testosterone (Hannes *et al.*, 1984), while subordinate male baboons have higher levels of stress hormones and are in poorer physical condition (Sapolsky & Ray, 1989). Dominant rhesus monkeys are more playful and confident at exploring new environments (Stevenson-Hinde, 1983). Status, which can only be produced as a result of specific aggressive or avoidance interactions, can in turn be generalised to a wider range of relationships, and ultimately influences the nature of the social structure. However, the conflict between co-operation and individual needs within a social context can be resolved in many ways, of which competitive dominance systems may be only one solution. The growing recognition that dominance operates within a more complex system of simultaneous interactions between a number of participants (see Harcourt & de Waal, 1992) provides a deeper insight into the expression of and variation in social structure. We can thus examine in greater detail the mechanisms for conflict resolution in animal societies, the available alternatives to conflict, and variation over an individual's lifespan as a function of its changing physical and social requirements.

A further fundamental concern in the origin and expression of social systems is that of parental investment (Trivers, 1972; Clutton-Brock, 1991). When both the production and survival of offspring are dependent on co-operative relationships between individuals, then the need for assistance in infant rearing will promote persisting sociality. The original models (e.g. Maynard Smith, 1977) described the necessary conditions for both sexes to co-operate. Subsequent work has emphasised the potential for help to be gained from non-reproductive group members of either sex (see Stacey & Koenig, 1990) or as a result of reciprocal help among simultaneous breeders (Gittleman, 1985; Lee, 1989). If enhancing infant survival is a critical component in an individual's lifetime reproductive success, then a number of options are available for gaining this additional help. Which of these is taken depends on the capacity of the resource base to sustain groups, on litter size, and on the mortality risks to the infants (Promislow & Harvey, 1990; Packer *et al.*, 1992). Sociality which depends on helping relationships may be most common when infant mortality can be reduced through the presence or actions of others. If mortality is high and is caused by random environmental variation, no amount of help can be effective. If, however, the mortality is due to specific threats (e.g. food shortage, predators, disturbance), then the presence and actions of others can effectively minimise the consequences of those threats. As mentioned above, the presence of helpers may also improve the mother's energy balance such that she can care for her young more effectively. Except for the case of co-operatively breeding birds, few general models have yet been devised for predicting the form of sociality favoured by the costs of infant rearing.

One important effect of parental investment on sociality is that, once two or more individuals participate in infant care, and if the care is prolonged relative to lifespan, relationships between those individuals must be persistant and reliable. Most species with some form of communal care have low reproductive rates and overlap of generations (Gittleman, 1985). Thus stable, differentiated relationships can develop between care-giver and parent. Two primary social structures emerge. The first is that of monogamy, where contributions of both parents are critical to offspring survival. Monogamous systems have the potential to be poly-androus when additional energetic help is required and shared mating opportunities secure this help, as in dunnocks (Davies, 1990) or tamarin monkeys (Goldizen, 1987), or to use help from the previous year's offspring who remain as non-reproductive helpers in a large 'family'. Another common option is seen in many co-operative care systems which are based on female kinship (reliable, known social partners with shared interests) and where the mating system is of minor importance as there is little or no paternal help (Lee, 1989).

The social specialists tend to have a suite of traits which mark them out from the less socially complex species. Body size, and life-history traits correlated with size, follow the patterns described for classical 'K-strategists' (Stearns, 1976; Horn, 1978). These species are larger, with slower rates of development and reduced reproductive rates associated with prolonged, higher levels of parental care. Many have complex means of vocal or olfactory communication, allowing for individual recognition and discrimination over considerable distances. Communication between individuals about internal states and external events enhances decision making and takes the mode of interaction from the opportunistic and infrequent to the repeated, discriminatory and patterned events necessary for the existence of relationships (see Cheney & Seyfarth, 1990). Processing complex information about the social environment requires expanded brain size, particularly in those structures associated with integration and perception (Eisenberg, 1981; Western & Ssemakula, 1982; Dunbar, 1992). Neurological and endocrinological functioning of the brain are linked to the ability to solve the problems faced by social species of individual recognition, memory, assessment, prediction and learning. Thus behavioural tactics move from the level of stimulus and response to that of 'modelling' outcomes, and potentially forming cognitive theories as to the next acts of self and others (Cheney & Seyfarth, 1990). Behavioural sophistication, flexible responses and marked learning capacities over relatively long time periods are necessary for the development and maintenance of complex, long-term relationships, which characterise the social specialists.

9.5 Social options

While interactions and relationships vary depending on the internal characteristics of the society and of the individuals composing it, the social forms that can be expressed are in theory and practice limited. Once sociality has evolved as a solution to the problems of survival and reproduction, its form is constrained by the number of ways in which conflict between co-operation and competition can be resolved. The costs of reproduction among most mammals and at least some birds are

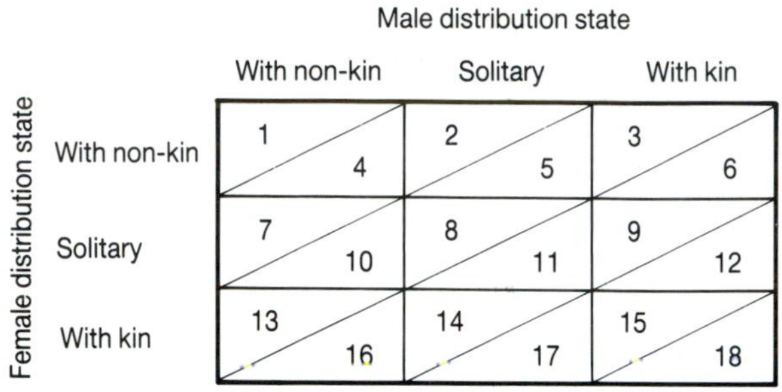

Male distribution state

Fig. 9.3. Core social states based on distributions defined for males and females in relation to same-sexed individuals. Associations between the sexes are indicated as either stable (above the diagonal) or transitory (below the diagonal). Each box is numbered to allow reference to the pathways between states (see Fig. 9.5). (Modified from Foley & Lee, 1989.)

imposed differentially on females due to their basic reproductive physiology. While in most cases, food limits the female's ability to produce young, males are limited primarily by access to females (Trivers, 1972; Emlen & Oring, 1977; Bradbury & Vehrencamp, 1977). As long as food supply affects the reproductive potential of females, the distribution of females in space will be influenced by the nature of that food supply (in terms of its quality, distribution and abundance), while that of the males is related to the presence and numbers of females relative to the number of other males (Wrangham, 1980, 1982).

Using these principles, it thus becomes possible to specify a limited number of ways in which females can be found in time and space. These female distributions then influence the limited number of ways in which males can be distributed relative to each other and to females (Foley & Lee, 1989). In Figure 9.3, three modes of 'distribution' for each sex are proposed, and for each potential 'distribution state' a distinction can be made between the existence of temporary or transitory relations, or alternatively, stable associations between males and females. In this model, there is theoretically a maximum of 17 different modes of sociality, if the category of both sexes 'solitary, with transitory relations between sexes' is excluded as non-social. Each of these different types of distribution is termed a 'social state', which is defined by the nature of the distributions of members of the same sex with each other on two dimensions and the associations between the sexes on the third dimension (Foley & Lee, 1989). The model is not limited to simple associations, but can be extended to distributions which are removed in time and space but nonetheless

(a) Primates

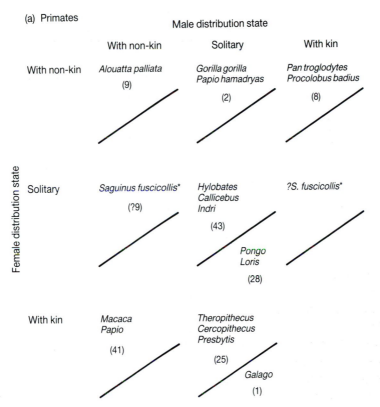

Fig. 9.4. Examples of species or genera of (a) primates, (b) African ungulates and (c) carnivores (excluding vivarids and mustelids) which fall into the social states defined in Figure 9.3. The number of species in each box is given to indicate the relative frequencies, and is presented as a rough guide. Species indicated with a (*) are known to have social states that shift as a function of ecology or (+) non-migratory phases (*continued*).

cognitively recognised by individuals, such as lineages among humans or associations between stable reproductive or foraging units into some form of more complex community. However, the basic principle remains: there is a finite number of options for sociality.

Figure 9.4 presents examples of species of primates (A), African ungulates (B) and carnivores (C) that can be categorised into the different social states. With sufficient information, it should prove possible to add a larger number of animal species to such a framework. It becomes clear that not all options for the different social states are viable; some are rare, and some, such as 'female kin' states, are relatively common. This leads us to ask how the social states that are observed have come to exist,

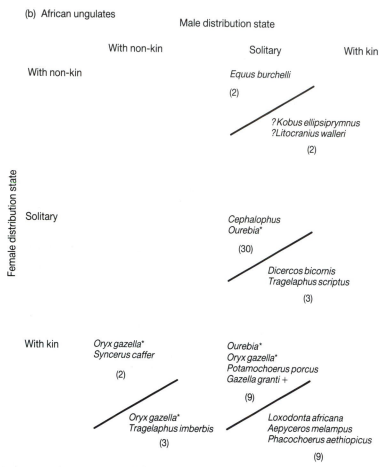

(b) African ungulates

Male distribution state

Fig. 9.4(b) (*continued*).

whether species can shift between states during evolution, and if so, how frequently can such shifts occur?

Among the primates, the vast majority of species fall into the distribution states of female kin and stable associations with the male (either one male or several non-kin males as a function of the size of the group) (Foley & Lee, 1989). In part, this is due to phylogenetic effects, in that most extant species are found among the superfamily of Old World monkeys, the Cercopithecoidea. These species radiated relatively recently compared to the apes and the New World monkeys or platyrrhines. Many species are frugivores, able to digest relatively less palatable and more toxic or poisonous unripe fruits. Such fruits tend to occur in patches large enough

(c) Carnivores

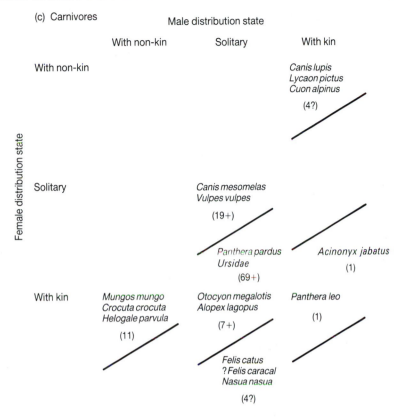

Fig. 9.4(c) (*continued*).

to support several females, who then group with kin to limit access by non-group members (see below). Among the specialist folivores (the Colobines), with fermentation chambers in the stomach, resources tend to be more evenly distributed and consequently more defensible. Female kin distribution states in colobines are slightly less common than in the fruit-eaters and may also depend on high levels of co-operative infant rearing. Among the platyrrhines, a greater variety of states is observed, but as most are frugivorous, female kin states are still common. The interesting exceptions are in the phylogenetically older and least numerous group of the apes. One factor unites the extant ape species: none has a society based on female kinship. These large-bodied, selective feeders on scarce and patchy ripe fruits fall into a number of distribution states, each with specific characteristics. Gibbons tend to be monogamous, while organgutans fall into the state defined as 'solitary' on the basis of their distribution and associations. For chimpanzees, females tend to forage in a relatively solitary state, embedded in an interactive non-kin group of other females, which is surrounded by a stable kin unit of males. Among gorillas, males have determined the female distribution state: in this case, unrelated females are attracted to associate with a dominant mature breeding male, the 'silverback'. This strategy seems most viable when foods are evenly distributed and of relatively low quality, making female resource competition infrequent, and where there is no competitive advantage to female resource defence. Among the apes, females fall into either solitary states or states with non-kin, and male reproductive strategies play a major role in determining the female distributions.

Among the African ungulates, female kin states are again the most common form of stable sociality (Estes, 1991). Several interesting observations can be made about these ungulates. Firstly, there are few truly solitary species; the very large rhino and the relatively small bushbuck are examples. The need for defence against predators in a terrestrial environment has limited the possible exploitation of this option. Secondly, a large number of species (about 20) as diverse as giraffe, kob, Grevy's zebra and eland, show no stable associations between either females or males. The group composition depends on the nature of the activity (e.g. feeding, mating, migrating), on the season and habitat, and is unstable over time. The frequency of occurrence of unstable aggregations is related to the highly seasonal nature of the distribution over relatively large areas. A final point with respect to the male distributions among ungulates is that there are few species in which males associate with other males. Ungulate males appear to be highly intolerant of others at times of mating, but will aggregate seasonally outside this period. This pattern is probably related to the short duration of the female's oestrus, drastically limiting the opportunities for mating during most months of the year, and thus reducing the advantages of stable associations between males and females while increasing the degree of male–male competition during brief ruts or oestrus. This pattern is in marked contrast to the primate patterns of stable male–female associations in combination with ovulatory cycling rather than oestrus.

Among the carnivores, sociality is itself rare. Most felid species tend not to associate either with members of the same sex or with the opposite sex (Caro, 1989). Among the canids, the majority of species are monogamous, with offspring residing with their parents for varying periods of time (Moehlman, 1986). Among mongooses, 21 species are solitary, five live in pairs, and eight in groups (Rood, 1986). Daughters are more likely to remain in the groups, although some offspring of both sexes can disperse or remain resident. Rood (1986) characterises the group-living mongoose species as insectivorous, living under conditions where prey renewal is rapid enough to sustain groups in a foraging area, but predation pressure is thought to maintain the groups. Of the bears, all are relatively solitary, while other small to medium-sized species such as badgers (Kruuk, 1978) occasionally form groups when offspring remain resident for long periods. The coati is one of the few carnivores for which female units, lacking stable associations with males, exist (Kaufmann, 1962; Russell, 1983). When sociality is present among carnivores, male-kin states are somewhat more common than among the primates or ungulates, since large litter membership may enhance the opportunities for such permanent associations between brothers. Female kin states are also observed, for example among the co-operative hunters where communal infant care and resource defence are critical.

The conditions for the existence of the different forms of sociality can be examined by returning to the general cost–benefit analyses discussed above. However, once the social state has been established, it is possible to explore in greater detail the specific ecological and life-history variables at the basis of that state. Variables such as body size, reproductive rates, mortality patterns, energetic requirements and the nature of the food, as well as the interactions between these factors, are fundamental *constraints* on social options. When and why species occupy different social states are determined by this interaction between ecology and life history, within the limited set of options available.

9.5.1 *Constraints on options*

Ecological constraints arise from the need to extract energy from finite sources. Differences in the nature of the energy supply, in space and over time, and in an individual's requirements for energy (for maintenance, growth, reproduction, and locomotion) are the first constraint. Categorisation of food supplies is a notoriously difficult process, and most models take as their basis distribution (e.g. dispersed, even, random, clumped), quality (high, low), availability (ease of location, productivity), and predictability (seasonal, unpredictable, constant) (see Krebs, 1978). Other factors, such as the risks of losing food to competitors, not finding additional foods, or the costs of switching between foods in terms of energetic returns for energy outlay, can also be assessed (at least in theory). But, why should these variables constrain social options?

If we start with the state of 'female solitary', this is predicted to occur when food patches are too small for more than one female to exploit them at the same time. The small prey taken by carnivores such as foxes and jackals (relative to their own body size and nutritional requirements) restrict opportunities for the formation of large groups. Large patches or prey items, which can sustain several females, promote female associations. In Wrangham's models (1980, 1982) this is referred to as interference mutualism, where the costs of reduced access are transferred to the excluded individuals as a result of co-operative defence of the patch. Since the reliability, consistency and predictability of partners acting together are enhanced, and shared costs of jointly exploiting a limited resource are reduced when the co-operating individuals are kin, the option of female-kin associations should arise. Indeed, since many foods for large-bodied species come in relatively large clumps, this option should be, and is, a frequent social state. 'Matrilineality' (where female kin remain together in the same group) is thus relatively common. In contrast, the 'female non-kin' state should only be observed when there are few or no advantages to co-operative defence (small, dispersed foods, low quality foods), and female groups form, if at all, in response to male ability to aggregate and control female distributions.

By establishing the ecological conditions for female states in relation to their energetic requirements for reproduction, the subsequent constraints on male options can be specified. If females are clumped, then males can either attempt to control that clump alone ('male solitary') or, if the female groups are large, may be forced into sharing mating access with others (either 'male non-kin' or 'male kin'). Shared access should actually promote the male kin state if competition for access to mates is to be minimised, but a further constraint can be added to the model in the form of the need for outbreeding. If the female state is kin, and the male state is kin, then matings between genetic relatives and the consequent reduction in fertility through inbreeding become more likely, especially when male tenure in the group is long relative to lifespan (Clutton-Brock, 1989). Although inbreeding should, at least potentially, be better than not breeding at all for an individual, the 'female kin/male kin' state remains sensitive to local population pressures. It will probably only be occupied under conditions of population decline, and should be easily invaded by unrelated, outbreeding males with higher reproductive success. Thus, when female kin associate, males tend to be unrelated to each other. However, related males can associate with units of related females when siblings of the same sex disperse from their group of birth together. This may be a rare state associated specifically with large and mixed sex litters. The switch between 'solitary male' to 'non-kin males' associating with female groups depends on the average patch size, and hence female group size. An analysis of uni-male–multi-male groups among primates suggested a threshold size of up to five females for groups that were consistently uni-male, whereas when group size increased to 10 or more females, several males were present (Andelman, 1986). Comparisons between populations suggest that these thresholds explain local demographic variations in the numbers of males present in a group within species as well (Dunbar, 1988).

When ecological factors are relatively unimportant in determining female associations, as under conditions of dispersed or small patches, or the even distribution of resources, then the need for co-operative infant rearing can become the critical influence on female associations. The mechanisms of kin selection again increase the likelihood of female kin states forming. Reciprocity in infant care, provisioning of young, and defence against predators are all important influences on the tendency for female kin to associate in order to promote success in infant rearing. If, on the other hand, predation or infanticide is the major influence on survival and reproductive success, then non-kin states also could be viable, especially where the role of males in reducing such risks is critical.

The associations between males and females are a function of the initial distribution of females, of the male's ability to control females or the resources necessary for females and infants, to enhance infant survival, and to minimise the energetic costs of infant rearing. Depending on the factors with the most influence on the lifetime reproductive success of each sex, a number of social states can be occupied. A detailed knowledge of the ecological conditions, of the importance of the three components of lifetime reproductive success (breeding lifespan, fecundity, and infant survival: Clutton-Brock, 1988), of dispersal prior to breeding, and the life-history variables of the species is necessary before defining which states are likely to be occupied.

The influence of life-history variables on social states within this model can be demonstrated clearly by the differences between carnivores and primates in their occupation of State 15 – male kin/female kin. While rare (or non-existent) among the primates, it is typical of lions (Packer *et al.*, 1991). Its availability as a viable option for lions may be due to the large litter size of lions compared to primates. Only when a cohort of brothers is present can kin of the same generation move together into a new pride. Among wild dogs, with litters potentially as large as 16, sibling cohorts of sisters potentially can disperse and form new groups with their same-sex kin, while males remain resident with their brothers (Frame *et al.*, 1979). The effect of large litter sizes in social evolution is further demonstrated in comparative studies of communal infant care (Packer *et al.*, 1992). A large litter size within small units of closely related females explains the majority of observations of non-offspring nursing in mammals.

The nutritional problems for ruminant ungulates, of digesting low quality foods in large enough amounts to sustain relatively large body sizes (Jarman, 1974; Demment & van Soest, 1985), suggest that female foraging needs will generally promote loose female-kin units for larger species, similar to the patterns observed among primates. However, the reliance on seasonally variable primary productivity (grass, herbs and leaves) may place constraints on the stability of female units in space, which then makes permanent bonding between males and females more problematic. Many of the ungulates fall into states with transitory associations between males and females, unlike the primates (Estes, 1991). The added effect of a restricted oestrus period limits further the advantages to males of stable male–female associations, and promotes male–male intolerance. As was originally proposed by Jarman (1974), the small-bodied selective feeders tend towards high quality foods in small ranges in a more solitary mode. The opportunity for monogamy develops with a stabilising of transitory associations between males and females. An interesting exception for this pattern appears to be the African buffalo, which forms stable female-kin units with several to many males associated with the group for varying lengths of time (see Estes, 1991). As large-bodied animals, able to sustain intake over large but stable areas, the need for female defence of large resource patches, and their vulnerability to large predators could cause group size to increase. With a large female group size in stable areas, more permanent associations between males and females may exist, and larger numbers of males will be associated with large female groups, again as in the primates (Andelman, 1986).

9.6 Social evolution

Each social state can be initially proposed as equally likely to exist in the simplest model for social evolution. Alternatively, the model can be used to assess changes in the nature and expression of social options during evolution. If it is specified that change between states (state shift) involves stepwise movement between adjacent states (Foley & Lee, 1989), then it becomes clear that each state can be reached by a number of pathways (Fig. 9.5). In this way, changes between states, the direction of the change and the probability of such change occurring can be modelled over evolutionary time. The point of interest is to determine the likelihood of the existence of different social states and the probabilities of changes between states in the light of conditions operating in the biological world.

The constraints on pathways are many: as mentioned above, the size of the animal, its life history and demography; energetic requirements and ecological conditions are critical. Furthermore, some social states may be more stable than others, while some are expressed only temporarily during the transition between more stable states. The probability distributions for stable and unstable social states have yet to be determined quantitatively. A final problem is that of phylogeny. Examination of Figure 9.4 suggests that once a state has been attained within a phylogenetic group, it can be maintained across a number of closely related species. One reason for presenting three taxonomic groups with diverse adaptations is to make apparent the effects of phylogeny on social evolution.

What, then, determines state shifts? The first factor is that of adjacency. In some models of social evolution, the initial shift is that where 'both sexes solitary' stabilises male–female associations to produce the classic monogamous pattern. Monogamy, while common among birds, is relatively rare among mammals (Kleiman, 1977; Goldizen, 1987). One reason for its rarity is that many species are either solitary (as in the carnivores) or are not yet known to form stable social units (as in the ungulates). If the definition of sociality presented above is used, and only the permanently social species are considered, then monogamy is relatively frequent among the three taxonomic groups examined (a total of 92 species). Female kin states are also relatively common (Pusey, 1987). Of the species examined, 113 appear to have some female kin structure to their social states. Male kin units are rare (14), as are the non-kin states (24). It is also interesting to note that stable distributions of one sex tend to promote a stabilising of male–female associations, with only a small number of species falling into the transitory male–female states. Indeed, some states, especially those with transitory relations between males and females, are not observed in the taxa considered here.

If it is assumed that social states arose from animals who were initially solitary, the movement between adjacent states can be used to model the different pathways for change in social states over time. A state shift from 'solitary' to 'female kin' is as probable as that to monogamy (female solitary/male solitary but stable male–female associations), with both

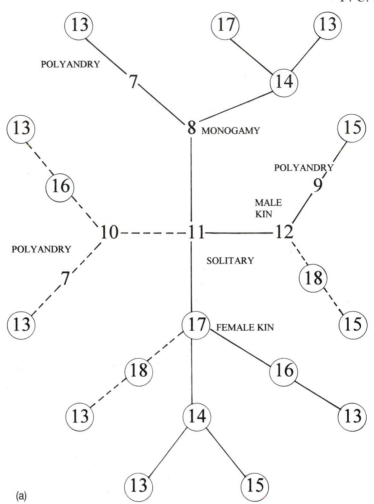

(a)

Fig. 9.5. Pathways of social evolution. All state shifts to (a) 'female kin' distributions (States 13, 14, 17 common; 15, 16 rare; 18 not observed) and (b) 'male kin' distributions (State 3 common; 9, 12, 15 rare; 6, 18 not observed) are illustrated, using as a starting state that of both sexes solitary (state 11). Numbers refer to the social states defined in Figure 9.3, and only a maximum of three state shifts is presented. Movement between states is constrained to shifts between adjacent states. Dark lines indicate pathways through states known to be stable, and dotted lines represent pathways through infrequent or non-occurring states (from Fig. 9.4). Circled states refer to those with a kin association.

(*continued*)

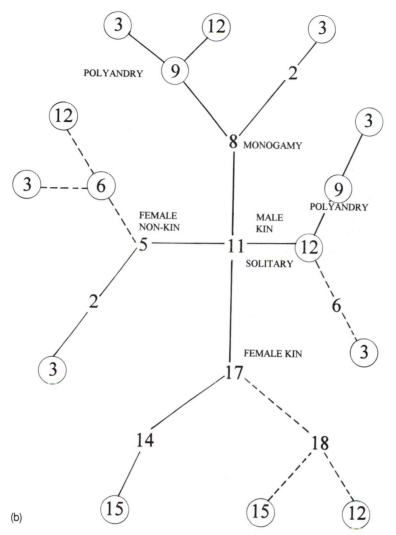

Fig. 9.5. (*continued*) (b) 'male kin' distributions.

involving a single state change as long as associations between males and females are transitory (Fig. 9.5a). In this way, the evolution of states such as those for elephants or coatis can occur. For the case of a shift to female kin associations out of monogamous families, as proposed by Rood (1986) for mongoose, the social route consists of two steps, where specific ecological and life-history variables must first fix the state of monogamy and then produce the next shift to female kin units. An equally possible alternative is that most likely among the primates, where the initial shift from solitary to female kin occurred in the context of female–female co-operation, with the second shift to stabilise male–female relations. The final shift between 'male solitary' and 'male non-kin' can be locally determined rather than typical of a species and requires: (a) an additional state change, and (b) a large female group size, as suggested above. Seven routes can be proposed for the shift from 'solitary' to 'female kin' distributions, all of which pass through states that are known to exist. While one route shifts through male kin and polyandry, the others all stem from either monogamy or other female kin distributions. Since the conditions operating to produce female kin distributions are more commonly observed than those to produce male kin states, it is to be expected that the female kin states are more frequent.

The male kin states are also a single step from the state of both sexes solitary, with the movement in the hypothetical evolutionary space away from female kin states (at least initially). What becomes evident is that there are few species where male kin co-reside in the absence of females. One exception is the cheetah (Caro, 1989), demonstrating the important effect of access to females on male distributions. However, at least seasonally, there can be male non-kin units not associated with females, as in kudu (Estes, 1991). How, then, can the male kin units associated with females arise? In terms of the model's parameters, the shift from 'solitary' to either 'female kin' or to 'male kin' states can be made in the same number of ways. However, the rarity of the 'male kin' state suggests that the route to male kin involves shifts through states that are either ecologically or physiologically unavailable (see Fig. 9.5b). Only five of the proposed routes are through occupied (or potentially viable) states (Foley & Lee, 1989), and three of these involve an initial shift to monogamy (see Fig. 9.5). When the conditions promoting monogamy result in greater state stability, it is only to be expected that subsequent shifts will be relatively rare and will involve specific ecological conditions or events. The routes through other states may be less likely since one is through polyandry, another through female non-kin groups in the absence of males, and the final one through female kin groups. However, rapid 'movement' through some unoccupied states, theoretically at least, should be possible. For example, a shift through polyandry might be possible, although the addition of more females would require a change in the allocation of parental investment to promote polyandry. Other shifts require more complex states, such as the stabilising of male–male relations throughout the breeding period.

As long as there is the potential for shifts between states at the level of a species, intraspecific variation should also be expected. There is no theoretical fixation at a specific taxonomic level in this model, and indeed, using the model's parameters, it should be possible to predict when and where intraspecific differences in social states will exist. Thus, within a species, it is possible for several states to be occupied, depending on local group dynamics and ecological conditions. For example, while the model state for saddle-back tamarins is that of polyandry, it is as yet unclear if the two males are kin or non-kin. Under conditions of large, rich food patches, polygynous groups have been observed, while monogamy is also frequent (Terborgh & Goldizen, 1985). A single group could, potentially, occupy several states over time, such as seasonally or, in the longer term, as a function of demographic changes and altering levels of food availability. The critical factor is which states will be stable under which ecological and demographic conditions, and how the transitions occur. A state shift from monogamy to polyandry is indeed a single transition between adjacent states, as is that from polyandry to 'female kin/male non-kin' (see Figs 9.3 & 9.5). Of interest here is that each state appears to be stable for specific conditions. The seasonal shifts of ungulates as a function of migrations or mating seasons is one example of short-term and local changes in state. An example of longer term changes can be seen in the shift from 'female non-kin' to 'female kin' seen in a group of mountain gorillas (Stewart & Harcourt, 1987). This appeared to be a demographic phenomenon resulting from the long tenure of the dominant silverback male, in combination with relatively few dispersal opportunities for females. Although this state change was effected in reality without the intervening steps (Fig. 9.5) proposed in the model, it was in fact an unstable state which broke down upon the death of the silverback (Harcourt, personal communication).

State shifts can also be proposed for oryx (Estes, 1991). Here, the female state of 'with kin' is maintained, but the distribution of males shifts from one male in stable associations with females, to several males (probably non-kin) in transitory association with females, and then a further shift to stabilise the male–female associations. The state shifts appear to be related to the oryx's general adaptation to an arid environment, where sparse, unpredictable and patchy resources limit the numbers of individuals which can associate. When, however, food and water are more reliable as a result of locally rich habitats, then the distributions of males and females can overlap and stabilise, imposed on the initial female stable state.

These examples highlight the differences in the model's predictions for state changes which become fixed and those which are a short-term response to specific and varying external conditions. The intention of the model is not to restrict a species to a single state. It allows for an examination of the potential for variation, and gives a clear rationale for when and why the variation exists. The point of the model is to establish theoretical limits to the options for sociality. The limits then allow for the determination of which costs and benefits are important for each specific social state. Rather than arguing at an all-encompassing level, with untestable competing hypotheses, only a limited number of factors emerge as conditions for each different state. The model specifies alternative routes to different stable states, which can be defined in relation to the life history or ecology of the animal, and the nature of the stable or unstable states can be examined. Finally, this perspective can be used to speculate about social states among extinct animals, incorporating knowledge of phylogeny and evolutionary change, since the total number of states available remains constant regardless of evolutionary time.

9.7 Conclusions

True sociality is surprisingly common in the animal kingom, and thus attracts attention for its diversity and patterning. Descriptions of sociality need to rely on a useful definition – how do we know when to call a species 'social'? The definition used here is relatively restricted, and relies on persistence in associations between individuals over time and space. Under such conditions, it then becomes possible to categorise a limited set of social options, to examine their frequency and distribution in different taxonomic groups, and to identify more precisely the costs and benefits involved in the establishment of each social option, rather than to speculate on these when the range of options is lumped into an arbitrary category of 'social'.

Having presented a reductionist model for sociality, it remains to be emphasised that, whatever category of sociality emerges as dominant for a particular species, the proximate mechanism for maintenance and change is that of repeated interactions between individuals. Thus, inter-actions, relationships and social structure are dynamic properties of individual participants embedded within their mode of society. The social processes influence the stability and potential for change in a social system. However, the initial expression of sociality and changes between social modes are both limited in the range of options available. The question of the evolution of sociality needs to address the proximate mechanisms determining social variation and its dynamics, as well as the constraints affecting those social options that can arise.

A number of factors influence the costs and benefits of sociality. As emphasised above, ecological conditions, life history and phylogeny all contribute to the observed patterns. Mating, rearing infants, finding food, and avoiding predators are the critical and complex tasks occurring in stable groups. The distribution of each activity through time and space and the conflicts of interest between these activities for each sex, for individuals of different ages, and in different environments, determine which social option is viable and limits others. Sociality and social variability are linked, but our knowledge is not yet sufficient to specify the nature of these links for all but a few well-studied species. The area thus remains one with great potential for further exploration.

References

Adams, E. S. (1990). Boundary disputes in the territorial ant *Azteca trigona*: effects of asymmetries in colony size. *Animal Behaviour*, **39**, 321–8.

Ahnesjo, I., Vincent, A., Alatalo, R., Halliday, T. R. & Sutherland, B. (1993). The role of females in influencing mating patterns. *Behavioral Ecology*, **4**, 187–9.

Alatalo, R. V., Carlson, A., Lundberg, A. & Ulfstrand, S. (1981). The conflict between male polygamy and female monogamy: the case of the pied flycatcher *Ficedula hypoleuca*. *American Naturalist*, **117**, 738–53.

Alcock, J. (1979). *Animal Behavior: An Evolutionary Approach*. Sunderland, Mass.: Sinauer Associates.

Alcock, J., Jones, C. E. & Buchmann, S. L. (1977). Male mating strategies in the bee *Centris pallida* Fox (Anthrophoridae: Hymenoptera). *American Naturalist*, **111**, 145–55.

Alexander, R. D. (1962a). The role of behavioral study in cricket classification. *Systematic Zoology*, **11**, 53–72.

Alexander, R. D. (1962b). Evolutionary change in cricket acoustical communication. *Evolution*, **16**, 433–67.

Alexander, R. D. (1974). The evolution of social behaviour. *Annual Review of Ecological Systems*, **5**, 325–83.

Altmann, S. A. (1979). Altruistic behaviour: the fallacy of kin deployment. *Animal Behaviour*, **27**, 958–9.

Altmann, J. (1980). *Baboon Mothers and Infants*. Chicago: University of Chicago Press.

Andelman, S. J. (1986). Ecological and social determinants of cercopithecine mating patterns. In *Ecological Aspects of Social Evolution*, ed. D. I. Rubinstein & R. W. Wrangham, pp. 201–16. Princeton, Princeton University Press.

Anderson, J. R. (1984). Monkeys with mirrors: some questions for primate psychology. *International Journal of Primatology*, **5**, 81–89.

Anderson, R. M. (1982). Coevolution of hosts and parasites. *Parasitology*, **85**, 411–26.

Andersson, M. (1982). Sexual selection, natural selection and quality advertisement. *Biological Journal of the Linnean Society*, **17**, 375–93.

Andersson, M. (1984). Brood parasitism within species. In *Producers and Scroungers*, ed. C. J. Barnard, pp. 195–228. London: Chapman & Hall.

Andersson, M. (1986). Evolution of condition-dependent sex ornaments and mating preferences: sexual selection based on viability differences. *Evolution*, **40**, 804–16.

Andrew, R. J. (1956). Intention movements of flight in certain passerines, and their use in systematics. *British Journal of Animal Behaviour*, **4**, 85–91.

Armstrong, E. (1985). Allometric considerations of the adult mammalian brain with special emphasis on primates. In *Size and Scaling in Primate Biology*, ed. W. L. Jungers, pp. 115–46. New York: Plenum Press.

Arnold, S. J. (1980). The microevolution of feeding behavior. In *Foraging Behavior: Ecological, Ethological and Psychological Perspectives*, ed. A. Kamil & T. Sargent, pp. 409–53. New York: Garland Press.

Arnold, S. J. (1981). Behavioral variation in natural populations. II. The inheritance of a feeding response in crosses between geographic races of the garter snake, *Thamnophis elegans*. *Evolution*, **35**, 510–15.

Arnold, S. J. & Wade, M. J. (1984a). On the measurement of natural and sexual selection: theory. *Evolution*, **38**, 709–19.

Arnold, S. J. & Wade, M. J. (1984b). On the measurement of natural and sexual selection: applications. *Evolution*, **38**, 720–34.

Ashlock, P. D. (1971). Monophyly and associated terms. *Systematic Zoology*, **20**, 63–9.

Ashlock, P. D. (1979). An evolutionary systematist's view of classification. *Systematic Zoology*, **28**, 441–50.

Atmar, W. (1991). On the role of males. *Animal Behaviour*, **41**, 195–205.

Atz, J. W. (1970). The application of the idea of homology to behavior. In *Development and Evolution of Behavior*, ed. L. R. Aronson, E. Tobach, D. S. Lehrman, &. J. S. Rosenblatt, pp. 53–74. San Francisco: W. H. Freeman.

Axelrod, R. (1984). *The Evolution of Cooperation*. New York: Basic Books.

Axelrod, R. & Hamilton, W. D. (1981). The evolution of cooperation. *Science*, **211**, 1390–96.

Bakker, T. C. M. (1986). Aggressiveness in sticklebacks (*Gasterosteus aculeatus* L.): a behavior-genetic study. *Behaviour*, **98**, 1–144.

Balmford, A. & Turyaho, M. (1992). Predation risk and lek-breeding in Uganda kob. *Animal Behaviour*, **44**, 117–28.

Barnard, C. J. (1990). Kin recognition: problems, prospects, and the evolution of discrimination systems. *Advances in the Study of Behavior*, **19**, 29–81.

Barnard, C. J. & Fitzsimons, J. (1989). Kin recognition and mate choice in mice: fitness consequences of mating with kin. *Animal Behaviour*, **38**, 35–40.

Barnes, R. F. W. (1982). Mate searching behaviour of elephant bulls in a semi-arid environment. *Animal Behaviour*, **30**, 1217–23.

Bart, J. & Tornes, A. (1989). Importance of monogamous male birds in determining reproductive success. Evidence for house wrens and a review of male-removal studies. *Behavioral Ecology and Sociobiology*, **24**, 109–16.

Barton, N. H. & Hewitt, G. M. (1989). Adaptation, speciation and hybrid zones. *Nature*, **341**, 497–503.

Barton, N. H. & Turelli, M. (1989). Evolutionary quantitative genetics: how little do we know? *Annual Review of Genetics*, **23**, 337–70.

Bartz, S. H. (1979). Evolution of eusociality in termites. *Proceedings of the National Academy of Sciences USA*, **76**, 5764–8.

Bateman, A. J. (1948). Intra-sexual selection in *Drosophila*. *Heredity*, **2**, 349–68.

Bateson, P. P. G. (1973). Internal influences on early learning in birds. In *Constraints on Learning*, ed. R. A. Hinde & J. Stevenson-Hinde, pp. 101–16. London: Academic Press.

Bateson, P. (1978). Sexual imprinting and optimal outbreeding. *Nature*, **273**, 659–60.

Bateson, P. (1982). Preferences for cousins in Japanese quail. *Nature*, **295**, 236–7.

Bateson, P. P. G. (1983a). Genes, environment and the development of behaviour. In *Animal Behaviour*, Vol. 3, *Genes, Development and Learning*, ed. T. R. Halliday & P. J. B. Slater, pp. 52–81. Oxford, Blackwell Scientific.

Bateson, P. P. G. (1983b). Optimal outbreeding. In *Mate Choice*, ed. P. P. G. Bateson, pp. 257–77. Cambridge: Cambridge University Press.

Bateson, P. (1988a). The active role of behaviour in evolution. In *Process and Metaphors in the New Evolutionary Paradigm*, ed. M.-W. Ho & S. Fox, pp. 191–207. Chichester: Wiley.

Bateson, P. (1988b). Preferences for close relations in Japanese quail. In *Acta XIX Congressus Internationalis Ornithologici*, ed. H. Ouellet, pp. 961–72. Ottawa: University of Ottawa Press.

Bauer, P. & Mandler, J. (1989). One thing follows another: effects of temporal structure on 1- and 2-year-olds' recall of events. *Developmental Psychology*, **25**, 197–206.

Bayliss, J. R. (1981). The evolution of parental care in fishes, with reference to Darwin's rule of male sexual selection. *Environmental Biology of Fishes*, **6**, 223–51.

Beauchamp, G. K., Yamazaki, K., Bard, J. & Boyse, E. A. (1988). Preweaning experience in the control of mating preferences by genes in the major histocompatibility complex of the mouse. *Behavior Genetics*, **18**, 537–48.

Beauchamp, G. K., Yamazaki, K. & Boyse, E. A. (1985). The chemosensory recognition of genetic individuality. *Scientific American*, **253**, 66–72.

Beck, B. B. (1980). *Animal Tool Behavior*. New York: Garland Press.

Beehler, B. M. & Foster, M. S. (1988). Hotshots, hotspots and female preferences in the organisation of mating systems. *American Naturalist*, **131**, 203–19.

Bell, G. (1982). *The Masterpiece of Nature: the Evolution and Genetics of Sexuality*. London: Croom Helm.

Bellman, H. (1988). *A Field Guide to the Grasshoppers and Crickets of Britain and Northern Europe*. London: Collins.

Bellman, R. (1957). *Dynamic Programming*. Princeton: Princeton University Press.

Bennet-Clark, H. C. & Ewing, A. W. (1969). Pulse interval as a critical parameter in the courtship song of *Drosophila melanogaster*. *Animal Behaviour*, **17**, 755–9.

Bercovitch, F. B. (1988). Coalitions, cooperation and reproductive tactics among adult male baboons. *Animal Behaviour*, **36**, 1198–209.

van den Berghe, E. P., Wernerus, F. & Warner, R. R. (1989). Female choice and the mating cost of peripheral males. *Animal Behaviour*, **38**, 875–84.

Bertram, B. C. R. (1978). Living in groups: predators and prey. In *Behavioural Ecology*, ed. J. R. Krebs & N. B. Davies, pp. 64–96. Oxford: Blackwell Scientific.

Bertram, B. C. R. (1980). Vigilance and group size in ostriches. *Animal Behaviour*, **28**, 278–86.

Binet, A. & Simon, T. H. (1915). *Method of Measuring the Development of the Intelligence of Young Children*. Chicago: Chicago Medical Book Co.

Birch, H. G. (1945). The relation of previous experience to insightful problem-solving. *Journal of Comparative and Physiological Psychology*, **38**, 367–83.

Birkhead, T. R. (1977). The effect of habitat and density on breeding success in common guillemots, *Uria aalge*. *Journal of Animal Ecology*, **46**, 751–64.

Birkhead, T. R. (1988). Behavioral aspects of sperm competition in birds. *Advances in the Study of Behavior*, **18**, 35–72.

Birkhead, T. R., Burke, T., Zann, R., Hunter, F. M. & Krupa, A. P. (1990). Extra-pair paternity and intraspecific brood parasitism in wild zebra finches *Taeniopygia guttata*, revealed by DNA fingerprinting. *Behavioral Ecology and Sociology*, **27**, 315–24.

Birkhead, T. R. & Møller, A. P. (1991). *Sperm Competition in Birds*. London: Academic Press.

Björklund, M. (1990). A phylogenetic interpretation of sexual dimorphism in body size and ornament in relation to mating system in birds. *Journal of Evolutionary Biology*, **3**, 171–83.

Blackburn, T. M. (1991). A comparative examination of life-span and fecundity in parasitoid Hymenoptera. *Journal of Animal Ecology*, **60**, 151–64.

Blaustein, A. R., Bekoff, M., Byers, J. A. & Daniels, T. J. (1991). Kin recognition in vertebrates: what do we really know about adaptive value? *Animal Behaviour*, **41**, 1079–83.

Blaustein, A. R. & O'Hara, R. K. (1986). An investigation of kin recognition in red-legged frog (*Rana aurora*) tadpoles. *Journal of Zoology* (*London*), **209**, 347–53.

Blaustein, A. R. & Waldman, B. (1992). Kin recognition in anuran amphibians. *Animal Behaviour*, **44**, 207–21.

Boesch, C. (1991). Teaching in wild chimpanzees. *Animal Behaviour*, **41**, 530–32.

Boesch, C. & Boesch, H. (1984). Mental map in wild chimpanzees: an analysis of hammer transports for nut cracking. *Primates*, **25**, 160–70.

Boyd, R. & Lorberbaum, J. P. (1987). No pure strategy is evolutionarily stable in the repeated Prisoner's Dilemma game. *Nature*, **327**, 58–9.

Boyd, R. & Richerson, P. J. (1989). The evolution of reciprocity in sizeable groups. *Journal of Theoretical Biology*, **132**, 337–56.

Bradbury, J. W. & Davies, N. B. (1987). Relative roles of intra- and intersexual selection. In *Sexual Selection: Testing the Alternatives*, ed. J. W. Bradbury & M. B. Andersson, pp. 143–63. Chichester: John Wiley & Sons.

Bradbury, J. W. & Gibson, R. (1983). Leks and mate choice. In *Mate Choice*, ed. P. P. G. Bateson, pp. 109–38. Cambridge: Cambridge University Press.

Bradbury, J. W., Gibson, R. M. & Tsai, I. M. (1986). Hotspots and the evolution of leks. *Animal Behaviour*, **34**, 1694–1709.

Bradbury, J. W. & Vehrencamp, S. L. (1977). Social organization and foraging in emballonurid rats. III. Mating systems. *Behavioral Ecology and Sociology*, **2**, 1–17.

Breder, C. M. & Rosen, D. E. (1966). *Modes of Reproduction in Fishes*. Neptune City: T. F. H. Publications.

Broadhurst, P. L. (1960). Experiments in psychogenetics: applications of biometrical genetics to the inheritance of behaviour. In *Experiments in Personality*, Vol. 1, *Psychogenetics and Psychopharmacology*, ed. H. J. Eysenck, pp. 1–102. London: Routledge.

Broadhurst, P. L. (1979). The experimental approach to behavioural evolution. In *Theoretical Advances in Behaviour Genetics*, ed. J. R. Royce & L. P. Moss, pp. 43–95. Alphen aan de Rijn: Sijthoff and Noordhoff.

Brooks, D. R. & McLennan, D. A. (1991). *Phylogeny, Ecology, and Behavior*. Chicago: University of Chicago Press.

Brower, L. P., Brower, J. V. Z. & Cranston, F. P. (1965). Courtship behavior of the queen butterfly. *Danaus gilippus berenice* (Cramer). *Zoologica*, **50**, 1–39.

Brown, C. R., Brown, M. B. & Shaffer, M. L. (1991). Food sharing signals among socially foraging cliff swallows. *Animal Behaviour*, **42**, 551–64.

Brown, J. L. (1983). Intersexual selection. *Nature*, **302**, 472.

Bull, J. J. (1981). Coevolution of haplo-diploidy and sex determination in the Hymenoptera. *Evolution*, **35**, 568–80.

Bull, J. J. (1983). *Evolution of Sex Determining Mechanisms*. Menlo Park, California: Benjamin Cummings.

Buechner, H. K. & Roth, H. D. (1974). The lek system in Uganda kob with special reference to behavior. *American Zoologist*, **14**, 145–62.

Burgess, J. W. (1976). Social spiders. *Scientific American*, **234**, 100–106.

Burghardt, G. M. & Gittleman, J. L. (1990). Comparative behavior and phylogenetic analysis: New wine, old bottles. In *Interpretation and Explanation in the Study of Animal Behavior*, ed. M. Bekoff & D. Jamieson, pp. 192–225. Boulder: Westview Press.

Burke, T., Davies, N. B., Bruford, M. W. & Hatchwell, B. J. (1989). Parental care and mating behaviour of polyandrous dunnocks *Prunella modularis* related to paternity by DNA fingerprinting. *Nature*, **338**, 249–51.

Burley, N. (1986). Sex-ratio manipulation in color-banded populations of zebra finches. *Evolution*, **40**, 1191–206.

Bush, G. (1974). The mechanism of sympatric host race formation in the true fruit flies (Tephritidae). In *Genetic Mechanisms of Speciation in Insects*, ed. M. J. D. White, pp. 3–23. Melbourne: Australia and New Zealand Book Co.

Butlin, R. K. (1987). Speciation by reinforcement. *Trends in Ecology and Evolution*, **2**, 8–13.

Butlin, R. K. & Ritchie, M. G. (1989). Genetic coupling in mate recognition systems: What is the evidence? *Biological Journal of the Linnean Society*, **37**, 237–46.

Butlin, R. K. & Ritchie, M. G. (1991). Variation in female mate preference across a grasshopper hybrid zone. *Journal of Evolutionary Biology*, **4**, 227–40.

Byrne, R. W. (1990). Having the imagination to suffer, and to avoid suffering. Commentary on M. S. Dawkins 'From an animal's point of view: consumer demand theory and animal welfare.' *The Behavioral and Brain Sciences*, **13**, 15–16.

Byrne, R. W. (1993). Do larger brains mean greater intelligence? *The Behavioral and Brain Sciences*, **16**, 696–7.

Byrne, R. W. (in press). Using anecdotes to distinguish psychological mechanisms in primate tactical deception. In *Anthropomorphism, Anecdotes, and Animals: The Emperor's New Clothes?*, ed. R. W. Mitchell & N. S. Thompson, Lincoln, Nebraska: University of Nebraska Press.

Byrne, R. W. & Byrne, J. M. E. (1991). Hand preference in the skilled gathering tasks of mountain gorillas (*Gorilla g. beringei*). *Cortex*, **27**, 521–46.

Byrne, R. W. & Byrne, J. M. E. (1993). The complex leaf-gathering skills of mountain gorillas (*Gorilla g. beringei*): variability and standardization. *American Journal of Primatology*, **31**, 241–61.

Byrne, R. W. & Whiten, A. (1988). *Machiavellian Intelligence: Social Expertise and the Evolution of Intellect in Monkeys, Apes and Humans*. Oxford: Clarendon Press.

Byrne, R. W. & Whiten, A. (1990). Tactical deception in primates: the 1990 database. *Primate Report*, **27**, 1–101.

Byrne, R. W. & Whiten, A. (1991). Computation and mindreading in primate tactical deception. In *Natural Theories of Mind*, ed. A. Whiten, pp. 127–41. Oxford: Basil Blackwell.

Byrne, R. W. & Whiten, A. (1992). Cognitive evolution in primates: evidence from tactical deception. *Man.* **27**, 609–27.

Cade, W. (1975). Acoustically orientating parasitoids: fly phonotaxis to cricket song. *Science*, **190**, 1312–13.

Cade, W. H. (1979). The evolution of alternative male reproductive strategies in field crickets. In *Sexual Selection and Reproductive Competition in Insects*, ed. M. Blum & N. A. Blum, pp. 343–79. London: Academic Press.

Cambefort, J. P. (1981). A comparative study of culturally transmitted patterns of feeding habits in the chacma baboon *Papio ursinus* and the vervet monkey *Cercopithecus aethiops*. *Folia primatologica*, **36**, 243–63.

Caro, T. M. (1989). Determinants of asociality in felids. In *Comparative Socioecology*, ed. V. Standen & R. Foley, pp. 41–74. Oxford: Blackwell Scientific.

Caro, T. M. & Collins, D. A. (1987). Male cheetah social organization and territoriality. *Ethology*, **74**, 52–64.

Caro, T. M. & Bateson, P. (1986). Organization and ontogeny of alternative tactics. *Animal Behaviour*, **34**, 1483–99.

Carson, H. L. (1982). Evolution of *Drosophila* on the newer Hawaiian volcanoes. *Heredity*, **48**, 3–25.

Carson, H. L. & Templeton, A. R. (1984). Genetic revolutions in relation to speciation phenomena: The founding of new populations. *Annual Review of Ecology and Systematics*, **15**, 97–131.

Cavalli-Sforza, L. L. Bodmer, W. F. (1971). *The Genetics of Human Populations*. San Francisco: W. H. Freeman.

Chapman, C. A. & Lefebvre, L. (1990). Manipulating foraging group size: spider monkey food calls at fruiting trees. *Animal Behaviour*, **39**, 891–6.

Charles-Dominique, P. (1977). *Ecology and Behaviour of Nocturnal Prosimians*. London: Duckworth.

Charnov, E. L., Los-den Hartogh, R. L., Jones, W. T. & van den Assem, J. (1981). Sex ratio evolution in a variable environment. *Nature*, **289**, 27–33.

Charnov, E. L. & Skinner, S. W. (1985). Complementary approaches to the understanding of parasitoid oviposition decisions. *Environmental Entomology*, **14**, 383–91.

Cheney, D. L. & Seyfarth, R. M. (1990). *How Monkeys See the World: Inside the Mind of Another Species*. Chicago: University of Chicago Press.

Chesler, P. (1969). Maternal influence in learning by observation in kittens. *Science*, **166**, 901–3.

Cheverud, J. M., Dow, M. M. & Leutenegger, W. (1985). The quantitative assessment of phylogenetic constraints in comparative analyses: sexual dimorphism in body weight among primates. *Evolution*, **39**, 1335–51.

Chowdhury, S. H., Corbet, P. S. & Harvey, I. F. (1989). Feeding and prey selection by larvae of *Enallagma cyathigerum* (Charpentier) (Zygoptera: Coenagrionidae) in relation to size and density of prey. *Odonatologica*, **18**, 1–13.

Claridge, M. F., Den Hollander, J. & Morgan, J. C. (1984). Specificity of acoustic signals and mate choice in the brown planthopper *Nilaparvata lugens*. *Entomologia experimentalis et applicata*, **35**, 221–6.

Claridge, M. F., Den Hollander, J. & Morgan, J. C. (1985). Variation in courtship signals and hybridization between geographically definable populations of the rice brown planthopper, *Niliparvata lugens* (Stal). *Biological Journal of the Linnean Society*, **24**, 35–49.

Clutton-Brock, T. H. (ed.) (1988). *Reproductive Success.* Chicago: University of Chicago Press.

Clutton-Brock, T. H. (1989). Female transfer, male tenure and inbreeding avoidance in social mammals. *Nature*, **337**, 70–72.

Clutton-Brock, T. H. (1991). *The Evolution of Parental Care.* Princeton: Princeton University Press.

Clutton-Brock, T. H., Albon, S. D. & Guinness, F. E. (1986). Great expectations: dominance, breeding success and offspring sex ratio in red deer. *Animal Behaviour*, **34**, 460–71.

Clutton-Brock, T. H. & Godfrey, C. (1991). Parental investment. In *Behavioural Ecology: An Evolutionary Approach*, 3rd edn, ed. J. R. Krebs & N. B. Davies, pp. 234–62. Oxford: Blackwell Scientific.

Clutton-Brock, T. H., Guinness, F. E. & Albon, S. D. (1982). *Red Deer: Behaviour and Ecology of Two Sexes.* Edinburgh: Edinburgh University Press.

Clutton-Brock, T. H. & Harvey, P. H. (1980). Primates, brains and ecology. *Journal of Zoology, London*, **190**, 309–23.

Clutton-Brock, T. H. & Harvey, P. H. (1984). Comparative approaches to investigating adaptation. In *Behavioural Ecology*, 2nd edn, ed. J. R. Krebs & N. B. Davies, pp. 7–29. Oxford: Blackwell Scientific.

Cockburn, A. (1991). *An Introduction to Evolutionary Ecology.* Oxford: Blackwell Scientific.

Connor, R. C., Smolker, R. A. & Richards, A. F. (1992). Dolphin alliances and coalitions. In *Coalitions and Alliances in Humans and Other Animals*, ed. A. H. Harcourt & F. B. M. de Waal, pp. 415–44. Oxford: Oxford University Press.

Convey, P. (1989). Influences on the choice between territorial and satellite behaviour in male *Libellula quadrimaculata* (Odonata: Libellulidae). *Behaviour*, **109**, 125–41.

Cook, R. M. & Cockrell, B. J. (1978). Predator ingestion rate and its bearing on feeding time and the theory of optimal diets. *Journal of Animal Ecology*, **47**, 529–47.

Cords, M. (1987). *Mixed-Species Association of Cercopithecus Monkeys in the Kakamega Forest, Kenya. University of California Publications in Zoology*, Vol. 117. Berkeley: University of California Press.

Coulson, J. C. (1966). The influence of the pair-bond and age on the breeding biology of the kittiwake gull *Rissa tridactyla. Journal of Animal Ecology*, **35**, 269–79.

Cox, C. R. & Le Boeuf, B. J. (1977). Female incitation of male competition: A mechanism of mate selection. *American Naturalist*, **111**, 317–35.

Coyne, J. A. & Orr, H. A. (1989). Patterns of speciation in *Drosophila. Evolution*, **43**, 362–81.

Coyne, J. A., Orr, H. A. & Futuyma, D. J. (1988). Do we need a new species concept? *Systematic Zoology*, **37**, 190–200.

Crews, D. (1987). Diversity and evolution of behavioral controlling mechanisms. In *Psychobiology of Reproductive Behavior*, ed. D. Crews, pp. 88–119. New Jersey: Prentice-Hall.

Crews, D. (1988). The problem with gender. *Psychobiology*, **16**, 321–34.

Crews, D. & Williams, E. E. (1977). Hormones, reproductive behaviour, and speciation. *American Zoologist*, **17**, 271–86.

Cronin, E. W. & Sherman, P. W. (1976). A resource-based mating system: the orange-rumped honeyguide. *Living Birds*, **15**, 5–32.

Crook, J. H. & Gartlan, J. S. (1966). Evolution of primate societies. *Nature*, **210**, 1200–3.

Cullen, J. M. (1959). Behaviour as a help in taxonomy. In *Function and Taxonomic Importance*, ed. A. J. Cain, pp. 131–40. London: The Systematics Association.

Darwin, C. (1859). *On the Origin of Species by Means of Natural Selection or the Preservation of Favoured Races in the Struggle for Life*. London: John Murray.

Darwin, C. (1871). *The Descent of Man and Selection in Relation to Sex*. London: John Murray.

Davies, N. B. (1978). Territorial defence in the speckled wood butterfly (*Pararge aegeria*): the resident always wins. *Animal Behaviour*, **26**, 138–47.

Davies, N. B. (1983). Polyandry, cloaca pecking and sperm competition in dunnocks. *Nature*, **302**, 334–6.

Davies, N. B. (1985). Cooperation and conflict among dunnocks. *Prunella modularis*, in a variable mating system. *Animal Behaviour*, **33**, 628–48.

Davies, N. B. (1986). Reproductive success of dunnocks. *Prunella modularis*, in a variable mating system. I. Factors influencing provisioning rate, nestling weight and fledgling success. *Journal of Animal Ecology*, **55**, 123–38.

Davies, N. B. (1989). Sexual conflict and the polygyny threshold. *Animal Behaviour*, **38**, 226–34.

Davies, N. B. (1990). Dunnocks: cooperation and conflict among males and females in a variable mating system. In *Cooperative Breeding in Birds*, ed. P. B. Stacey & W. D. Koenig, pp. 455–86. Cambridge: Cambridge University Press.

Davies, N. B. (1991). Mating systems. In *Behavioural Ecology: An Evolutionary Approach*, 3rd edn, ed. J. R. Krebs & N. B. Davies, pp. 263–94. Oxford: Blackwell Scientific.

Davies, N. B. (1992). *Dunnock Behaviour and Social Evolution*. Oxford: Oxford University Press.

Davies, N. B. & Houston, A. (1986). Reproductive success of dunnocks, *Prunella modularis*, in a variable mating system. II. Conflicts of interest among breeding adults. *Journal of Animal Ecology*, **55**, 139–54.

Davies, N. B. & Lundberg, A. (1984). Food distribution and a variable mating system in the dunnock, *Prunella modularis*. *Journal of Animal Ecology*, **53**, 895–912.

Davis, R. L. & Dauwalder, B. (1991). The *Drosophila* dunce locus. *Trends in Genetics*, **7**, 224–9.

Dawkins, M. S. (1986). *Unravelling Animal Behaviour*. Harlow: Longman.

Dawkins, R. (1976a). *The Selfish Gene*. Oxford: Oxford University Press.

Dawkins, R. (1976b). Hierarchical organisation: a candidate principle for ethology. In *Growing Points in Ethology*, ed. P. P. G. Bateson & R. A. Hinde, pp. 7–54. Cambridge: Cambridge University Press.

Dawkins, R. (1979). Twelve misunderstandings of kin selection. *Zeitschrift für Tierpsychologie*, **51**, 184–200.

Dawkins, R. (1982). *The Extended Phenotype*. Oxford: W. H. Freeman.

Dawkins, R. & Carlisle, T. R. (1976). Parental investment, mate desertion and a fallacy. *Nature*, **262**, 131–3.

Dawson, B. V. & Foss, B. M. (1965). Observational learning in budgerigars. *Animal Behaviour*, **13**, 470–74.

Dawson, W. D., Lake, C. E. & Schumpert, S. S. (1988). Inheritance of burrow building in *Peromyscus. Behavior Genetics*, **18**, 371–82.

Deacon, T. W. (1990). Fallacies of progression in theories of brain-size evolution. *International Journal of Primatology*, **11**, 193–236.

de Belle, J. S., Hilliker, A. J. & Sokolowski, M. B. (1989). Genetic localization of foraging (*for*): a major gene for larval behavior in *Drosophila melanogaster. Genetics*, **123**, 157–68.

de Belle, J. S. & Sokolowski, M. B. (1987). Heredity of rover/sitter: alternative foraging strategies of *Drosophila melanogaster* larvae. *Heredity*, **59**, 73–83.

Demment, M. W. & van Soest, P. J. (1985). A nutritional explanation for body size patterns of ruminant and nonruminant herbivores. *American Naturalist*, **125**, 641–72.

Denenberg, V. H., Hudgens, G. A. & Zarrow, M. X. (1964). Mice reared with rats: modification of behavior by early experience with another species. *Science*, **143**, 380–81.

Dickinson, A. (1980). *Contemporary Animal Learning Theory*. Cambridge: Cambridge University Press.

Diehl, S. R. & Bush, G. L. (1989). The role of habitat preference in adaptation and speciation. In *Speciation and its Consequences*, ed. D. Otte & J. A. Endler, pp. 345–65. Sunderland, Mass.: Sinauer Associates.

Dingle, H. (1981). Geographic variation and behavioral flexibility in milkweed bug life histories. In *Insect Life History Patterns: Habitat and Geographical Variation*, ed. R. F. Denno & H. Dingle, pp. 57–73. New York: Springer-Verlag.

Dittus, W. P. J. (1979). The evolution of behaviours regulating density and age-specific sex ratios in a primate population. *Behaviour*, **69**, 265–302.

Dobzhansky, T. (1951). *Genetics and the Origin of Species*, 3rd edn. New York: Columbia University Press.

Downhower, J. F., Brown, J. L., Pedersen, R. & Staples, G. (1983). Sexual selection and sexual dimorphism in mottle sculpins. *Evolution*, **37**, 96–103.

Dowsett-Lemaire, F. (1979). The imitative range of the song of the marsh warbler, *Acrocephalus palustris*, with special reference to imitations of African birds. *Ibis*, **121**, 453–68.

Drent, R. H. & Daan, S. (1980). The prudent parent: Energetic adjustments in avian breeding. *Ardea*, **68**, 225–52.

Dublin, H. T. (1983). Cooperation and reproductive competition among female African elephants. In *Social Behaviour of Female Vertebrates*, ed. S. K. Wasser, pp. 291–313. London: Academic Press.

Dudai, Y. (1988). Neurogenetic dissection of learning and short-term memory in *Drosophila. Annual Review of Neurosciences*, **11**, 537–63.

Dugatkin, L. A. (1988). Do guppies play TIT FOR TAT during predator inspection visits? *Behavioral Ecology and Sociobiology*, **25**, 395–9.

Dugatkin, L. A. (1991). Dynamics of the TIT FOR TAT strategy during predator inspection visits in the guppy (*Poecilia reticulata*). *Behavioral Ecology and Sociology*, **29**, 127–32.

Dugatkin, L. A. & Alfieri, M. (1991). Guppies and the TIT FOR TAT strategy: preference based on past interaction. *Behavioral Ecology and Sociology*, **28**, 243–6.

Dunbar, R. I. M. (1988). *Primate Social Systems*. London: Croom Helm.

Dunbar, R. I. M. (1989). Reproductive strategies of female gelada baboons. In *Sociology of Sexual and Reproductive Strategies*. ed. A. E. Rasa, C. Vogel & E. Voland, pp. 74–92. London: Chapman & Hall.

Dunbar, R. I. M. (1992). Neocortex size as a constraint on group size in primates. *Journal of Human Evolution*, **20**, 469–93.

Duncan, J. R. & Bird, D. M. (1989). The influence of relatedness and display effort on the mate choice of captive female American kestrels. *Animal Behaviour*, **37**, 112–17.

Dunn, J. A. (1959). The biology of the lettuce root aphid. *Annals of Applied Biology*, **47**, 475–91.

Eberhard, W. G. (1990). Inadvertent machismo? *Trends in Ecology and Evolution*, **5**, 263.

Edwards, S. V. & Naeem, S. (1993). The phylogenetic component of cooperative breeding in perching birds. *American Naturalist*, **141**, 754–89.

Egeland, J. A., Gerhard, D. S., Pauls, D. L., Sussex, J. N., Kidd, K. K., Allen, C. L., Hostetter, A. M. & Housman, D. E. (1987). Bipolar affective disorders linked to DNA markers on chromosome 11. *Nature*, **325**, 783–7.

Egid, K. & Brown, J. L. (1989). The major histocompatibility complex and female mating preferences in mice. *Animal Behaviour*, **38**, 548–9.

Ehrman, L. & Parsons, P. A. (1981). *Behavior Genetics and Evolution*. New York: McGraw-Hill.

Eibl-Eibesfeldt, I. (1975). *Ethology: The Biology of Behavior*. New York: Holt, Rinehart and Winston.

Eisenberg, J. E. (1981). *The Mammalian Radiations*. Chicago: Chicago University Press.

Elgar, M. A. (1987). Food intake rate and resource availability: flocking decisions in house sparrows. *Animal Behaviour*, **35**, 1168–76.

Elsner, N. & Popov, A. V. (1978). Neuroethology of acoustic communication. *Advances in Insect Physiology*, **13**, 229–335.

Elwood, R. W. & Ostermeyer, M. C. (1984). Does copulation inhibit infanticide in male rodents? *Animal Behaviour*, **32**, 293–4.

Emerson, A. E. (1938). Termite nests – a study of the phylogeny of behavior. *Ecological Monographs*, **8**, 247–84.

Emlen, S. T. (1970). Celestial rotation: its importance in the development of migratory orientation. *Science*, **170**, 1198–202.

Emlen, S. T. (1990). White fronted bee-eaters: helping in a colonially nesting species. In *Cooperative Breeding in Birds*, ed. P. B. Stacey & W. D. Koenig, pp. 487–526. Cambridge: Cambridge University Press.

Emlen, S. T. & Oring, L. W. (1977). Ecology, sexual selection, and the evolution of mating systems. *Science*, **197**, 215–23.

Emlen, S. T. & Wrege, P. H. (1988). The role of kinship in helping decisions among white-fronted bee-eaters. *Behavioral Ecology and Sociobiology*, **23**, 305–15.

Endler, J. A. (1986). *Natural Selection in the Wild*. Princeton: Princeton University Press.

Endler, J. A. (1992). Signals, signal conditions, and the direction of evolution. *American Naturalist*, **139**, s125–53.

Estes, R. D. (1991). *The Behavior Guide to African Mammals*. Berkeley: University of California Press.

Estes, R. D. & Estes, R. K. (1979). The birth and survival of wildebeest calves. *Zeitschrift für Tierpsychologie*, **50**, 45–95.

Ewald, P. W. & Rohwer, S. (1982). Effects of supplemental feeding on timing of breeding, clutch size and polygyny in red-winged blackbirds *Agelaius phoeniceus*. *Journal of Animal Ecology*, **51**, 429–50.

Fairbairn, D. J. & Roff, D. A. (1990). Genetic correlations among traits determining migratory tendency in the sand cricket, *Gryllus firmus*. *Evolution*, **44**, 1787–95.

Falconer, D. S. (1989). *Introduction to Quantitative Genetics*, 3rd edn. Harlow: Longman.

Felsenstein, J. (1981). Skepticism towards Santa Rosalia or why are there so few kinds of animals? *Evolution*, **35**, 124–38.

Felsenstein, J. (1985). Phylogenies and the comparative method. *American Naturalist*, **125**, 1–15.

Ferguson, M. W. J. & Joanen, T. (1983). Temperature-dependent sex determination in *Alligator mississippiensis*. *Journal of Zoology, London*, **200**, 143–77.

Fischer, E. A. (1981). Sexual allocation in a simultaneously hermaphroditic reef fish. *American Naturalist*, **117**, 64–82.

Fisher, J. & Hinde, R. A. (1949). The opening of milk bottles by birds. *British Birds*, **42**, 347–57.

Fisher, R. A. (1930). *The Genetical Theory of Natural Selection*. Oxford: Clarendon Press.

FitzGibbon, C. D. (1990). Why do hunting cheetahs prefer male gazelles? *Animal Behaviour*, **40**, 837–45.

FitzGibbon, C. D. & Fanshawe, J. H. (1989). The condition and age of Thomson's gazelles killed by cheetahs and wild dogs. *Journal of Zoology, London*, **218**, 99–107.

Foley, R. A. & Lee, P. C. (1989). Finite social space, evolutionary pathways, and reconstructing hominid behaviour. *Science*, **243**, 901–6.

Forsyth, A. & Montgomerie, R. D. (1987). Alternative reproductive tactics in the territorial damselfly Calopteryx maculata: Sneaking by older males. *Behavioral Ecology and Sociobiology*, **21**, 73–81.

Fouts, R. S., Fouts, D. H. & Van Cantfort, T. E. (1989). The infant Loulis learns signs from cross fostered chimpanzees. In *Teaching Sign Language to Chimpanzees*, ed. R. A. Gardner, B. T. Gardner & T. E. Van Cantfort, pp. 280–92. New York: State University of New York Press.

Frame, L. H., Malcolm, J. R., Frame, G. W. & Van Lawick, H. (1979). Social organization of African wild dogs (*Lycaon pictus*) on the Serengeti plains, Tanzania. *Zeitschrift für Tierpsychologie*, **50**, 225–49.

Frank, L. G. (1986). Social organization of the spotted hyena (*Crocuta crocuta*). II: Dominance and reproduction. *Animal Behaviour*, **34**, 1500–09.

Freeman, M. C. & Grossman, G. D. (1992). Group foraging by a stream minnow: shoals or aggregations. *Animal Behaviour*, **44**, 393–404.

von Frisch, K. (1967). *The Dance Language and Orientation of Bees*. Cambridge, Mass.: Belknap Press.

Funk, V. A. & Brooks, D. R. (1990). Phylogenetic systematics as the basis of comparative biology. *Smithsonian Contributions to Botany*, **73**, 1–45.

Futuyma, D. J. (1986). *Evolutionary Biology*. Sunderland, Mass.: Sinauer Associates.

Futuyma, D. J. (1992). History and evolutionary processes. In *History and Evolution*, ed. M. H. Nitecki & D. V. Nitecki, pp. 103–30. Albany: State University of New York Press.

Galef, B. G. (1988). Imitation in animals: History, definition and interpretation of data from the psychological laboratory. In *Comparative Social Learning*, ed. T. Zentall & B. G. Galef Jr, pp. 3–28. Hillsdale, NJ: Lawrence Erlbaum.

Galef, B. G. (1991). Information centres of Norway rats: sites for information exchange and information parasitism. *Animal Behaviour*, **41**, 295–301.

Galef, B. G., Manzig, L. A. & Field, R. M. (1986). Imitation learning in budgerigars: Dawson and Foss (1965) revisited. *Behavioral Processes*, **13**, 191–202.

Gallup, G. G., Jr (1970). Chimpanzees: self-recognition. *Science*, **167**, 86–7.

Gallup, G. G., Jr (1982). Self-awareness and the emergence of mind in primates. *American Journal of Primatology*, **2**, 237–48.

Garcia, J., Ervin, F. R. & Koelling, R. A. (1966). Learning with prolonged delay of reinforcement. *Psychonomic Science*, **5**, 121–2.

Garcia, J. & Koelling, R. A. (1966). Relation of cue to consequence in avoidance learning. *Psychonomic Science*, **4**, 123–4.

Garland, T. Jr., Huey, R. B. & Bennett, A. F. (1991). Phylogeny and coadaptation of thermal physiology in lizards: a reanalysis. *Evolution*, **45**, 1969–75.

Geist, V. (1971). *Mountain Sheep*. Chicago: Chicago University Press.

Ghiselin, M. T. (1969). The evolution of hermaphroditism among animals. *Quarterly Review of Biology*, **44**, 189–208.

Ghiselin, M. T. (1974). *The Economy of Nature and the Evolution of Sex*. Berkeley, Calif.: University of California Press.

Gibson, R. M. & Bradbury, J. W. (1985). Sexual selection in the lekking sage grouse: phenotypic correlates of male mating success. *Behavioral Ecology and Sociobiology*, **18**, 117–23.

Gilder, P. M. & Slater, P. J. B. (1978). Interest of mice in conspecific male odours is influenced by degree of kinship. *Nature*, **274**, 364–5.

Gittleman, J. L. (1981). The phylogeny of parental care in fishes. *Animal Behaviour*, **29**, 936–41.

Gittleman, J. L. (1985). Functions of communal care in mammals. In *Evolution: Essays in Honour of John Maynard Smith*, ed. P. J. Greenwood, P. H. Harvey & M. Slatkin, pp. 187–205. Cambridge: Cambridge University Press.

Gittleman, J. L. (1986), Carnivore brain size, behavioral ecology, and phylogeny. *Journal of Mammalogy*, **67**, 540–54.

Gittleman, J. L. (1989). The comparative approach in ethology: Aims and limitations. In *Perspectives in Ethology*, Vol. 8, ed. P. P. G. Bateson & P. H. Klopfer, pp. 55–83. New York: Plenum Press.

Gittleman, J. L. (1991). Carnivore olfactory bulb size: allometry, phylogeny, and ecology. *Journal of Zoology*, **225**, 253–72.

Gittleman, J. L. (1993). Carnivore life histories: a reanalysis in light of new models. In *Mammals as Predators*, ed. N. Dunstone & M. Gorman, pp. 65–86. Oxford: Oxford University Press.

Gittleman, J. L. & Kot, M. (1990). Adaptation: statistics and a null model for estimating phylogenetic effects. *Systematic Zoology*, **39**, 227–41.

Gittleman, J. L. & Luh, H.-K. (1992). On comparing comparative methods. *Annual Review of Ecology and Systematics*, **23**, 383–404.

Gittleman, J. L. & Luh, H.-K. (1993), Phylogeny, evolutionary models, and comparative methods: a simulation study. In *Pattern and Process: Phylogenetic Approaches to Ecological Problems*, ed. P. Eggleton & R. Vane-Wright. London: Academic Press.

Gladikas, B. (1979). Orangutan adaptation at Tunjung Puting Reserve: mating and ecology. In *The Great Apes*, ed. D. Hamburg & E. McCown, pp. 194–223. Menlo Park: Benjamin Cummings.

Gladikas, B. (1985). Orangutan sociality at Tunjung Puting. *American Journal of Primatology*, **9**, 101–19.

Glatson, A. (1986). Communal nesting in mouse lemurs. In *Primate Ontogeny*,

Cognition and Social Behaviour, ed. J. G. Else & P. C. Lee, pp. 355–62. Cambridge: Cambridge University Press.

Goldizen, A. W. (1987). Tamarins and marmosets: communal care of offspring. In *Primate Societies*, ed. B. Smuts, D. Cheney, R. Seyfarth, R. Wrangham & T. Stuhsaker, pp. 34–43. Chicago: University of Chicago Press.

Gomez, J. C. (1991). Visual behaviour as a window for reading the mind of others in primates. In *Natural Theories of Mind: Evolution, Development and Simulation of Everyday Mindreading*, ed. A. Whiten, pp. 195–207. Oxford: Basil Blackwell.

Goodall, J. (1970). Tool using in primates and other vertebrates. *Advances in the Study of Behaviour*, **3**, 195–249.

Goodall, J. (1986). *The Chimpanzees of Gombe: Patterns of Behavior*. Cambridge, Mass.: Harvard University Press.

Gorman, M. L. & Trowbridge, B. J. (1989). The role of odor in the social lives of carnivores. In *Carnivore Behavior, Ecology, and Evolution*, ed. J. L. Gittleman, pp. 57–88. London: Chapman & Hall.

Gosling, L. M. (1986). The evolution of male mating strategies in antelopes. In *Ecological Aspects of Social Evolution*, ed. D. I. Rubenstein & R. W. Wrangham, pp. 241–81. Princeton: Princeton University Press.

Gosling, L. M. & Petrie, M. (1990). Lekking in topi: a consequence of satellite behaviour by small males at hot spots. *Animal Behaviour*, **40**, 272–87.

Gould, J. L. & Marler, P. (1984). Ethology and the natural history of learning. In *The Biology of Learning* (Report of Dahlem Workshop on the Biology of Learning, Berlin 1983, October 23–28), ed. P. Marler & H. S. Terrace, pp. 47–74. Berlin: Springer-Verlag.

Gould, S. J. (1986). Evolution and the triumph of homology, or why history matters. *American Scientist*, **74**, 60–69.

Gould, S. J. & Lewontin, R. C. (1979). The spandrels of San Marco and the Panglossian paradigm: a critique of the adaptationist programme. *Proceedings of the Royal Society of London* B, **205**, 581–98.

Gould, S. J. & Vrba, E. S. (1982). Exaptation – a missing term in the science of form. *Paleobiology*, **8**, 4–15.

Grafen, A. (1980). Opportunity costs, benefit and degree of relatedness. *Animal Behaviour*, **28**, 967–8.

Grafen, A. (1982). How not to measure inclusive fitness. *Nature*, **274**, 364–5.

Grafen, A. (1989). The phylogenetic regression. *Philosophical Transactions of the Royal Society of London*, **326**, 119–57.

Grafen, A. (1990). Do animals really recognise kin? *Animal Behaviour*, **39**, 42–54.

Graves, J. A., Hay, R. T., Scallan, M. & Rowe, S. (1992). Extra-pair paternity in the shag, *Phalacrocorax aristotelis*, as determined by DNA fingerprinting. *Journal of Zoology* (*London*), **226**, 399–408.

Green, S. (1975). Dialects in Japanese monkeys: vocal learning and cultural transmission of locale-specific vocal behaviour? *Zeitschrift für Tierpsychologie*, **38**, 304–14.

Greenacre, M., Ritchie, M. G., Byrne, B. C. & Kyriacou, C. P. (1993). Female song preference and the period gene of *Drosophila melanogaster*. *Behavior Genetics*, **23**, 85–90.

Greene, H. W. & Burghardt, G. M. (1978). Behavior and phylogeny: constriction in ancient and modern snakes. *Science*, **200**, 74–7.

Greenlaw, J. S. & Post, W. (1985). Evolution of monogamy in seaside sparrows, *Ammodramus maritimus*: tests of hypotheses. *Animal Behaviour*, **33**, 373–83.

Greenwood, P. J. (1980). Mating systems, philopatry and dispersal in birds and mammals. *Animal Behaviour*, **28**, 1140–62.

Greenwood, P. J., Harvey, P. H. & Perrins, C. M. (1978). Inbreeding and dispersal in the great tit. *Nature*, **271**, 52–54.

Grosberg, R. K. & Quinn, J. F. (1986). The genetic control and consequences of kin recognition by the larvae of a colonial marine invertebrate. *Nature*, **322**, 456–9.

Gross, M. R. (1985). Disruptive selection for alternative life histories in salmon. *Nature*, **313**, 47–8.

Gross, M. R. & Shine, R. (1981). Parental care and the mode of fertilisation in ectothermic vertebrates. *Evolution*, **35**, 775–93.

Grubb, T. C. (1987). Changes in flocking behaviour of wintering English titmice with time, weather and supplementary food. *Animal Behaviour*, **35**, 794–806.

Gwynne, D. T. (1981). Sexual difference theory: mormon crickets show role reversal in mate choice. *Science*, **213**, 779–90.

Gyllensten, U. B., Jakobsson, S. & Temrin, H. (1990). No evidence for illegitimate young in monogamous and polygynous warblers. *Nature*, **343**, 168–70.

Hall, J. C., Greenspan, R. J. & Harris, W. A. (1982). *Genetic Neurobiology*. Cambridge: Mass.: MIT Press.

Halliday, T. R. (1978). Sexual selection and mate choice. In *Behavioural Ecology. An Evolutionary Approach*, ed. J. R. Krebs & N. B. Davies, pp. 180–213. Oxford: Blackwell Scientific.

Halliday, T. R. (1981). Sexual behaviour and group cohesion. In *Group Cohesion: Psychological and Sociological Reflections*, ed. H. Kellerman, pp. 171–89. New York: Grune & Stratton.

Halliday, T. R. (1987). Physiological constraints on sexual selection. In *Sexual Selection: Testing the Alternatives*, ed. J. W. Bradbury & M. B. Andersson, pp. 247–64. Chichester: John Wiley & Sons.

Halliday, T. R. (1990a). Morphology and sexual selection. *Atti VI Convegno Nazionale Ass. 'Alessandro Ghigi', Museo Regionale di Scienze Naturali*, Torino, 9–21.

Halliday, T. R. (1990b). The evolution of courtship behaviour in newts and salamanders. *Advances in the Study of Behavior*, **19**, 137–69.

Hamilton, W. D. (1964a). The genetical evolution of social behaviour. I. *Journal of Theoretical Biology*, **7**, 1–16.

Hamilton, W. D. (1964b). The genetical evolution of social behaviour. II. *Journal of Theoretical Biology*, **7**, 17–52.

Hamilton, W. D. (1967). Extraordinary sex ratios. *Science*, **156**, 477–88.

Hamilton, W. D. (1971). Geometry for the selfish herd. *Journal of Theoretical Biology*, **31**, 295–311.

Hamilton, W. D. (1982). Pathogens as causes of genetic diversity in their host organisms. In *Population Biology of Infectious Diseases*, ed. R. M. Anderson & R. M. May, pp. 269–96. New York: Springer-Verlag.

Hamilton, W. D., Axelrod, R. & Tanese, R. (1990). Sexual selection as an adaptation to resist parasites (a review). *Proceedings of the National Academy of Sciences USA*, **87**, 3566–73.

Hammerstein, P. & Parker, G. A. (1987). Sexual selection: games between the sexes. In *Sexual Selection: Testing the Alternatives*, ed. J. W. Bradbury & M. B. Andersson, pp. 119–42. Chichester: John Wiley & Sons.

Hannes, R. P., Rranck, D. & Lienmann, F. (1984). Effects of rank order fights on whole body and blood concentrations of androgens and corticosteroids in the male swordtail (*Xiphophorus belleri*). *Zeitschrift für Tierpsychologie*, **65**, 53–65.

Hannon, S. J., Mumme, R. L., Koenig, W. D. & Pitelka, F. A. (1985). Replacement of breeders and within-group conflict in the cooperatively breeding acorn woodpeckers. *Behavioral Ecology and Sociobiology*, **17**, 303–12.

Harcourt, A. H. & de Waal, F. B. M. (eds.) (1992). *Coalitions and Alliances in Humans and Other Animals*. Oxford: Oxford University Press.

Harlow, H. F. (1949). The formation of learning sets. *Psychological Review*, **56**, 51–65.

Harvey, I. F. & White, S. A. (1990). Prey selection by larvae of *Pyrrhosoma nymphula* (Sulzer) (Zygoptera: Coenagrionidae). *Odonatologica*, **16**, 17–25.

Harvey, P. H. & Krebs, J. R. (1990). Comparing brains. *Science*, **249**, 140–46.

Harvey, P. H. & Pagel, M. D. (1991). *The Comparative Method in Evolutionary Biology*. Oxford: Oxford University Press.

Harvey, P. H. & Purvis, A. (1991). Comparative methods for explaining adaptations. *Nature*, **351**, 619–24.

Hay, D. A. (1986). *Essentials of Behaviour Genetics*. Melbourne: Blackwell.

Hay, D. E. & McPhail, J. D. (1975). Mate selection in three-spined sticklebacks. *Canadian Journal of Zoology*, **53**, 441–50.

Hayes, K. J. & Hayes, C. (1952). Imitation in a home-raised chimpanzee. *Journal of Comparative Physiological Psychology*, **45**, 450–59.

Hayes, K. J., Thompson, R. & Hayes, C. (1953). Discrimination learning set in chimpanzees. *Journal of Comparative and Physiological Psychology*, **46**, 99–104.

Hazel, W. N. & Johnson, M. S. (1990). Microhabitat choice and polymorphism in the land snail *Theba pisana* (Muller). *Heredity*, **65**, 449–54.

Hedrick, A. V. (1988). Female choice and the heritability of attractive male traits: an empirical study. *American Naturalist*, **132**, 267–76.

Heim, A. (1970). *Intelligence and Personality*. Harmondsworth, Middlesex: Penguin.

Heinroth, O. (1911). Contributions to the biology, especially the ethology and psychology of the Anatidae. Reprinted and translated in *Foundations in Comparative Ethology*, ed. G. M. Burghardt, pp. 246–301. New York: Van Nostrand Reinhold.

Hepper, P. G. (1986). Kin recognition: functions and mechanisms. A review. *Biological Reviews*, **61**, 63–93.

Hepper, P. G. (ed.) (1991). *Kin Recognition*. Cambridge: Cambridge University Press.

Heyes, C. M., Dawson, G. R. & Nokes, T. (1992). Imitation in rats: initial responding and transfer evidence from a bidirectional control procedure. *Quarterly Journal of Experimental Psychology*. Section B: *Comparative and Physiological Psychology*, **45B**, 229–40.

Hinde, R. A. (1970). *Animal Behaviour*, 2nd edn. New York: McGraw-Hill.

Hinde, R. A. (1974). *Biological Bases of Human Social Behaviour*. New York: McGraw-Hill.

Hinde, R. A. (1976). Interactions, relationships and social structure. *Man*, **11**, 1–17.

Hinde, R. A. (ed.) (1983). *Primate Social Relationships: an Integrated Approach*. Oxford: Blackwell Scientific.

Hirsch, J. (1963). Behavior genetics and individuality understood. *Science*, **142**, 1436–42.

Hoekstra, R. F. (1987). The evolution of sexes. In *The Evolution of Sex and its Consequences*, ed. S. C. Stearns, pp. 59–91. Basel: Birkhäuser Verlag.

Hoffmann, A. A. (1988). Heritable variation for territorial success in two *Drosophila melanogaster* populations. *Animal Behaviour*, **36**, 1180–9.

Hoffmann, A. A. & Cacoyianni, Z. (1989). Selection for territoriality in *Drosophila melanogaster*: correlated responses in mating success and other fitness components. *Animal Behaviour*, **38**, 837–45.

Hofman, M. A. (1983a). Encephalization in hominids: evidence for the model of punctuationalism. *Brain, Behavior and Evolution*, **22**, 102–17.

Hofman, M. A. (1983b). Energy metabolism, brain size and longevity in mammals. *Quarterly Review of Biology*, **58**, 495–512.

Hoglund, J. (1989). Size and plumage dimorphism in lek-breeding birds: a comparative analysis. *American Naturalist*, **134**, 72–87.

Holland, P. C. & Straub, J. J. (1979). Differential effects of two ways of developing the unconditioned stimulus after Pavlovian appetitive conditioning. *Journal of Experimental Psychology: Animal Behavior Processes*, **5**, 65–8.

Holmes, W. G. & Sherman, P. W. (1983). Kin recognition in animals. *American Scientist*, **71**, 46–55.

Hoogland, J. L. (1979). Aggression, ectoparasitism and other possible costs of prairie dog (Sciuridae: *Cynomys* spp.) coloniality. *Behaviour*, **69**, 1–35.

Horn, H. S. (1978). Optimal tactics of reproduction and life history. In *Behavioural Ecology*, ed. J. R. Krebs & N. B. Davies, pp. 411–29. Oxford: Blackwell Scientific.

Houston, A. & Davies, N. B. (1985). The evolution of cooperation and life history in the dunnock, *Prunella modularis*. In *Behavioural Ecology: Ecological Consequences of Adaptive Behaviour*, ed. R. M. Sibly & R. H. Smith, pp. 471–87. Oxford: Blackwell Scientific.

Howard, R. D. (1979). Estimating reproductive success in natural populations. *American Naturalist*, **114**, 221–31.

Howe, H. F. (1977). Sex ratio adjustment in the common grackle. *Science*, **198**, 744–6.

Hrdy, S. B. (1974). Male–male competition and infanticide among the langurs (*Presbytis entellus*) of Abu, Rajasthan. *Folia primatologia*, **22**, 19–58.

Hull, D. (1988). *Science as a Process*. Chicago: Chicago University Press.

Humphrey, N. K. (1976). The social function of intellect. In *Growing Points in Ethology*, ed. P. P. G. Bateson & R. A. Hinde, pp. 303–17. Cambridge: Cambridge University Press.

Isbell, L. (1991). Contest and scramble competition: patterns of female aggression and ranging behaviour among primates. *Behavioral Ecology*, **2**, 143–55.

Janzen, F. J. & Paukstis, G. L. (1991). Environmental sex determination in reptiles: ecology, evolution, and experimental design. *Quarterly Review of Biology*, **66**, 149–79.

Jarman, P. J. (1974). The social organization of antelope in relation to their ecology. *Behaviour*, **48**, 215–67.

Jarman, P. J. (1982). Prospects for interspecific comparison in sociobiology. In *Current Problems in Sociobiology*, ed. King's College Sociobiology Group, pp. 323–42. Cambridge: Cambridge University Press.

Jarvis, J. U. M. (1981). Eusociality in a mammal: cooperative breeding in naked mole rat colonies. *Science*, **212**, 571–3.

Jeffreys, A. J., Wilson, V. & Thein, S. L. (1985). Individual-specific 'fingerprints' of human DNA. *Nature*, **316**, 76–9.

Jenni, D. A. (1974). Evolution of polyandry in birds. *American Zoologist*, **14**, 129–41.

Jerison, H. J. (1973). *Evolution of the Brain and Intelligence*. New York: Academic Press.

Jermy, T., Labos, E. & Molnar, I. (1990). Stenography of phytophagous insects – a result of constraints on the evolution of the nervous system. In *Organizational Constraints on the Dynamics of Evolution*, ed. J. Maynard Smith & G. Vida, pp. 157–66. Manchester: Manchester University Press.

John, E. R., Chesler, I. & Bartlett, F. (1968). Observation learning in cats. *Science*, **159**, 1489–91.

Johnson, M. S. (1982). Polymorphism for direction of coil in *Partula suturalis*: behavioural isolation and positive frequency dependent selection. *Heredity*, **49**, 145–51.

Jolly, A. (1966). Lemur social behavior and primate intelligence. *Science*, **153**, 501–06.

Jolly, A. (1991). Conscious chimpanzees? A review of recent literature. In *Cognitive Ethology: The Minds of Other Animals*, ed. C. Ristau, pp. 231–52. Hillsdale, NJ: Lawrence Erlbaum.

Jones, J. S. (1982). Genetic differences in individual behaviour associated with shell polymorphism in the snail *Cepaea nemoralis*. *Nature*, **298**, 749–50.

Jones, J. S., Leith, B. H. & Rawlings, P. (1977). Polymorphism in *Cepaea*: a problem with too many solutions? *Annual Review of Ecology and Systematics*, **9**, 109–43.

Kacser, H. & Burns, J. A. (1981). The molecular basis of dominance. *Genetics*, **97**, 639–66.

Kaneshiro, K. Y. (1989). The dynamics of sexual selection and founder effects in species formation. In *Genetics, Speciation and Founder Principle*, ed. L. V. Giddings, K. Y. Kaneshiro & W. W. Anderson, pp. 279–96. New York: Oxford University Press.

Kaneshiro, K. Y. & Boake, C. R. B. (1987). Sexual selection and speciation: Issues raised by Hawaiian *Drosophila*. *Trends in Ecology and Evolution*, **2**, 207–12.

Kareem, A. M. & Barnard, C. J. (1986). Kin recognition in mice: age, sex and parental effects. *Animal Behaviour*, **34**, 1814–24.

Karowe, D. N. (1990). Predicting host range evolution: colonization of *Coronilla varia* by *Colias philodice* (Lepidopter Pieridae). *Evolution*, **44**, 1637–47.

Kaufman, J. H. (1962). Ecology and social behaviour of the coati, *Naua narica*, on Barro Colorado Island, Panama. *University of California Publications in Zoology*, **60**, 95–222. Berkeley: University of California Press.

Kawai, M. (1965). Newly-acquired pre-cultural behavior of the natural troop of Japanese monkeys on Koshima Islet. *Primates*, **6**, 1–30.

Keane, B. (1990). The effect of relatedness on reproductive success and mate choice in the white-footed mouse, *Peromyscus leucopus*. *Animal Behaviour*, **39**, 264–73.

Kirkpatrick, M., Price, T. & Arnold, S. J. (1990). The Darwin–Fisher theory of sexual selection in monogamous birds. *Evolution*, **44**, 180–93.

Kirkpatrick, M. & Ryan, M. J. (1991). The evolution of mating preferences and the paradox of the lek. *Nature*, **350**, 33–8.

Kitahara-Frisch, J. & Norikoshi, K. (1982). Spontaneous sponge-making in captive chimpanzees. *Journal of Human Evolution*, **11**, 41–7.

Kitchell, J. A. (1986). The evolution of predator–prey behavior: naticid gastropods and their molluscan prey. In *Evolution of Animal Behavior, Paleontological and Field Approaches*, ed. M. H. Nitecki & J. A. Kitchell, pp. 88–110. New York: Oxford University Press.

Kitcher, P. (1985). *Vaulting Ambition: Sociobiology and the Quest for Human Nature*. Cambridge, Mass.: MIT Press.

Kleiman, D. G. (1977). Monogamy in mammals. *Quarterly Review of Biology*, **52**, 39–69.

Kleiman, D. G. (1985). Paternal care in New World primates. *American Zoologist*, **25**, 857–9.

Klump, G. M., Kretzschmar, E. & Curio, E. (1986). The hearing of an avian predator and its avian prey. *Behavioural Ecology and Sociobiology*, **18**, 317–23.

Kodric-Brown, A. & Brown, J. H. (1984). Truth in advertising: the kind of traits favored by sexual selection. *American Naturalist*, **124**, 309–23.

Koella, J. C. (1988). The tangled bank: the maintenance of sexual reproduction through competitive interactions. *Journal of Evolutionary Biology*, **1**, 95–116.

Koenig, W. D. & Albano, S. S. (1986). On the measurement of sexual selection. *American Naturalist*, **127**, 403–09.

Koenig, W. D. & Stacey, P. B. (1990). Acorn woodpeckers: group living and food storage under contrasting ecological conditions. In *Cooperative Breeding in Birds*, ed. P. B. Stacey & W. D. Koenig, pp. 413–54. Cambridge: Cambridge University Press.

Koford, R. R., Bowem, B. S. & Vehrencamp, S. L. (1990). Groove-billed anis: joint nesting in a tropical cuckoo. In *Cooperative Breeding in Birds*, ed. P. B. Stacey & W. D. Koenig, pp. 333–56. Cambridge: Cambridge University Press.

Kohler, W. (1925). *The Mentality of Apes*. London: Routledge & Kegan Paul.

Krebs, J. R. (1978). Optimal foraging: decision rules for predators. In *Behavioural Ecology: An Evolutionary Approach*, ed. J. R. Krebs & N. B. Davies, pp. 23–63. Oxford: Blackwell Scientific.

Krebs, J. R. (1990). Food-storing birds: adaptive specialization in brain and behaviour? *Philosophical Transactions of the Royal Society* B, **329**, 153–9.

Krebs, J. R. & Davies, N. B. (1981). *An Introduction to Behavioural Ecology*. Oxford: Blackwell Scientific.

Krebs, J. R. & Davies, N. B. (1987). *An Introduction to Behavioural Ecology*, 2nd edn. Cambridge, Mass.: Sinauer.

Kruuk, H. (1964). Predators and anti-predator behaviour of the black headed gull, *Larus ridibundus*. *Behaviour*, Suppl. **11**, 1–129.

Kruuk, H. (1975). Functional aspects of social hunting by carnivores. In *Function and Evolution in Behaviour*, ed. G. Baerends, C. Beer & A. Manning, pp. 119–41. Oxford: Oxford University Press.

Kruuk, H. (1978). Spatial organization and territorial behaviour of the European badger. *Meles meles*. *Journal of Zoology, London*, **184**, 1–19.

Kyriacou, C. P. & Hall, J. C. (1988). Interspecific genetic control of courtship song production and reception in *Drosophila*. *Science*, **232**, 494–7.

Lande, R. (1982). Rapid origin of sexual isolation and character divergence within a cline. *Evolution*, **36**, 213–23.

Lank, D. B., Mineau, P., Rockwell, R. F. & Cooke, F. (1989). Intraspecific nest parasitism and extra-pair copulation in lesser snow geese. *Animal Behaviour*, **37**, 74–89.

Lashley, K. S. (1951). The problem of serial order in behavior. In *Cerebral*

Mechanisms in Behavior: The Hixon Symposium, ed. L. A. Jeffress, pp.
112–36. New York: Wiley.

Lazarus, J. & Metcalfe, N. B. (1990). Tit-for-tat cooperation in sticklebacks: a
critique of Milinski. *Animal Behaviour*, **39**, 987–8.

Le Boeuf, B. J. (1974). Male–male competition and reproductive success in
elephant seals. *American Zoologist*, **14**, 163–76.

Lee, P. C. (1988). Ecological constraints and opportunities: interactions,
relationships and social organization of primates. In *Ecology and Behaviour
of Food-Enhanced Primate Groups*, ed. J. E. Fa & C. H. Southwick,
pp. 297–312. New York: Alan R. Liss.

Lee, P. C. (1989). Family structure, communal care and female reproductive
effort. In *Comparative Socioecology*, ed. V. Standen & R. Foley, pp. 323–40.
Oxford: Blackwell Scientific.

Leighton, D. R. (1987). Gibbons: territoriality and monogamy. In *Primate
Societies*, ed. B. Smuts, D. Cheney, R. Seyfarth, R. Wrangham & T.
Struhsaker, pp. 135–45. Chicago: University of Chicago Press.

Leonard, J. L. & Lukowiak, K. (1991). Sex and the simultaneous hermaphrodite:
testing models of male–female conflict in a sea slug *Navanax inermis*
(Opisthobranchia). *Animal Behaviour*, **41**, 255–66.

Lewis, W. M. (1987). The cost of sex. In *The Evolution of Sex and its
Consequences*, ed. S. C. Stearns, pp. 33–57. Basel: Birkhäuser Verlag.

Ligon, J. D. & Ligon, S. H. (1978). Communal breeding in green woodhoopoes
as a case for reciprocity. *Nature*, **276**, 496–8.

Lill, A. (1974). Social organisation and space utilisation in the lek-forming
white-bearded manakin, *M. manacus trinitatis*. *Zeitschrift für
Tierpsychologie*, **36**, 513–30.

Littlejohn, M. J. (1965). Premating isolation in the *Hyla ewingi* complex
(Anura: Hylidae). *Evolution*, **19**, 234–43.

Littlejohn, M. J. & Watson, G. F. (1985). Hybrid zones and homogamy in
Australian frogs, *Annual Review of Ecology and Systematics*, **16**, 85–112.

Lively, C. M. (1987). Evidence from a New Zealand snail for the maintenance of
sex by parasitism. *Nature*, **328**, 519–21.

Lloyd, J. E. (1966). Studies on the flash communication system in *Photuris*
fireflies. *Miscellaneous Publications of the Museum of Zoology, University
of Michigan*, **130**, 1–195.

Lloyd, J. E. (1984). On deception, a way of all flesh, and firefly signalling and
systematics. *Oxford Surveys in Evolutionary Biology*, **1**, 48–84.

Lorenz, K. (1939). The comparative study of behavior. In *Motivation of Animal
and Human Behavior*, ed. K. Lorenz & P. Leyhausen, pp. 1–31. New York:
Van Nostrand Reinhold.

Lorenz, K. Z. (1941a). Vergleichende Bewegungsstudien am Anatinen. *Journal
für Ornithologie* (Supplement), **89**, 194–294.

Lorenz, K. (1941b). Comparative studies of the motor patterns of Anatinae. In
Studies in Animal and Human Behavior, ed. R. D. Martin, pp. 14–114.
Cambridge, Mass.: Harvard University Press.

Lorenz, K. (1966). *On Aggression*. London: Methuen.

Losos, J. B. (1990). The evolution of form and function: morphology and
locomotor performance in West Indian *Anolis* lizards. *Evolution*, **44**,
1189–203.

Lynch, C. B. (1980). Response to divergent selection for nesting behavior in
Mus musculus. *Genetics*, **96**, 757–65.

Lynch, C. B. & Hegmann, J. P. (1973). Genetic differences influencing

behavioral temperature regulation in small mammals. II.
Genotype – environment interactions. *Behavior Genetics*, **3**, 145–54.

Lynch, C. B. & Sulzbach, D. S. (1984). Quantitative genetic analysis of temperature regulation in *Mus musculus*. II. Diallel analysis of individual traits. *Evolution*, **38**, 527–40.

Maddison, W. P. (1990). A method for testing the correlated evolution of two binary characters: are gains or losses concentrated on certain branches of phylogenetic tree? *Evolution*, **44**, 539–57.

Maddison, W. P. (1992). *MacClade* (*Version 3.0*). Sunderland, Mass.: Sinauer Associates.

Maddison, W. P. & Maddison, D. R. (1992). *MacClade: Analysis of Phylogeny and Character Evolution*. Version 3.0. Sunderland, Mass: Sinauer.

Majiluf, M. P. J. (1987). *The Reproductive Ecology of Female South American Fur Seals at Punta San Juan, Peru*. PhD Thesis, University of Cambridge.

Major, P. F. (1978). Predator–prey interactions in two schooling fishes, *Caranx ignobilis* and *Stolephorus purpureus*. *Animal Behaviour*, **26**, 760–77.

Mangel, M. & Clark, C. W. (1988). *Dynamic Modeling in Behavioral Ecology*. Princeton: Princeton University Press.

Martin, A. P. & Simon, C. (1988). Anomalous distribution of nuclear and mitochondrial DNA markers in periodical cicadas. *Nature*, **336**, 237–9.

Martins, E. & Garland, T. Jr (1991). Phylogenetic analysis of the correlated evolution of continuous characters: a simulated study. *Evolution*, **45**, 534–57.

Mason, J. R. & Reidinger, R. F. (1981). Effects of social facilitation and observational learning on feeding behaviour of the red-winged blackbird (*Agelaius phoeniceus*). *Auk*, **98**, 778–84.

Mason, J. R. & Reidinger, R. F. (1982). Observational learning of food aversions in red-winged blackbirds (*Agelaius phoeniceus*). *Auk*, **99**, 548–54.

Masters, J. C. & Spencer, H. G. (1989). Why we need a new genetic species concept. *Systematic Zoology*, **38**, 270–79.

Masters, W. M. & Waite, T. A. (1990). Tit-for-tat during predator inspection or shoaling. *Animal Behaviour*, **39**, 603–04.

Mather, K. & Jinks, J. L. (1971). *Biometrical Genetics:the Study of Continuous Variation*. London: Chapman & Hall.

May, R. M. & Anderson, R. (1987). The transmission dynamics of HIV infection. *Nature*, **326**, 137–42.

Maynard Smith, J. (1974). The theory of games and the evolution of animal conflict. *Journal of theoretical Biology*, **47**, 209–21.

Maynard Smith, J. (1976a). Group selection. *Quarterly Review of Biology*, **54**, 277–83.

Maynard Smith, J. (1976b). A short-term advantage for sex recombination through sib competition. *Journal of theoretical Biology*, **63**, 245–58.

Maynard Smith, J. (1977). Parental investment: a prospective analysis. *Animal Behaviour*, **25**, 1–9.

Maynard Smith, J. (1978a). Optimization theory in evolution. *Annual Review of Ecology and Systematics*, **9**, 13–56.

Maynard Smith, (1978b). *The Evolution of Sex*. Cambridge: Cambridge University Press.

Maynard Smith, J. (1982a). The evolution of social behaviour: a classification of models. In *Current Problems in Sociobiology*, ed. King's College Sociobiology Group, pp. 29–44. Cambridge: Cambridge University Press.

Maynard Smith, J. (1982b). *Evolution and the Theory of Games*. Cambridge: Cambridge University Press.

Maynard Smith, J. (1984). The ecology of sex. In *Behavioural Ecology. An Evolutionary Approach*, 2nd edn, ed. J. R. Krebs & N. B. Davies, pp. 201–21. Oxford: Blackwell Scientific.

Maynard Smith, J. (1989). *Evolutionary Genetics*. Oxford: Oxford University Press.

Maynard Smith, J. & Parker, G. A. (1976). The logic of asymmetric contests. *Animal Behaviour*, **24**, 159–75.

Maynard Smith, J. & Price, G. R. (1973). The logic of animal conflict. *Nature, London*, **246**, 15–18.

Maynard Smith, J. & Riechert, S. E. (1984). A conflict tendency model of spider agonistic behaviour: hybrid-pure population line comparisons. *Animal Behaviour*, **32**, 564–78.

Mayr, E. (1942). *Systematics and the Origin of Species*. New York: Columbia University Press.

Mayr, E. (1958). Behavior and systematics. In *Behavior and Evolution*, ed. A. Roe & G. G. Simpson, pp. 341–62. New Haven: Yale University Press.

Mayr, E. (1963). *Animal Species and Evolution*. Cambridge, Mass.: Belknap Press.

Mayr, E. (1972). Sexual selection and natural selection. In *Sexual Selection and the Descent of Man, 1871–1971*, ed. B. Campbell, pp. 87–104. London: Heinemann.

Mayr, E. (1983). How to carry out the adaptationist program? *American Naturalist*, **121**, 324–34.

McClure, P. A. (1981). Sex-biased litter reduction in food-restricted wood rats (*Neotoma floridana*). *Science*, **211**, 1058–60.

McComb, K. E. (1991). Female choice for high roaring rate in red deer *Cervus elaphus*. *Animal Behaviour*, **41**, 79–88.

McCracken, G. F. & Brussard, P. F. (1980). Self-fertilisation in the white-lipped snail *Tridodopsis albolabris*. *Biological Journal of the Linnean Society*, **14**, 429–34.

MacDonald, D. W. (1983). The ecology of carnivore social behaviour. *Nature*, **301**, 379–84.

MacDonald, D. W. & Carr, G. M. (1989). Food security and the rewards of tolerance. In *Comparative Socioecology*, ed. V. Standen & R. Foley, pp. 75–100. Oxford: Blackwell Scientific.

MacDonald, D. W. & Moehlman, P. D. (1982). Cooperation, altruism and restraint in the reproduction of carnivores. In *Perspectives in Ethology*, Vol. 5, ed. P. Bateson & P. H. Klopfer, pp. 433–67. New York: Plenum Press.

McGregor, P. K. (1989). Bird song and kin recognition: potential, constraints and evidence. *Ethology, Ecology and Evolution*, **1**, 124–7.

McGregor, P. K. & Krebs, J. R. (1982). Mating and song types in the great tit. *Nature*, **297**, 60–61.

McGrew, W. C. (in press). *Chimpanzee Material Culture: Implications for Human Evolution*. Cambridge: Cambridge University Press.

Mackinnon, J. (1978). *The Ape Within Us*. London: Collins.

McLennan, D. A., Brooks, D. R. & McPhail, J. D. (1988). The benefits of communication between comparative ethology and phylogenetic systematics: a case study using gasterosteid fishes. *Canadian Journal of Zoology*, **66**, 2177–90.

McNamara, J. M. & Houston, A. T. (1985). Optimal foraging and learning. *Journal of theoretical Biology*, **117**, 231–49.

Macphail, E. M. (1985). Vertebrate intelligence: the null hypothesis. In *Animal Intelligence*, ed. L. Weiskrantz, pp. 37–51. Oxford: Clarendon Press.

Meltzoff, A. N. (1988). The human infant as *Homo imitans*. In *Comparative Social Learning*, ed. T. Zentall & B. G. Galef Jr, pp. 319–41. Hillsdale, NJ: Lawrence Erlbaum.

Meltzoff, A. N. & Moore, M. K. (1977). Imitation of facial and manual gestures by human neonates. *Science*, **198**, 75–8.

Menken, S. B. J., Herrebout, W. M. & Wiebas, J. T. (1992). Small ermine moths (Yponomeuta): their host relations and evolution. *Annual Review of Entomology*, **37**, 41–66.

Menzel, E. W. (1974). A group of chimpanzees in a 1-acre field: leadership and communication. In *Behavior of Nonhuman Primates*, ed. A. M. Schier & F. Stollnitz, pp. 83–153. New York: Academic Press.

Milinski, M. (1987). Tit for tat in sticklebacks and the evolution of cooperation. *Nature*, **325**, 433–5.

Milinski, M. (1992). Predator inspection: cooperation or 'safety in numbers'? *Animal Behaviour*, **43**, 679–80.

Milinski, M., Pfluger, D., Kulling, D. & Kettler, R. (1990). Do sticklebacks cooperate repeatedly in reciprocal pairs? *Behavioral Ecology and Sociobiology*, **27**, 17–21.

Milton, K. (1981). Distribution patterns of tropical plant foods as a stimulus to primate mental development. *American Anthropologist*, **83**, 534–48.

Mineka, S. & Cook, M. (1988). Social learning and the acquisition of snake fear in monkeys. In *Social Learning: Psychological and Biological Perspectives*, ed. T. Zentall & B. Galef, pp. 51–71. Hillsdale, NJ: Lawrence Erlbaum.

Mineka, R. W., Davidson, M., Cook, M. & Keir, R. (1984). Observational conditioning of snake fear in rhesus monkeys. *Journal of Abnormal Psychology*, **93**, 355–72.

Mitani, J. C., Gether, G. F., Rodman, P. S. & Priatna, D. (1991). Associations among wild orangutans: sociality, passive aggregations or chance? *Animal Behaviour*, **42**, 33–46.

Mitchell, R. W. (1987). A comparative-development approach to understanding imitation. In *Perspectives in Ethology*, Vol. 7, Alternatives, ed. P. P. G. Bateson & P. H. Klopfer, pp. 183–215. New York: Plenum Press.

Mock, D. W. & Fijioka, M. (1990). Monogamy and long-term pair bonding in vertebrates. *Trends in Ecology and Evolution*, **5**, 39–43.

Moehlman, P. D. (1979). Jackal helpers and pup survival. *Nature*, **277**, 382–3.

Moehlman, P. D. (1986). The ecology of cooperation in canids. In *Ecological Aspects of Social Evolution*, ed. D. I. Rubenstein & R. W. Wrangham, pp. 64–86. Princeton: Princeton University Press.

Møller, A. P. (1987). Variation in badge size in male house sparrows *Passer domesticus*: evidence for status signalling. *Animal Behaviour*, **35**, 1637–44.

Møller, A. P. (1988). Badge size in the house sparrow *Passer domesticus*. Effects of intra- and intersexual selection. *Behavioral Ecology and Sociobiology*, **22**, 373–8.

Møller, A. P. (1991). Sperm competition, sperm depletion, paternal care, and relative testis size in birds. *American Naturalist*, **137**, 882–906.

Monaghan, P. & Metcalfe, N. B. (1985). Group foraging in wild brown hares: effects of resource distribution and social status. *Animal Behaviour*, **33**, 993–9.

Montgomerie, R. & Thornhill, R. (1989). Fertility advertisement in birds: a means of enticing male–male competition? *Ethology*, **81**, 209–20.

Montgomery, G. G. & Sunquist, M. E. (1978). Habitat selection and use by two-toed and three-toed sloths. In *The Ecology of Arboreal Folivores*, ed. G. G. Montgomery, pp. 329–69. Washington DC: Smithsonian Institute Press.

Moore, W. S. (1981). Assortative mating genes selected along a gradient. *Heredity*, **46**, 191–5.

Morton, E. S. (1975). Ecological sources of selection on avian sounds. *American Naturalist*, **109**, 17–34.

Morton, N. E., Crow, J. F. & Muller, H. J. (1956). An estimate of the mutational damage in man from data on consanguineous marriages. *Proceedings of the National Academy of Sciences USA*, **42**, 855–63.

Moss, C. J. (1988) *Elephant Memories*. New York: Morrow.

Moss, C. J. & Poole, J. H. (1983). Relationships and social structure of African elephants. In *Primate Social Relationships: An Integrated Approach*, ed. R. A. Hinde, pp. 315–25. Oxford: Blackwell Scientific.

Muller, H. J. (1964). The relation of recombination to mutational advance. *Mutation Research*, **1**, 2–9.

Newton, I. (1989). *Lifetime Reproduction in Birds*. London: Academic Press.

Nishida, T. & Uehara, S. (1980). Natural diet of chimpanzees (*Pan troglodytes schweinfurhii*): long term record from the Mahale Mountains. *African Study Monographs*, **3**, 109–30.

Noe, R. (1990). A Veto game played by baboons: a challenge to the use of the Prisoner's Dilemma as a paradigm for reciprocity and cooperation. *Animal Behaviour*, **39**, 79–90.

Noordwijk, A. J. van & Scharloo, W. (1981). Inbreeding in an island population of the great tit. *Evolution*, **35**, 674–88.

Nowak, M. A. & Sigmund, K. (1992). Tit for tat in heterogeneous populations. *Nature*, **355**, 250–53.

Nowak, M. & Sigmund, K. (1993). A strategy of win–stay, lose–shift that outperforms tit-for-tat in the Prisoner's Dilemma game. *Nature*, **364**, 56–8.

Odendaal, F. J., Bull, C. M. & Telford, S. R. (1986). Influence of the acoustic environment on the distribution of the frog *Ranidella riparia*. *Animal Behaviour*, **34**, 1836–43.

Oftedal, O. (1984). Milk composition, milk yield and energy output at peak lactation: a comparative review. *Symposia of the Zoological Society, London*, **51**, 33–85.

Ollason, J. G. (1980). Learning to forage – optimally? *Theoretical Population Biology*, **18**, 44–56.

Orians, G. H. (1969). On the evolution of mating systems in birds and mammals. *American Naturalist*, **103**, 589–603.

Otte, D. (1989). Speciation in Hawaiian crickets. In *Speciation and its Consequences*, ed. D. Otte & J. A. Endler, pp. 482–526. Sunderland, Mass.: Sinauer Associates.

Owen, M. J. & Mullan, M. J. (1990). Molecular genetic studies of manic-depression and schizophrenia. *Trends in Neuroscience*, **13**, 29–31.

Packer, C. (1978). Reciprocal altruism in *Papio anubis*. *Nature*, **265**, 441–3.

Packer, C., Gilbert, D. A., Pusey, A. E. & O'Brien, S. J. (1991). A molecular genetic analysis of kinship and cooperation in African lions. *Nature*, **351**, 562–5.

Packer, C., Herbst, L., Pusey, A. E., Bygott, J. D., Hanby, J. P., Cairns, S. J. & Bogerhoff Mulder, M. (1988). Reproductive success in lions. In *Reproductive Success*, ed. T. H. Clutton-Brock, pp. 419–35. Chicago: University of Chicago Press.

Packer, C., Lewis, S. & Pusey, A. (1992). A comparative analysis of non-offspring nursing. *Animal Behaviour*, **43**, 265–81.

Pagel, M. D. (1991). *Program: Evolutionary covariance regression*. Private distribution. Department of Zoology, University of Oxford, England.

Pagel, M. D. (1992). A method for the analysis of comparative data. *Journal of theoretical Biology*, **156**, 431–42.

Pagel, M. D. & Harvey, P. H. (1989). Comparative methods for examining adaptation depend on evolutionary models. *Folia Primatologica*, **53**, 203–20.

Palameta, B. (1989). The importance of socially transmitted information in the acquisition of novel foraging skills by pigeons and canaries. PhD Thesis, University of Cambridge.

Parker, G. A. (1970a). Sperm competition and its evolutionary consequences in the insects. *Biological Reviews*, **45**, 525–68.

Parker, G. A. (1970b). The reproductive behaviour and nature of sexual selection in *Scatophaga stercoraria* L, II. The fertilization rate and the spatial and temporal relationships of each sex around the site of mating and oviposition. *Journal of Animal Ecology*, **39**, 205–28.

Parker, G. A. (1970c). Sperm competition and its evolutionary effect on copula duration in the fly *Scatophaga stercoraria*. *Journal of Insect Physiology*, **16**, 1301–28.

Parker, G. A. (1974). Assessment strategy and the evolution of fighting behaviour. *Journal of theoretical Biology*, **47**, 223–43.

Parker, G. A. (1978). Searching for mates. In *Behavioural Ecology: An Evolutionary Approach*, 1st edn, ed. J. R. Krebs & N. B. Davies, pp. 214–44.

Parker, G. A. (1984). Evolutionarily stable strategies. In *Behavioural Ecology: An Evolutionary Approach*, 2nd edn, ed. J. R. Krebs & N. B. Davies, pp. 30–61. Oxford: Blackwell Scientific.

Parker, G. A. (1992). The marginal value theorem with exploitation time costs: sperm, diet, and optimal copula duration in dung flies. *American Naturalist*, **139**, 1237–56.

Parker, G. A., Baker, R. R. & Smith, V. G. F. (1972). The origin and evolution of gamete dimorphism and the male–female phenomenon. *Journal of theoretical Biology*, **36**, 529–53.

Parker, G. A., Simmons, L. W. & Ward, P. I. (1993). Optimal copula duration in dung flies: effects of frequency dependence and female mating status. *Behavioral Ecology and Sociobiology*, **32**, 157–66.

Parker, G. A. & Stuart, R. A. (1976). Animal behavior as a strategy optimizer: evolution of resource assessment strategies and optimal emigration thresholds. *American Naturalist*, **110**, 1055–76.

Parker, S. T. & Gibson, K. R. (1979). A developmental model for the evolution of language and intelligence in early hominids. *Behavioral and Brain Sciences*, **2**, 367–408.

Partridge, L. & Halliday, T. (1984). Mating patterns and mate choice. In *Behavioural Ecology. An Evolutionary Approach*, 2nd edn, ed. J. R. Krebs & N. B. Davies, pp. 222–50. Oxford: Blackwell Scientific.

Passingham, R. E. (1981). Primate specializations in brain and intelligence. *Symposia of the Zoological Society of London*, **46**, 361–88.

Passingham, R. E. (1982). *The Human Primate*. Oxford: W. H. Freeman.

Passingham, R. E. & Ettlinger, G. (1974). A comparison of cortical function in man and other primates. *International Review of Neurobiology*, **16**, 233–99.

Paterson, H. E. H. (1982). Perspective on speciation by reinforcement. *South African Journal of Science*, **78**, 53–7.

Paterson, H. E. H. (1985). The recognition concept of species. In *Species and Speciation*, ed. E. S. Vrba, pp. 21–9. Pretoria: Transvaal Museum.

Patterson, F. & Cohn, R. (in press). Self-recognition and self-awareness in lowland gorilla. In *Comparative Reflections on Self-awareness in Animals and Humans*, ed. S. Parker, M. Boccia & R. Mitchell. Cambridge: Cambridge University Press.

Payne, R. B. & Payne, L. L. (1989). Heritability estimates and behaviour observations: extra-pair matings in indigo buntings. *Animal Behaviour*, **38**, 457–67.

Perdeck, A. C. (1958). The isolating value of specific song patterns in two species of grasshoppers (*Chorthippus brunneus* Thunb. and *Ch. bigittulus* L.). *Behaviour*, **12**, 1–75.

Peres, C. A. (1991). *Ecology of Mixed-Species Groups of Tamarins in Amazonian terra firme Forests*. PhD. Thesis, University of Cambridge.

Perrett, D. I., Harris, M. H., Bevan, R., Thomas, S., Benson, P. J., Mistlin, A. J., Chitty, A. J., Hietanen, J. K. & Ortega, J. E. (1989). Frameworks of analysis for the neural representations of animate objects and actions. *Journal of Experimental Biology*, **146**, 87–113.

Petrie, M. (1983a). Female moorhens compete for small fat males. *Science*, **220**, 413–15.

Petrie, M. (1983b). Mate choice in role-reversed species. In *Mate Choice*, ed. P. Bateson, pp. 167–79. Cambridge: Cambridge University Press.

Petrie, M., Halliday, T. & Sanders, C. (1991). Peahens prefer peacocks with elaborate trains. *Animal Behaviour*, **41**, 323–31.

Pierce, G. J. & Ollason, J. G. (1987). Eight reasons why optimal foraging theory is a complete waste of time. *Oikos*, **49**, 111–18.

Plomin, R. DeFries, J. C. & McClearn, G. E. (1990). *Behavioral Genetics: a Primer*. New York: Freeman.

Poole, J. H. (1989). Mate guarding, reproductive success and female choice in African elephants. *Animal Behaviour*, **37**, 842–9.

Porter, R. H. & Blaustein, A. R. (1989). Mechanisms and ecological correlates of kin recognition. *Science Progress*. **73**, 53–66.

Porter, R. H., Wyrick, M. & Pankey, J. (1978). Sibling recognition and spiny mice (*Acomys cahirinus*). *Behavioral Ecology and Sociobiology*, **3**, 61–8.

Potts, W. K., Manning, C. J. & Wakeland, E. K. (1991). Mating patterns in seminatural populations of mice influenced by MHC genotype. *Nature*, **352**, 619–21.

Povinelli, D. J. (1989). Failure to find self-recognition in Asian elephants (*Elephas maximus*) in contrast to their use of mirror cues to discover hidden food. *Journal of Comparative Psychology*, **103**, 122–31.

Povinelli, D. J. (1991). *Social intelligence in monkeys and apes*. PhD Thesis, New Haven, Connecticut: Yale University.

Povinelli, D. J., Nelson, K. E. & Boysen, S. T. (1990). Inferences about guessing and knowing in chimpanzees (*Pan troglodytes*). *Journal of Comparative Psychology*, **104**, 203–10.

Povinelli, D. J., Nelson, K. E. & Boysen, S. T. (1992). Comprehension of role reversal in chimpanzees: evidence of empathy? *Animal Behaviour*, **43**, 633–40.

Povinelli, D. J., Parks, K. A. & Novack, M. A. (1991). Do rhesus monkeys (*Macaca mulatta*) attribute knowledge and ignorance to others? *Journal of Comparative Psychology*, **105**, 318–25.

Povinelli, D. J., Parks, K. A. & Novak, M. A. (1992). Role reversal by rhesus monkeys but no evidence of empathy. *Animal Behaviour*, **43**, 269–81.

Powell, J. R. (1978). The founder-flush speciation theory: An experimental approach. *Evolution*, **32**, 465–74.

Premack, D. (1988). 'Does the chimpanzee have a theory of mind?' revisited. In *Machiavellian Intelligence: Social Expertise and the Evolution of Intellect in Monkeys, Apes and Humans*, ed. R. W. Byrne & A. Whiten, pp. 94–110. Oxford: Clarendon Press.

Premack, D. & Woodruff, G. (1978). Does the chimpanzee have a theory of mind? *The Behavioral and Brain Sciences*, **4**, 515–26.

Price, M. V. & Waser, N. M. (1979). Pollen dispersal and optimal outcrossing in *Delphinium nelsoni*. *Nature*, **277**, 294–7.

Price, P. W. (1980). *Evolutionary Biology of Parasites*. Princeton: Princeton University Press.

Prins, H. H., Ydenberg, R. C. & Drent, R. H. (1980). The interaction of brent geese *Branta bernicla* and sea plantain *Plantago maritima* during spring staging: field observations and experiments. *Acta Botanica Neerlandii*, **29**, 585–96.

Proctor, H. C. (1991). Courtship in the water mite *Neumania papillator*: males capitalize on female adaptations for predation. *Animal Behaviour*, **42**, 589–98.

Promislow, D. E. L. & Harvey, P. H. (1990). Living fast and dying young: a comparative analysis of life-history variation among mammals. *Journal of Zoology, London*, **220**, 417–37.

Provine, R. R. (1989). Contagious yawning and infant imitation. *Bulletin of the Psychonomic Society*, **27**, 125–6.

Prum, R. O. (1990). Phylogenetic analysis of the evolution of display behavior in the Neotropical manakins (Aves: Pipiridae). *Ethology*, **84**, 202–31.

Purvis, A. (1991). Comparative analysis by independent contrasts (C.A.I.C.). A statistical package for the Apple Macintosh Version TEST. Private distribution. Department of Zoology, University of Oxford, Oxford, England.

Pusey, A. E. (1987). Sex-biased dispersal and inbreeding avoidance in birds and mammals. *Trends in Ecology and Evolution*, **2**, 295–9.

Queiroz, A. de & Wimberger, P. H. (1993). The usefulness of behavior for phylogeny estimation: levels of homoplasy in behavioral and morphological characters. *Evolution*, **47**, 46–60.

Rasa, A. E. (1989). Helping in dwarf mongoose societies: an alternative reproductive strategy. In *The Sociobiology of Sexual and Reproductive Stragegies*, ed. A. E. Rasa, C. Vogel & E. Voland, pp. 61–73. London: Chapman & Hall.

Readhead, C. & Hood, L. (1990). The dysmyelinating mouse mutations shiverer (*shi*) and myelin deficient (*shi-mld*). *Behavior Genetics*, **20**, 213–34.

Readhead, C., Popko, B., Takahashi, N., Shine, H. D., Saavedra, R. A., Sidman, R. L. & Hood, L. (1987). Expression of a myelin basic protein gene in transgenic shiverer mice: correction of the dysmyelinating phenotype. *Cell*, **48**, 703–12.

Reeve, H. K., Westneat, D. F., Noon, W. A., Sherman, P. W. & Aquadro, C. F. (1990). DNA 'fingerprinting' reveals high levels of inbreeding in colonies of the eusocial naked mole-rat. *Proceedings of the National Academy of Sciences USA*, **87**, 2496–500.

Ridley, M. (1983). *The Explanation of Organic Diversity*. Oxford: Clarendon Press.

Ridley, M. (1986). *Evolution and Classification: the Reformation of Cladism*. London: Longman.

Ridley, M. (1989). Why not to use species in comparative tests. *Journal of theoretical Biology*, **136**, 361–4.

Ridlye, M. (1990). The control and frequency of mating in insects. *Functional Ecology*, **4**, 75–84.

Riechart, S. E. & Maynard Smith, J. (1989). Genetic analyses of two behavioural traits linked to individual fitness in the desert spider *Agelenopsis aperta*. *Animal Behaviour*, **37**, 624–37.

Ritchie, M. G., Butlin, R. K. & Hewitt, G. M. (1992). Fitness consequences of potential assortative mating inside and outside a hybrid zone in *Chorthippus parallelus* (Orthoptera: Acrididae): implications for reinforcement and sexual selection theory. *Biological Journal of the Linnean Society*, **45**, 219–34.

Robinson, J. G. (1988). Group size in wedge-capped capuchin monkeys *Cebus olivaceus* and the reproductive success of males and females. *Behavioral Ecology and Sociobiology*, **23**, 187–97.

Rockwell, R. F. & Seiger, M. B. (1973). Phototaxis in *Drosophila*: a critical evaluation. *American Scientist*, **61**, 339–45.

Roelofs, W. L. & Brown, R. L. (1982). Pheromones and evolutionary relationships of *Tortricidae*. *Annual Review of Ecology and Systematics*, **13**, 395–422.

Rogers, D. J. (1972). Random search and insect population models. *Journal of Animal Ecology*, **41**, 369–83.

Rood, J. P. (1986). Ecology and social evolution in the mongooses. In *Ecological Aspects of Social Evolution*, ed. D. I. Rubenstein & R. W. Wrangham, pp. 131–52. Princeton: Princeton University Press.

Roper, T. J. (1983). Learning as a biological phenomenon. In *Animal Behaviour*, Vol. 3, *Genes, Development and Learning*, ed. T. R. Halliday & P. J. B. Slater, pp. 178–212. Oxford: Blackwell Scientific.

Roughgarden, J. (1991). The evolution of sex. *American Naturalist*, **138**, 934–53.

Rubenstein, D. I. & Wrangham, R. W. (eds.) (1986). *Ecological Aspects of Social Evolution*. Princeton: Princeton University Press.

Russell, J. K. (1983). Altruism in coati bands: nepotism or reciprocity? In *Social Behaviour of Female Vertebrates*, ed. S. K. Wasser, pp. 263–90. London: Academic Press.

Russon, A. E. & Galdikas, B. M. F. (1993). Imitation in ex-captive orangutans. *Journal of Comparative Psychology*, **107**, 147–61.

Ryan, M. J. (1991). Sexual selection and communication in frogs. *Trends in Ecology and Evolution*, **6**, 351–5.

Ryan, M. J., Fox, J. H., Wilczynski, W. & Rand, A. S. (1990). Sexual selection for sensory exploitation in the frog. *Physalaemus pustolosus*. *Nature*, **343**, 66–7.

Ryan, M. J. & Wilczynski, W. (1991). Evolution of intraspecific variation in the advertisement call of a cricket frog (*Acris crepitans*, Hylidae). *Biological Journal of the Linnean Society*, **44**, 249–71.

Sapolsky, R. M. & Ray, J. C. (1989). Styles of dominance and their endocrine correlates among individual olive baboons *Papio anubis*. *American Journal of Primatology*, **18**, 1–13.

Sawaguchi, T. & Kudo, H. (1990). Neocortical development and social structure in primates. *Primates*, **31**, 283–9.

van Schaik, C. P. (1983). Why are diurnal primates living in groups? *Behaviour*, **87**, 120–44.

van Schaik, C. P. & Dunbar, R. I. M. (1990). The evolution of monogamy in large primates: a new hypothesis and some crucial tests. *Behaviour*, **115**, 30–62.

Scheel, D. & Packer, C. (1991). Group hunting behaviour of lions: a search for cooperation. *Animal Behaviour*, **41**, 697–709

Schiller, P. H. (1952). Innate constituents of complex responses in primates. *Psychological Review*, **59**, 177–91.

Schultz, R. J. (1971). Special adaptive problems associated with unisexual fishes. *American Zoologist*, **11**, 351–60.

Schwagmeyer, P. L. & Foltz, D. W. (1990). Factors affecting the outcome of sperm competition in thirteen-lined ground squirrels. *Animal Behaviour*, **39**, 156–62.

Scott, D. K. (1988). Reproductive success in Bewick's swans. In *Reproductive Success*, ed. T. H. Clutton-Brock, pp. 220–36. Chicago: Chicago University Press.

Searcy, W. A. & Yasukawa, K. (1989). Alternative models of territorial polygyny in birds. *American Naturalist*, **134**, 323–43.

Seilacher, A. (1967). Fossil behavior. *Scientific American*, **217**, 72–80.

Seilacher, A. (1986). Evolution of behavior as expressed in marine trace fossils. In *Evolution of Animal Behavior, Paleontological and Field Approaches*, ed. M. H. Nitecki & J. A. Kitchell, pp. 62–87. New York: Oxford University Press.

Sella, G. (1985). Reciprocal egg trading and brood care in a hermaphrodite polychaete worm. *Animal Behaviour*, **33**, 938–44.

Sella, G. (1988). Reciprocation, reproductive success, and safeguards against cheating in a hermaphrodite polychaete worm, *Ophryotrocha diadema*. *Biological Bulletin*, **175**, 212–17.

Seyfarth, R. M. & Cheney, D. L. (1984). Grooming, alliances and reciprocal altruism in vervet monkeys. *Nature*, **308**, 541–3.

Shea, B. T. (1983). Phyletic size change and brain/body allometry: a consideration based on the African pongids and other primates. *International Journal of Primatology*, **4**, 33–61.

Sherman, P. (1977). Nepotism and the evolution of alarm calls. *Science*, **197**, 1246–53.

Sherman, P. (1980). The limits of ground squirrel nepotism. In *Beyond Nature/Nurture*, ed. G. W. Barlow & J. Silverberg, pp. 505–44. Boulder: Westview Press.

Sherman, P. W. (1981). Reproductive competition and infanticide in Belding's ground squirrels and other animals. In *Natural Selection and Social Behaviour: Recent Research and New Theory*, ed. R. D. Alexander & D. W. Tinkle, pp. 311–31. New York: Chiron Press.

Sherry, D. F. & Galef, B. G., Jr (1984). Cultural transmission without imitation: milk bottle opening by birds. *Animal Behaviour*, **32**, 937.

Shettleworth, S.J. (1975). Reinforcement and the organisation behavior in golden hamsters: hunger, environment and food reinforcement. *Journal of Experimental Psychology: Animal Behaviour Processes*, **1**, 56–87.

Shields, W. M. (1984). *Philopatry, Inbreeding and the Evolution of Sex*. Albany: State University of New York Press.

Shuster, S. M. (1989). Male alternative reproductive strategies in a marine isopod crustacean (*Paracerceis sculpta*): the use of genetic markers to measure differences in fertilization success among α-, β-, and γ-males. *Evolution*, **43**, 1683–98.

Shuster, S. M. & Wade, M. J. (1991). Equal mating success among male reproductive strategies in a marine isopod. *Nature,* **350,** 608–10.

Sibly, R. M. (1983). Optimal group size is unstable. *Animal Behaviour,* **31,** 947–8.

Sigg, H. & Stolba, A. (1981). Home range and daily march in a hamadryas baboon troop. *Folia Primatologica,* **36,** 40–75.

Sillen-Tullberg, B. (1988). Evolution of gregariousness in aposematic butterfly larvae: A phylogenetic analysis. *Evolution,* **42,** 293–305.

Simon, C. (1979). Debut of the seventeen-year-old cicada. *Natural History,* **88,** 38–45.

Simpson, G. G. (1958). Behavior and evolution. In *Behavior and Evolution,* ed. A. Roe & G. G. Simpson, pp. 507–35. New Haven: Yale University Press.

Singer, M. C. & Thomas, C. D. (1988). Heritability of oviposition preference and its relationship to offspring performance within a single insect population. *Evolution,* **42,** 977–85.

Singh, P., Brown, R. E. & Roser, B. (1987). MHC antigens in the urine as olfactory recognition cues. *Nature,* **327,** 161–4.

Skelton, P. (ed.) (1993). *Evolution. A Biological and Palaeontological Approach.* Wokingham: Addison-Wesley.

Skinner, B. F. (1981). Selection by consequences. *Sciences,* **213,** 501–4.

Slater, P. J. B. (1983). The development of individual behaviour. In *Animal Behaviour,* Vol. 3, *Genes, Development and Learning,* ed. T. R. Halliday & P. J. B. Slater, pp. 82–113. Oxford: Blackwell Scientific.

Slater, P. J. B. (1986). The cultural transmission of bird song. *Trends in Ecology and Evolution,* **1,** 94–7.

Slater, P. J. B., Ince, S. A. & Colgan, P. W. (1980). Chaffinch song types: their frequencies in the population and distribution between the repertoires of different individuals. *Behaviour,* **75,** 207–18.

Smith, D. C. (1988). Heritable divergence of *Rhagoletis pomonella* host races by seasonal asynchrony. *Nature,* **336,** 66–7.

Smith, P. K. (1988). The cognitive demands of children's social interactions with peers. In *Machiavellian Intelligence: Social Expertise and the Evolution of Intellect in Monkeys, Apes and Humans,* ed. R. W. Byrne & A. Whiten, pp. 94–110. Oxford: Clarendon Press.

Smith, R. L. (ed.) (1984). *Sperm Competition and the Evolution of Animal Mating Systems.* New York: Academic Press.

Snow, D. W. (1962). A field study of the black-and-white manakin, *Manacus manacus,* in Trinidad. *Zoologica, NY,* **47,** 65–104.

Sokolowski, M. B. (1980). Foraging strategies of *Drosophila melanogaster*: a chromosomal analysis. *Behaviour Genetics,* **10,** 291–302.

Sokolowski, M. B., Bauer, S. J., Wai-Ping, V., Rodriguez, L., Wong, J. L. & Kent, C. (1986). Ecological genetics and behaviour of *Drosophila melanogaster* larvae in nature. *Animal Behaviour,* **32,** 403–8.

Spence, K. W. (1937). Experimental studies of learning and higher mental processes in infra-human primates. *Psychological Bulletin,* **34,** 806–50.

Stacey, P. B. & Koenig, W. D. (eds.) (1990). *Cooperative Breeding in Birds.* Cambridge: Cambridge University Press.

Stander, P. E. (1992). Foraging dynamics of lions in a semi-arid environment. *Canadian Journal of Zoology,* **70,** 8–21.

Stearns, S. C. (1976). Life history tactics: a review of ideas. *Quarterly Review of Biology,* **51,** 3–47.

Stearns, S. C. & Schmid-Hempel, P. (1987). Evolutionary insights should not be wasted. *Oikos,* **49,** 118–25.

Stephens, D. W. & Krebs, J. R. (1986). *Foraging Theory*. Princeton: Princeton University Press.

Stevens, P. F. (1980). Evolutionary polarity of character states. *Annual Review of Ecology and Systematics*, **11**, 333–58.

Stevenson-Hinde, J. (1983). Individual characteristics and the social situation. In *Primate Social Relationships*, ed. R. A. Hinde, pp. 28–35. Oxford: Blackwell Scientific.

Stewart, K. & Harcourt, A. H. (1987). Gorillas: variation in female relationships. In *Primate Societies*, ed. B. B. Smuts, D. L. Cheney, R. M. Seyfarth, R. W. Wrangham & T. T. Struhsaker, pp. 155–64. Chicago: Chicago University Press.

Stratton, G. E. & Uetz, G. W. (1986). Inheritance of courtship behavior in *Schizocoas* wolf spiders (Araneae: Lycosidae). In *Evolutionary Genetics of Invertebrate Behavior*, ed. M. D. Huettel, pp. 63–78. New York: Plenum Press.

Stuart, R. J. (1991). Kin recognition as a functional concept. *Animal Behaviour*, **41**, 1093–4.

Suarez, S. & Gallup, G. G. (1981). Self-recognition in chimpanzees and orangutans, but not gorillas. *Journal of Human Evolution*, **10**, 175–88.

Summers, K. (1992). Mating strategies in two species of dart-poison frogs: a comparative study. *Animal Behaviour*, **43**, 907–19.

Svare, B. & Mann, M. (1981). Infanticide: genetics, developmental and hormonal influences in mice. *Physiology & Behaviour*, **27**, 921–7.

Swofford, D. (1991). *Phylogenetic Analysis Using Parsimony (PAUP)*. Champaign, Ill.: Illinois Natural History Survey.

Taylor, C. K. & Saayman, G. S. (1973). Imitative behaviour by Indian Ocean bottlenose dolphins (*Tursiops aduncus*) in captivity. *Behaviour*, **44**, 286–98.

Terborgh, J. & Goldizen, A. W. (1985). On the mating system of the cooperatively breeding saddle-backed tamarin (*Saguinus fuscicollis*). *Behavioral Ecology and Sociobiology*, **16**, 293–9.

Thomas, J. H. (1990). Genetic analysis of defecation in *Caenorhabditis elegans*. *Genetics*, **124**, 855–72.

Thompson, J. N. (1988). Evolutionary genetics of oviposition preference in swallowtail butterflies. *Evolution*, **42**, 1223–34.

Thornhill, R. (1988). The jungle fowl hen's cackle incites male competition. *Verhandlungen der Deutschen zoologischen Gesellschaft*, **81**, 145–54.

Thornhill, R. & Alcock, J. (1983). *The Evolution of Insect Mating Systems*. Cambridge, Mass.: Harvard University Press.

Thorpe, W. H. (1956). *Learning and Instinct in Animals*. London: Methuen.

Thorpe, W. H. (1963). *Learning and Instinct in Animals*, 2nd edn. London: Methuen.

Thorpe, W. H. (1967). Vocal imitation and antiphonal song and its implications. *Proceedings of the International Ornithological Congress*, **14**, 245–63.

Tinbergen, N. (1951). *The Study of Instinct*. Oxford: Oxford University Press.

Tinbergen, N. (1959). Comparative studies of the behaviour of gulls (Laridae): a progress report. *Behaviour*, **15**, 1–70.

Tomasello, M. (1990). Cultural transmission in the tool use and communicatory signaling of chimpanzees? In *'Language' and Intelligence in Monkeys and Apes: Comparative Developmental Perspectives*, ed. S. Parker & K. Gibson, pp. 274–311. Cambridge: Cambridge University Press.

Tomasello, M., Davis-Dasilva, M., Camak, L. & Bard, K. (1987). Observational learning of tool-use by young chimpanzees. *Human Evolution*, **2**, 175–83.

Tomasello, M., Kruger, A. C. & Rarner, H. H. (1991). Cultural learning. *Emory Cognition Project, Report 21.*

Torchio, P. F. & Tepedino, V. J. (1980). Sex ratio, body size and seasonality in a solitary bee, *Osmia lignaria propinqua* Cresson (Hymenoptera: Megachilidae). *Evolution*, **34**, 993–1003.

Trivers, R. L. (1971). The evolution of reciprocal altruism. *Quarterly Review of Biology*, **46**, 35–57.

Trivers, R. L. (1972). Parental investment and sexual selection. In *Sexual Selection and the Descent of Man*, ed. B. Campbell, pp. 136–79. Chicago: Aldine.

Trivers, R. L. (1974). Parent-offspring conflict. *American Zoologist*, **14**, 249–64.

Trivers, R. (1985). *Social Evolution*. Menlo Park: Benjamin Cummings.

Trivers, R. L. & Willard, D. E. (1973). Natural selection of parental ability to vary the sex ratio of offspring. *Science*, **191**, 249–63.

Trune, D. R. & Slobodchikoff, C. N. (1976). Social effects of roosting on the metabolism of the pallid bat, *Antrozous pallidus*. *Journal of Mammology*, **57**, 656–63.

Tully, T. & Quinn, W. G. (1985). Classical conditioning and retention in normal and mutant *Drosophila melanogaster*. *Journal of Comparative Physiology*, **157**, 263–77.

Turner, G. F. & Robinsons, R. L. (1992). Milinski's Tit-for-tat hypothesis: do fish preferentially inspect in pairs? *Animal Behaviour*, **43**, 677–8.

Tutin, C. E. G., McGrew, W. C. & Baldwin, P. J. (1983). Social organisation of savanna-dwelling chimpanzees, *Pan troglodytes verus*, at Mt. Assirik, Senegal. *Primates*, **24**, 154–73.

Van Valen, L. (1973). A new evolutionary law. *Evolutionary Theory*, **1**, 1–30.

Vehrencamp, S. L. (1983). A model for the evolution of despotic versus egalitarian societies. *Animal Behaviour*, **31**, 667–82.

Vehrencamp, S. L. & Bradbury, J. W. (1984). Mating systems and ecology. In *Behavioural Ecology. An Evolutionary Approach*, 2nd edn, ed. J. Krebs & N. B. Davies, pp. 251–78. Oxford: Blackwell Scientific.

Verner, J. & Willson, M. F. (1966). The influence of habitats on the mating systems of North American passerine birds. *Ecology*, **47**, 143–7.

Vincent, A. C. J. (1990). A seahorse father makes a good mother. *Natural History*, December, pp. 4–43.

Visalberghi, E. & Fragaszy, D. (1990a). Food washing behaviour in tufted capuchin monkeys, *Cebus apella*, and crabeating macaques, *Macaca fascicularis*. *Animal Behaviour*, **40**, 829–36.

Visalberghi, E. & Fragaszy, D. (1990b). Do monkeys ape? In *'Language' and Intelligence in Monkeys and Apes: Comparative Developmental Perspectives*, ed. S. Parker & K. Gibson, pp. 247–73. Cambridge: Cambridge University Press.

Visalberghi, E. & Trinca, L. (1987). Tool use in capuchin monkeys: distinguishing between performing and understanding. *Primates*, **30**, 511–21.

Von Neumann, J. & Morgenstern, O. (1944). *The Theory of Games and Economic Behaviour*. Princeton: Princeton University Press.

de Waal, F. (1982). *Chimpanzee Politics*. London: Jonathan Cape.

Wade, M. J. & Arnold, S. J. (1980). The intensity of sexual selection in relation to male sexual behaviour, female choice and sperm precedence. *Animal Behaviour*, **28**, 446–61.

Waldman, B. (1981). Sibling recognition in toad tadpoles: the role of experience. *Zeitschrift für Tierpsychologie*, **56**, 341–58.

Waldman, B. (1985a). Olfactory basis of kin recognition in toad tadpoles. *Journal of Comparative Physiology* A, **156**, 565–77.

Waldman, B. (1985b). Sibling recognition in toad tadpoles: are kinship labels transferred among individuals? *Zeitschrift für Tierpsychologie*, **68**, 41–57.

Waldman, B., Frumhoff, P. C. & Sherman, P. W. (1988). Problems of kin recognition. *Trends in Ecology and Evolution*, **3**, 8–13.

Ward, P. & Zahavi, A. (1973). The importance of certain assemblages of birds as 'information-centres' for food finding. *Ibis*, **115**, 517–34.

Warden, C. J. & Jackson, T. A. (1935). Imitative behavior of rhesus monkeys. *Journal of Genetical Psychology*, **46**, 103–25.

Warner, R. R. (1984). Mating behaviour and hermaphroditism in coral reef fish. *American Scientist*, **72**, 128–36.

Warner, R. R., Robertson, D. R. & Leigh, E. G. (1975). Sex change and sexual selection. *Science*, **190**, 633–8.

Warren, J. M. (1973). Learning in vertebrates. In *Comparative Psychology: a Modern Survey*, ed. D. A. Dewsbury & D. A. Rethlingshafer, pp. 471–509. New York: McGraw-Hill.

Watson, P. J. (1990). Female-enhanced male competition determines the first mate and principal sire in the spider *Linyphia litigiosa* (Linyphiidae). *Behavioral Ecology and Sociobiology*, **26**, 77–90.

Watson, P. J. (1991). Multiple paternity and first mate sperm precedence in the sierra dome spider. *Linyphia litigiosa* Keyserling (Linyphiidae). *Animal Behaviour*, **41**, 135–48.

Wayne, R. K., Benveniste, R. E., Janczewski, D. N. & O'Brien, S. J. (1989). Molecular and biochemical evolution of the Carnivora. In *Carnivore Behavior, Ecology, and Evolution*, ed. J. L. Gittleman, pp. 465–95. Ithaca: Cornell University Press.

Wcislo, W. T. (1989). Behavioral environments and evolutionary change. *Annual Review of Ecology and Systematics*, **20**, 137–69.

Wechsler, D. (1944). *The Measurement of Adult Intelligence*. Baltimore: Williams and Wilkins.

Weigl, P. D. & Hanson, E. V. (1980). Observational learning and the feeding behaviour of the red squirrel *Tamiasciurus hudsonicus*: the ontogeny of optimization. *Ecology*, **61**, 213–18.

Welbergen, P. (in press). Sexual isolation between *Drosophila melanogaster* and *D. simulans*: Experimental assessment of behavioral mechanisms. In *Genetic Disorders Related to Brain Functions*, ed. J.-M. Guastavino, J.-M. Jallon & T. Archer. Hillsdale, NJ: Lawrence Erlbaum.

West-Eberhard, M. J. (1983). Sexual selection, social competition and speciation. *Quarterly Review of Biology*, **58**, 155–83.

Western, D. & Ssemakula, J. (1982). Life history patterns in birds and mammals and their evolutionary interpretation. *Oecologia*, **54**, 281–90.

Westneat, D. F. (1990). Genetic parentage in the indigo bunting: a study using DNA fingerprinting. *Behavioral Ecology and Sociobiology*, **27**, 67–76.

Whillans, K. V. & Falls, J. B. (1990). Effects of male removal on parental care of female white-throated sparrows, *Zonotrichia albicollis*. *Animal Behaviour*, **39**, 869–78.

Whitehead, J. M. (1987). Vocally mediated reciprocity between neighbouring groups of mantled howling monkeys. *Animal Behaviour*, **35**, 1615–27.

Whitehead, H., Waters, S. & Lyrholm, T. (1991). Social organization of female sperm whales and their offspring: constant companions, casual acquaintances. *Behavioral Ecology and Sociobiology*, **29**, 385–9.

Whiten, A. & Byrne, R. W. (1991). The emergence of meta-representation in human ontogeny and primate phylogeny. In *Natural Theories of Mind*, ed. A. Whiten, pp. 267–81. Oxford: Basil Blackwell.

Whiten, A. & Ham, R. (1992). On the nature and evolution of imitation in the animal kingdom: Reappraisal of a century of research. *Advances in the Study of Behavior*, **21**, 239–83.

Whitman, C. O. (1898). Animal behavior. *Woods Hole Marine Biological Laboratories Biological Lectures*, **6**, 285–338. Reprinted in Burghardt, G. M. (ed.) (1985). *Foundations of Comparative Ethology*, pp. 141–205. New York: Van Nostrand Reinhold.

Wichman, H. A. & Lynch, C. B. (1991). Genetic variation for seasonal adaptation in *Peromyscus leucopus*: nonreciprocal breakdown in a population cross. *Journal of Heredity*, **82**, 197–204.

Wiklund, C. (1975). The evolutionary relationship between adult oviposition preferences and larval host plant range in *Papilio machaon* L. *Oecologia*, **18**, 185–97.

Wilkinson, G. S. (1984). Reciprocal food sharing in the vampire bat. *Nature*, **308**, 181–4.

Williams, G. C. (1966). *Adaptation and Natural Selection*. Princeton: Princeton University Press.

Williams, G. C. (1975). *Sex and Evolution*. Princeton: Princeton University Press.

Wilson, D. S. (1983). The group selection controversy: history and current status. *Annual Reviews of Ecology and Systematics*, **14**, 159–87.

Wilson, E. O. (1975). *Sociobiology: The New Synthesis*. Cambridge, Mass.: Harvard University Press.

Wittenberger, J. F. (1976). The ecological factors selecting for polygyny in altricial birds. *American Naturalist*, **110**, 779–99.

Wood, D. (1989). Social interaction as tutoring. In *Interaction in Human Development*, ed. M. H. Bornstein & J. S. Bruner, pp. 59–80. Hillsdale, NJ: Lawrence Erlbaum.

Wood, T. K. (1993). Speciation in the *Enchenopa binotata* complex (Homoptera: Membracidae). In *Evolutionary Patterns and Processes*, ed. D. R. Lees & D. Edwards, pp. 299–317. London: Academic Press for the Linnean Society of London.

Woolfenden, G. E. & Fitzpatrick, J. W. (1990). Florida scrub jays: a synopsis after 18 years of study. In *Cooperative Breeding in Birds*, ed. P. B. Stacey & W. D. Koenig, pp. 239–66. Cambridge: Cambridge University Press.

Wrangham, R. W. (1980). An ecological model of female bonded primate groups. *Behaviour*, **75**, 262–300.

Wrangham, R. W. (1981). Drinking competition in vervet monkeys. *Animal Behaviour*, **29**, 904–10.

Wrangham, R. W. (1982). Mutualism, kinship and social evolution. In *Current Problems in Sociobiology*, ed. King's College Sociobiology Group, pp. 269–90. Cambridge: Cambridge University Press.

Wrangham, R. W. & Rubenstein, D. I. (1986). Social evolution in birds and mammals. In *Ecological Aspects of Social Evolution*, ed. D. I. Rubenstein & R. W. Wrangham, pp. 452–70. Princeton: Princeton University Press.

Wright, J. & Cuthill, I. (1989). Manipulation of sex differences in parental care. *Behavioral Ecology and Sociobiology*, **25**, 171–81.

Wynne-Edwards, V. C. (1961). *Animal Dispersion in Relation to Social Behaviour*. Edinburgh: Oliver & Boyd.

Yamazaki, K., Boyse, E. A., Mike, V., Thaler, H. T., Mathieson, B. J., Abbott, J., Boyse, J., Zayas, Z. A. & Thomas, L. (1976). Control of mating preferences in mice by genes in the major histocompatibility complex. *Journal of Experimental Medicine*, **144**, 1324–35.

Zajonc, R. B. (1965). Social facilitation. *Science*, **149**, 269–74.

Zimmerer, E. J. & Kallman, K. D. (1989). Genetic basis for alternative reproductive tactics in the pygmy swordtail, *Xiphophorus nigrensis*. *Evolution*, **43**, 1298–307.

Index

References to authors' names are selective.